Biotechnology in Agriculture and Forestry

Edited by
T. Nagata (Managing Editor)
H. Lörz
J. M. Widholm

Springer-Verlag Berlin Heidelberg GmbH

Volumes already published

Volume 1: Trees I (1986)
Volume 2: Crops I (1986)
Volume 3: Potato (1987)
Volume 4: Medicinal and Aromatic Plants I (1988)
Volume 5: Trees II (1989)
Volume 6: Crops II (1988)
Volume 7: Medicinal and Aromatic Plants II (1989)
Volume 8: Plant Protoplasts and Genetic Engineering I (1989)
Volume 9: Plant Protoplasts and Genetic Engineering II (1989)
Volume 10: Legumes and Oilseed Crops I (1990)
Volume 11: Somaclonal Variation in Crop Improvement I (1990)
Volume 12: Haploids in Crop Improvement I (1990)
Volume 13: Wheat (1990)
Volume 14: Rice (1991)
Volume 15: Medicinal and Aromatic Plants III (1991)
Volume 16: Trees III (1991)
Volume 17: High-Tech and Micropropagation I (1991)
Volume 18: High-Tech and Micropropagation II (1992)
Volume 19: High-Tech and Micropropagation III (1992)
Volume 20: High-Tech and Micropropagation IV (1992)
Volume 21: Medicinal and Aromatic Plants IV (1993)
Volume 22: Plant Protoplasts and Genetic Engineering III (1993)
Volume 23: Plant Protoplasts and Genetic Engineering IV (1993)
Volume 24: Medicinal and Aromatic Plants V (1993)
Volume 25: Maize (1994)
Volume 26: Medicinal and Aromatic Plants VI (1994)
Volume 27: Somatic Hybridization in Crop Improvement I (1994)
Volume 28: Medicinal and Aromatic Plants VII (1994)
Volume 29: Plant Protoplasts and Genetic Engineering V (1994)
Volume 30: Somatic Embryogenesis and Synthetic Seed I (1995)
Volume 31: Somatic Embryogenesis and Synthetic Seed II (1995)
Volume 32: Cryopreservation of Plant Germplasm I (1995)
Volume 33: Medicinal and Aromatic Plants VIII (1995)
Volume 34: Plant Protoplasts and Genetic Engineering VI (1995)
Volume 35: Trees IV (1996)
Volume 36: Somaclonal Variation in Crop Improvement II (1996)
Volume 37: Medicinal and Aromatic Plants IX (1996)
Volume 38: Plant Protoplasts and Genetic Engineering VII (1996)
Volume 39: High-Tech and Microprogation V (1997)
Volume 40: High-Tech and Microprogation VI (1997)
Volume 41: Medicinal and Aromatic Plants X (1998)
Volume 42: Cotton (1998)
Volume 43: Medicinal and Aromatic Plants XI (1999)
Volume 44: Transgenic Trees (1999)
Volume 45: Transgenic Medicinal Plants (1999)
Volume 46: Transgenic Crops I (1999)
Volume 47: Transgenic Crops II (2001)
Volume 48: Transgenic Crops III (2001)
Volume 49: Somatic Hybridization in Crop Improvement II (2001)
Volume 50: Cryopreservation of Plant Germplasm II (2002)
Volume 51: Medicinal and Aromatic Plants XII (2002)
Volume 52: Brassicas and Legumes: From Genome Structure to Breeding (2003)

Volumes in preparation

Tobacco BY-2 Cells
Brassica Biotechnology
Haploids in Crop Improvement II
Molecular Marker Systems in Plant Breeding and Crop Improvement

Biotechnology in Agriculture and Forestry 52

Brassicas and Legumes
From Genome Structure to Breeding

Edited by T. Nagata and S. Tabata

With 59 Figures, 8 in Color, and 19 Tables

Springer

Series Editors

Professor Dr. Toshiyuki Nagata
University of Tokyo
Graduate School of Science
Department of Biological Sciences
7-3-1 Hongo, Bunkyo-ku
Tokyo 113-0033, Japan

Professor Dr. Horst Lörz
Universität Hamburg
Institut für Allgemeine Botanik
und Botanischer Garten
Angewandte Molekularbiologie der Pflanzen II
Ohnhorststr. 18
22609 Hamburg, Germany

Prof. Dr. Jack M. Widholm
University of Illinois
285A E.R. Madigan Laboratory
Department of Crop Sciences
1201 W. Gregory
Urbana, IL 61801, USA

Volume Editors

Professor Dr. Toshiyuki Nagata
(*address see above*)

Professor Dr. Satoshi Tabata
Kazusa DNA Research Institute
2-6-7 Kazusa-kamatari, Kisarazu
Chiba 292-0818, Japan

ISSN 0934-943-X
ISBN 978-3-642-07662-6 ISBN 978-3-662-05036-1 (eBook)
DOI 10.1007/978-3-662-05036-1

Library of Congress Cataloging-in-Publication Data.

Brassicas and legumes: from genome structure to breeding / editors, Toshiyuki Nagata, Satoshi Tabata.
p. cm. – (Biotechnology in agriculture and forestry; 52)
Includes bibliographical references and index.

1. Brassica – Genome mapping. 2. Legumes – Genome mapping. 3. Brassica – Breeding.
4. Legumes – Breeding. I. Nagata, T. (Toshiyuki) II. Tabata, Satoshi, 1954-III. Series.

SB317.B65 B74 2003
635'.34233 – dc21

http://www.springer.de

Originally published by Springer-Verlag Berlin Heidelberg New York in 2003.
Softcover reprint of the hardcover 1st edition 2003

Production: PRO EDIT GmbH, Heidelberg, Germany
Cover design: design & production GmbH, Heidelberg, Germany
Typesetting: SNP Best-set Typesetter Ltd., Hong Kong
31/315050 5 4 3 2 1 0 Printed on acid free paper

This volume is dedicated to the late Professor Itaru Takebe, with whom I worked with great pleasure during the cradle stage of plant molecular biology.

Preface

This volume is the 52nd volume of the Series, *Biotechnology in Agriculture and Forestry* (BAF), founded by the late Professor Y.P.S. Bajaj in 1986. However, the present volume represents more than that, as it is the first of the new extension of the BAF Series edited by the editorial team of Professors Horst Lörz (Hamburg, Germany), Jack M. Widholm (Champaign/Urbana, USA), and Toshiyuki Nagata (Tokyo, Japan), which came into being in 2000 after Professor Bajaj passed away. One of us (T.N.) had already been involved in editing Volumes 49 and 51, undertakings partially initiated by Prof. Bajaj, and so the job of this editor should be considered as a kind of complementary work to the previous ones. In between, Volume 50 was published through the efforts of Dr. Leigh E. Towill (Fort Collins, CO, USA). Thus this Volume, *Brassicas and Legumes: From Genome Structure to Breeding*, is the first result in the new direction of the BAF Series and its new editorial team.

For editing this volume, it should be particularly noted that Dr. Satoshi Tabata of the Kazusa DNA Research Institute at Chiba, Japan, kindly joined this undertaking. He is one of the key persons from "The Arabidopsis Genome Initiative (AGI) 2000", and this volume could not have been completed without his tremendous efforts. This volume also includes another important aspect concerning legumes. In this regard, Dr. Tabata also played a key role, one of his activities being the determination of the genome sequence of the legume *Lotus japonicus*. Here, again, his contribution has been exemplary. An important feature of this volume in its entirety is its intent to form a bridge between genome sequence studies and practical breeding.

As noted above, the activity of the new editorial team has been launched. Two volumes, *Brassica Biotechnology* and *Tobacco BY-2 Cells*, are currently being processed. The former is being edited by Professors Eng Chong Pua (Singapore) and Carl Douglas (British Columbia, Canada) and overseen by T. Nagata, while the latter is being edited by Professors Seiichoro Hasezawa (Kashiwa, Japan), Dirk Inze (Ghent, Belgium), and Toshiyuki Nagata. In parallel with these two, the volume, *Molecular Markers in Plant Breeding and Crop Improvement*, is being planned by Professors Lörz and Gerd Wenzel (Munich, Germany). In addition, another volume, *Haploids and Plant Breeding*, has been proposed by Professors Widholm and Ken Katha (Alberta, Canada). Needless to say, other colleagues are also helping us in these activities.

Indeed, the new editorial team will attempt to cover the broad area of plant biotechnology. Furthermore, as this field is steadily expanding, the new editorial team is keen to hear the views of the general readers. It is our wish that these series of volumes will supplement original papers and annual reviews as another useful source for all readers. Therefore, all comments are welcome, sent either to us directly (Nagata, Lörz, and Widholm) or through Springer-Verlag.

Toshiyuki Nagata
Tokyo, January 2003

Contents

8 *Brassica* and Legume Chromosomes
K. Fukui

9 Sequence Analysis of the *Lotus japonicus* Genome
S. Sato, E. Asamizu, and S. Tabata

List of Contributors

ACARKAN, A., Bayer AG, Landwirschaftszentrum, PF-F-NWF, Geb. 6240, Alfred-Nobel-Strasse 50, 40789 Monheim, Germany

ASAMIZU, E., Kazusa DNA Research Institute, 2-6-7 Kazusa-kamatari, Kisarazu, Chiba 292-0818, Japan

BAUM, M., International Centre for Agricultural Research in the Dry Areas (ICARDA), Aleppo, Syria

BOIVIN, K., INRA-URGV, 2, rue Gaston Crémieux – CP5708, 91057 Évry cedex, France

CLARENZ, O., Max-Planck-Institut für Züchtungsforschung, Carl-von-Linné-Weg 10, 50829 Köln, Germany

DAS, S., Centre for Biotechnology, Faculty of Science, Hamdard University, Hamdard Nagar, New Delhi 110 062, India

ENDRE, G., Institute of Genetics, Biological Research Center, Hungarian Academy of Sciences, Szeged, Hungary

FELFÖLDI, K., Institute of Genetics, Biological Research Center, Hungarian Academy of Sciences, Szeged, Hungary

FUKUI, K., Department of Biotechnology, Graduate School of Engineering, Osaka University, Suita 565-0871, Osaka, Japan

HARADA, K., Faculty of Horticulture, Chiba University, 648 Matsudo, Matsudo, Chiba 271-8510, Japan

HAYASHI, M., Department of Biotechnology, Graduate School of Engineering, Osaka University, 2-1 Yamadakami, Suita, Osaka 565-0871, Japan

HAYASHI, M., Faculty of Horticulture, Chiba University, 648 Matsudo, Matsudo, Chiba 271-8510, Japan

HUGUET, T., Laboratoire de Biologie Moléculaire des Relations Plantes-Microorganismes, CNRS-INRA, BP27, 31326 Castanet-Tolosan Cedex, France

IMAIZUMI-ANRAKU, H., Plant Physiology Department, National Institute of Agrobiological Sciences, Kannon-dai 2-1-2, Tsukuba, Ibaraki 305-8602 Japan

JONGERIUS, M.C., Plant Research International B.V., P.O. Box 16, 6700 PB Wageningen, The Netherlands

KAHL, G., Plant Molecular Biology, Biocentre, University of Frankfurt/Main, Marie-Curie-Strasse 9, 60439 Frankfurt/M., Germany

KANEKO, T., Kazusa DNA Research Institute, 2-6-7 Kazusa-kamatari, Kisarazu, Chiba 292-0818, Japan

KANNE, H.J., Plant Research International B.V., P.O. Box 16, 6700 PB Wageningen, The Netherlands

KALÓ, P., Institute of Genetics, Biological Research Center, Hungarian Academy of Sciences, Szeged, Hungary

KATO, T., Kazusa DNA Research Institute, 2-6-7 Kazusa-kamatari, Kisarazu, Chiba 292-0818, Japan

KAWAGUCHI, M., Department of Environmental Sciences, Faculty of Science, Niigata University, 8050 Ikarashi, Niigaya City, 950-2181, Japan

KAWASAKI, S., Plant Physiology Department, National Institute of Agrobiological Sciences, Kannon-dai 2-1-2, Tsukuba, Ibaraki 305-8602 Japan

KERESZT, A., Laboratoire de Biologie Moléculaire des Relations Plantes-Microorganismes, CNRS-INRA, BP27, 31326 Castanet-Tolosan Cedex, France

KISS, G.B., Institute of Genetics, Biological Research Center, Hungarian Academy of Sciences, Szeged, Hungary

LAKSHMIKUMARAN, M., Bioresources and Biotechnology Division, TERI, Habitat Place, Lodhi Road, New Delhi 110 003, India

LUKENS, L., Department of Agronomy, University of Wisconsin, 1575 Linden Drive, Madison, Wisconsin 53706, USA

MIKAMI, I., Plant Physiology Department, National Institute of Agrobiological Sciences, Kannon-dai 2-1-2, Tsukuba, Ibaraki 305-8602 Japan

MINAMISAWA, K., Graduate School of Life Sciences, Tohoku University, Katahira, Aoba-ku, Sendai 980-8577, Japan

MURAKAMI, Y., Plant Physiology Department, National Institute of Agrobiological Sciences, Kannon-dai 2-1-2, Tsukuba, Ibaraki 305-8602 Japan

Nagata, T., The University of Tokyo, Graduate School of Science, Department of Biological Sciences, 7-3-1 Hongo, Bunkyo-ku, Tokyo 156-0042, Japan

Nakamura, Y., Kazusa DNA Research Institute, 2-6-7 Kazusa-kamatari, Kisarazu, Chiba 292-0818, Japan

Nishimura, R., Department of Life Sciences, Graduate School of Arts and Sciences, University of Tokyo, 3-8-1 Komaba, Meguro-ku, Tokyo, 153-8902 Japan

Nishio, T., Graduate School of Agricultural Science, Tohoku University, 1-1 Tsutsumidori-Amemiyamachi, Aoba-ku, Sendai 981-8555, Japan

Osborn, T.C., Department of Agronomy, University of Wisconsin, 1575 Linden Drive, Madison, Wisconsin 53706, USA

Prosperi, J.M., Station de Génétique et d'Amélioration des Plantes, INRA, Domaine de Melgueil, 34130 Mauguio, France

Rakshit, S., Indian Institute for Pulses Research, Kanpur, India

Rossberg, M., AMGEN GmbH, Riesstrasse 25, 80992 Munich, Germany

Saeki, K., Department of Biology, Graduate School of Science, Osaka University, Toyonaka, Osaka 560-0043, Japan

Sato, K., Graduate School of Agricultural Science, Tohoku University, 1-1 Tsutsumidori-Amamiyamachi, Aoba-ku, Sendai 981-8555, Japan

Sato, S., Kazusa DNA Research Institute, 2-6-7 Kazusa-kamatari, Kisarazu, Chiba 292-0818, Japan

Schmidt, R., Max-Planck-Institut für Molekulare Pflanzenphysiologie, 14424 Golm, Germany

Seres, A., Institute of Genetics, Biological Research Center, Hungarian Academy of Sciences, Szeged, Hungary

Shimizu, A., Plant Physiology Department, National Institute of Agrobiological Sciences, Kannon-dai 2-1-2, Tsukuba, Ibaraki 305-8602 Japan

Srivastava, P.S., Centre for Biotechnology, Faculty of Science, Hamdard University, Hamdard Nagar, New Delhi 110 062, India

Tabata, S., Kazusa DNA Research Institute, 2-6-7 Kazusa-kamatari, Kisarazu, Chiba 292-0818, Japan

THOQUET, P., Laboratoire de Biologie Moléculaire des Relations Plantes-Microorganismes, CNRS-INRA, BP27, 31326 Castanet-Tolosan Cedex, France

VOORRIPS, R.E., Plant Research International B.V., P.O. Box 16, 6700 PB Wageningen, The Netherlands

WINTER, P., Plant Molecular Biology, Biocentre, University of Frankfurt/Main, Marie-Curie-Strasse 9, 60439 Frankfurt/Main, Germany

1 Introductory Remarks to Brassicas and Legumes: From Genome Structure to Breeding

T. NAGATA[1] and S. TABATA[2]

Here, at the beginning of the 21st century, we have seen the unfolding of the complete genome sequence of *Arabidopsis thaliana,* product of the international collaborative work of "The Arabidopsis Genome Initiative 2000" (AGI 2000). This herbaceous plant has the smallest genome size known among the flowering plants, as previously demonstrated by Meyerowitz and Pruitt (1985). As might be expected, the genome complexity of this plant was not found to be great, rendering its sequencing relatively simple. Thus the whole genome sequence of this plant has been determined. In addition, the short generation time of this plant helped ease the analysis of its genetic traits. *A. thaliana* is considered to be a model plant among angiosperms.

More than a year has passed since the publication of the genome sequence of *A. thaliana*, time enough to recognize that it has brought us a revolutionary means of searching for genes for specific genetic traits. If one obtains any mutants by treatment with chemical mutagens, one is able to identify a genetic locus. Subsequently, by crossing these mutants with T-DNA-tagged lines, isolation of respective genes can be accomplished. If one knows the portions of the plant DNA sequence surrounding the T-DNA, one may identify and isolate a corresponding gene in the database of the *A. thaliana* genome. Thus, in a short time, *in silico* screening has become a routine tool in the discipline of the identification and isolation of specific genes. One is also able, with the help of several genetic markers, to reach corresponding genes by means of map-based screening. The basic issues determined from the sequence determination in *A. thaliana* are dealt with by Kato et al. in Chapter 2. It should be understood that the knowledge acquired on the genome organization of *A. thaliana* is not confined to this plant, those in the field are pleased to learn, but can be applied to other crop species such as tomato (Ku et al. 2000). That the closely related species of brassicas directly enjoyed the wealth of genetic information on *A. thaliana* is illustrated by the finding by genetic analysis that *Capsella rubella,* which is very closely related to *A. thaliana*, has very similar genome structures. This similarity can be further applied to other species of Brassicaceae, as shown by Schmidt et al. in Chapter 3.

Genetic marker studies on brassicas are complemented by this knowledge of the whole genome sequence of *A. thaliana*, as discussed by

[1] The University of Tokyo, Graduate School of Science, Department of Biological Sciences, 7-3-1 Hongo, Bunkyo-ku, Tokyo 156-0042, Japan
[2] Kazusa DNA Research Institute, 2-6-7 Kazusa-kamatari, Kisarazu, Chiba 292-0818, Japan

Biotechnology in Agriculture and Forestry, Vol. 52
Nagata/Tabata (Eds.) Brassicas and Legumes

Lakshmikumaran et al. in Chapter 4. In fact, analysis of most of the cultivated brassicas, which have complex genome compositions, have benefited from the detailed information on *A. thaliana*.

Based on the information on flowering of *A. thaliana*, exploration of flowering-related genes in brassicas is made possible, as demonstrated by Osborn and Lukens in Chapter 5. Genetic loci that are related to disease resistance to nematodes (which cause the formation of club roots) are also handled with this basic knowledge, which is discussed by Voorrips et al. in Chapter 6. Incompatibilities of *Brassica* species are analyzed by Nishio and Sato in Chapter 7.

By using genetic probes that have been identified in the genome of *A. thaliana*, one can identify the location of respective genes at the chromosome level, which is demonstrated in several members of the *Brassica* family by Fukui in Chapter 8. It is seen that the basic knowledge gained on *A. thaliana* genome allows interpretation of genome features among all the angiosperms.

Meanwhile, the genome sequence of two model legumes is currently a hot issue, as the nitrogen fixation by the symbiosis of legumes with *Rhizobia* is undoubtedly important in preserving the sustainable natural environments on earth. To this aim, the genome sequence of *Lotus japonicus* has been probed and discussed by Sato et al. in Chapter 9.

Furthermore, the introduction of a new accession of "Miyakojima" which has been collected at Miyakojima, the southernmost island in Japan, into *L. japonicus*, is described and its benefits discussed by Kawaguchi and Nishimura in Chapter 10. Genetic linkage maps and genetic markers on this species are also described by Harada et al. in Chapter 11. Construction of high density genetic mapping of *L. japonicus* is described by Kawasaki et al. in Chapter 12. Characterization of the genome structure of the soil-living bacterium *Mesorhizobium loti*, the symbiotic partner to *L. japonicus*, is described by Kaneko et al. in Chapter 13.

In Chapter 14, the genome structure of the other model legume of *Medicago truncatula* is described as another representative of nitrogen fixation legume by Huguet. Kalo et al. present and characterize certain genetic features of *Medicago sativa* in Chapter 15. The determination of the whole genome sequence of the symbiotic partner to this species, *Sinorhizobium meliloti*, has recently been published by Galibert et al. (2001).

From a more practical point of view, Winter et al. describe the introduction of molecular marker-aided resistance of chickpea to diseases in Chapter 16. Again, with these leguminous species, characterization of genome organization using several molecular markers is included by Fukui in Chapter 7.

The overall intent of this volume, is to describe what has been learned in two categories of brassicas and legumes of their genome structures and to the application thereof in plant breeding. This work will be complemented by a similar treatment of other crop species in future volumes.

References

Galibert F, Finan TM, Long SR, Pühler A, Abola P, Ampe F, Barloy-Hubler F, Barnett MJ, Becker A, Boistard P, Bothe G, Boutry M, Bowser L, Buhrmester J, Cadieu E, Capela D, Chain P, Cowie A, Davis RW, Dréano S, Federspiel NA, Fisher RF, Gloux S, Godrie T, Goffeau A, Golding B, Gouzy J, Gurjal M, Hernandez-Lucas I, Hong A, Huizar L, Hyman RW, Jones T, Kahn D, Kahn ML, Kalman S, Keating DH, Kiss E, Komp C, Lelaure V, Masuy D, Palm C, Peck MC, Pohl TM, Portetelle D, Purnelle B, Ramsperger U, Surzycki R, Thébault P, Vandenbol M, Vorhölter FJ, Weidner S, Wells DH, Wong K, Yeh K-C, Batut J (2001) The composite genome of the legume symbiont *Sinorhizobium meliloti*. Science 293:668–672

Ku H-M, Vision T, Liu J, Tanksley SD (2000) Comparing sequenced segments of the tomato and *Arabidopsis* genome: large-scale duplication followed by selective gene loss creates a network of syntheny. Proc Natl Acad Sci USA 97:9170

Meyerowitz EM, Pruitt RE (1985) *Arabidopsis thaliana* and plant molecular genetics. Science 229:1214–1218

The Arabidopsis Genome Initiative (2000) Analysis of the genome sequence of the flowering plant *Arabidopsis thaliana*. Nature 408:796–815

2 Genome Analysis of a Flowering Plant, *Arabidopsis thaliana*

T. Kato, E. Asamizu, Y. Nakamura, and S. Tabata

1 Introduction

Advances in DNA cloning and sequencing technologies have allowed the performance of comprehensive analysis of genetic information in various flowering plants of biological and agronomical importance. Among them, *Arabidopsis thaliana*, a member of the *Brassica* family, was chosen as a plant most suitable for genomic sequencing (Goodman et al. 1995; Meinke et al. 1998), because the estimated genome size of 125–130 Mb is the smallest among known higher plants and the content of repetitive sequences was assumed to be low. Its short life cycle (average 60 days) and prodigious seed production are the characteristics which make this small plant an ideal model organism in which to analyze metabolism, development, stress responses, and disease resistance in all the flowering plants.

The initial phase of the genome analysis began in 1990, which included the collection of expressed sequence tags (ESTs) (Höfte et al. 1993; Newman et al. 1994; Cooke et al. 1996; Asamizu et al. 2000), generation of genomic libraries (Choi et al. 1995; Creusot et al. 1995; Liu et al. 1995, 1999; Mozo et al. 1998) and physical maps of five chromosomes (Zachgo et al. 1996; Schmidt et al. 1996, 1997; Kotani et al. 1997; Camilleri et al. 1998; Sato et al. 1998, 1999). As the second phase of the genome analysis, a multinational genome sequencing project was initiated in August 1996, aimed at determining the nucleotide sequence of the entire *Arabidopsis* genome within 8 years. The project was coordinated by the Arabidopsis Genome Initiative (AGI) formed by six sequencing groups from Japan, EU, and the United States. Improvement of sequencing technologies accelerated the project, and completion of *A. thaliana* genome sequencing was officially announced in December 2000 (The Arabidopsis Genome Initiative 2000).

A large quantity of information on the genome and gene structures, together with hundreds of thousands of ESTs and corresponding cDNA clones, have changed the strategy of genetic analysis in this plant. Identification and isolation of causative genes of a variety of mutants have been dramatically accelerated by utilizing the sequence information of the genes. In addition, generation of a large number of mutants by insertion of T-DNA or

Kazusa DNA Research Institute, 2-6-7 Kazusa-kamatari, Kisarazu, Chiba 292-0818, Japan

Biotechnology in Agriculture and Forestry, Vol. 52
Nagata/Tabata (Eds.) Brassicas and Legumes

transposons followed by systematic characterization has become the most popular and promising approach for assignment of gene functions. Moreover, new approaches with cutting edge technologies such as transcriptome and proteome analyses have become available.

In this chapter, details of the genome analysis of *A. thaliana* including EST collection and genome sequencing, as well as the characteristic features of the genome revealed by the genome analysis, are summarized. Status of a genome-wide analysis of gene function, so-called functional genomics, is also described.

2 EST Analyses in *Arabidopsis thaliana*

An EST analysis is performed to make a catalogue of expressed portions of the genome by developing anonymous partial cDNA sequences. In an EST analysis, synthesized cDNA is directionally cloned into a vector and a single-pass sequence is read either from the 5′-end or 3′-end of the cDNA. A 300–500-bp-long sequence thus obtained is sufficient to identify a gene by similarity search against the public databases. The EST information is useful in understanding the expression levels or expression properties of certain genes. Since this is obviously an efficient way to rapidly generate gene sequence data, it has been applied in many organisms. As of January 2002, the EST database in GenBank (http://www.ncbi.nlm.nih.gov/dbEST/) contains 10,067,985 entries from more than 380 different organisms, and the number is continuously growing.

As for *A. thaliana*, more than 110,000 EST sequences have been deposited in the database, but no notable increase in the number has been observed since the completion of its genome sequencing at the end of 2000 (The Arabidopsis Genome Initiative 2000). It has now become the fifth largest number among plant species, behind the rapidly increasing crop ESTs such as soybean, tomato, and maize. However, some of the early *A. thaliana* EST projects have no doubt played a leading role in accelerating the large-scale sequencing efforts in plants.

Early important projects are represented by those promoted by two consortia of laboratories, one in France and the other in the United States. In 1993, a consortium of research groups in France made the first report of an EST analysis in *A. thaliana* (Höfte et al. 1993). In this and the following paper (Cooke et al. 1996), they reported the generation of 5′- and 3′-end sequences of approximately 6,000 nonredundant clones. The US group produced 31,000 ESTs, mainly from a single library made from a mixture of mRNAs from four different tissues (Newman et al. 1994).

The EST sequences first served as a useful tool to identify gene-coding regions in the genome sequence. To facilitate this process by complementing thus far available EST data, another large-scale EST analysis was launched in Japan. The project by Kazusa DNA Research Institute generated 14,026 5′-

end ESTs and 39,207 3′-end ESTs from two types of cDNA libraries, normalized and size-selected libraries made from five different samples, aboveground organs, roots, flower buds, green siliques, and liquid-cultured seedlings (Asamizu et al. 2000). This project resulted in establishing 12,028 nonredundant groups revealed by a simple clustering of the 3′-ESTs. The EST sequence data and search results of individual cDNA clones are available at the web site: http://www.kazusa.or.jp/en/plant/arabi/EST/.

Toward understanding of the whole genetic system carried by this model plant, it is important to make an effective use of the sequence information as well as clones accumulated in the EST projects. The effort includes large-scale gene expression analysis using arrayed cDNA clones, which is now performed extensively as described later.

3 Genome Sequencing

Genome sequencing of *A. thaliana* was initially planned independently by research groups in Japan, the EU, and the United States, then coordinated by AGI after August 1996. Therefore, the strategy taken for the project was similar but not identical for each of six members. The clone-by-clone method rather than the whole-genome shotgun method was adopted for sequencing, which consequently allowed accumulation of high-quality sequence data. Bacterial artificial chromosomes (BACs), P1s, and transformation-competent artificial chromosomes (TACs) with large-inserts were used as the primary substrates for sequence determination. To access the relative positions of the clones and to select the clones to sequence, restriction fragment fingerprinting analysis of BAC clones, hybridization or polymerase chain reaction (PCR) of sequence tagged sites generated on BACs, P1s, and TACs were performed. End sequences of BACs, P1s, and TACs were also used to extend the contigs from the marker-anchored clones in both directions.

Random libraries were generated from each clone and sequence files were accumulated with the redundancy of 8 to 10. The random sequences were assembled using standard techniques with a computer program phred-phrap (Phil Green, University of Washington, Seattle, USA: http://www.phrap.org/). Accuracy of the finished sequences was estimated to be 99.99–99.999%, which is the highest among the genomes fully sequenced up to now. The total length of sequenced regions that extend from either the telomeres or ribosomal DNA to the 180-bp centromeric repeats was 115,409,949 bp. By adding 10 Mb ribosomal DNA repeat regions and 1–2 Mb each for unsequenced centromeric regions, the total genome size of *A. thaliana* can be estimated to be 130–135 Mb.

4 Characterization of Genomic Information in *A. thaliana*

Annotation is the description of structural and biological features of a sequence. The primary computational tools available for annotation of nucleotide sequences are database search programs and coding-region prediction algorithms. In the *A. thaliana* genome sequencing project, all the AGI sequencing teams adopted similar annotation procedures, that is, a combination of comparison with EST and known protein sequences, *de novo* gene-finding methods, and manual adjustment of the resultant data.

The following is an example of annotation procedure taken by our group. Nucleotide sequences of the *A. thaliana* genome were subjected to similarity searches against the nonredundant protein database, nr, with use of the BLASTX algorithm in the BLAST2 program (Altschul et al. 1997). To identify transcribed regions and structural RNA genes, the BLASTN was used to compare nucleotide sequences with the EST and RNA gene databases. Potential exons of protein genes were predicted by the computer programs GRAIL (Uberbacher and Mural 1991) and GENSCAN (Burge and Karlin 1997). For localization of exon-intron boundaries, donor/acceptor sites for splicing were predicted by NetGene2 (Hebsgaard et al. 1996) and SplicePredictor (Brendel and Kleffe 1998). Transfer RNA genes and their structures were predicted by the tRNAscan-SE program (Lowe and Eddy 1997) and manual inspection of the BLASTN similarity search result. All the results thus obtained were then parsed and loaded into a web-based display system called *Arabidopsis* Genome Displayer [http://www.kazusa.or.jp/kaos/]. The process mentioned above is generally called "first-pass annotation".

The entire *A. thaliana* genome was divided into three portions and first-pass annotation was performed separately by three information centers; Kazusa (Japan), TIGR (The Institute for Genomic Research, USA), and MIPS (Munich Information center for Protein Sequences, Germany). The pilot annotation project indicated that approximately 80% of the gene models generated by three centers were consistent. Deduced genes were annotated functionally by similarities to known sequences, then classified automatically using the PEDANT system (Frishman et al. 2001) into categories with respect to different biological roles.

As a consequence, a total of 25,498 potential protein-coding genes were deduced in the 115,409,949 bp regions of five chromosomes of *A thaliana* (The Arabidopsis Genome Initiative 2000). The average density of the presumptive protein-coding genes in the annotated regions was estimated to be 1 gene per 4.5 kb. The average number per gene and the average length of the exons and introns were 5.1 and 250 bp, 4.1 and 168 bp, respectively, as shown in Table 1. Overall features such as gene density, expression levels and repeat distribution are very consistent across the five chromosomes.

Sixty-nine percent of protein-coding genes deduced in the *A. thaliana* genome had sequence similarity to proteins of annotated functions in other organisms, whereas only 9% of the genes have been characterized experimentally. Approximately 30% of the predicted proteins, including those of

Table 1. Features of *Arabidopsis thaliana* genome

(a) The DNA molecules	
Length	115,409,949 bp
Number of genes	25,498
Gene density	4.5 kb per gene
Average gene length	2,011 bp
Average peptide length	434 aa
Exons	
Average size	250 bp
Introns	
Average size	168 bp
Average number per gene	4.1
(b) The proteome	
Cellular metabolism	22.5%
Transcription	16.9%
Plant defense	11.5%
Signaling	10.4%
Growth	11.7%
Protein fate	9.9%
Intracellular transport	8.3%
Transport	4.8%
Protein synthesis	4.1%
Total	69.9%

The features are based on the publication by the Arabidopsis Genome Initiative (2000). Automated classification was done by the PEDANT analysis system.

either plant-specific or similarity to genes of unknown function from other organisms, could not be assigned to functional categories.

Transposons account for at least 10% of the genome. Most of pseudogenes were related to proteins found in retrotransposons and were located near the centromeres. Five hundred and eighty-nine cytoplasmic transfer RNA genes and 27 organelle-derived tRNAs and 13 tRNA pseudogenes were identified by tRNAscan-SE and similarity search results.

Notable features of the *A. thaliana* genome are as follow:

1. 4,432 genes (17% of the total genes in the genome) form 1,661 tandem arrays each containing 2–23 adjacent members.
2. 808 proteins are more similar to proteins from cyanobacteria than from any other organisms.
3. 31 duplicated gene clusters longer than 0.5 Mb, which encompass 67.9 Mb, 60% of the genome, were identified. The duplicated regions contain 17,193 genes (66% of the genes).
4. Plastid DNA insertions into nuclei (17 insertions totaling 11 kb) containing entire and fragments of genes and intergenic regions were identified.

Mitochondrial DNA insertions into nuclei, a total of 13 insertions totaling 7 kb and a large insertion (270 kb, 74% of the mitochondrial genome) into chr 2, are also found. The results suggest exchange of genetic materials among three genomes in the cells.

5 Large-Scale Functional Analyses

Upon completion of the *A. thaliana* genome sequence, one of the next goals in plant biology is to elucidate the function of all of the assigned 25,000 genes. Because more than 30% of the genes show no sequence similarity to genes of known or hypothesized function, it is necessary to use different approaches for dissecting gene function or find new tools for investigating gene function. In this section, examples of large-scale analyses of gene function in the post-sequencing era are described.

5.1 Profiling of Transcription by Array Technologies

Microarrays are one of the effective methods for analyzing transcriptional profiles of thousands of genes at one time. DNA samples are first deposited on a grass slide at high density. These DNA samples are derived from either cDNA clones or oligonucleotides corresponding to gene sequences. Extracted mRNAs from two different samples are labeled with different fluorescent nucleotide derivatives, Cy3 and Cy5 for example, respectively, then hybridized with the DNAs on the grass slide. Intensity of the labeled probes is scanned by confocal microscopy and analyzed with computer programs. Oligonucleotide arrays cost a great deal, but allow discrimination among members of gene families if properly designed. On the other hand, cDNA arrays have the advantage of direct connection with the EST analysis, which is a common and popular approach, and the cost is low. Because transcriptional patterns of thousands of genes can be analyzed simultaneously with microarrays, clues for understanding functions of a large number of hypothetical genes revealed by genome sequencing projects can potentially be obtained.

Several papers using *A. thaliana* microarrays have been published to date (Table 2). To facilitate utilization of microarray systems, two publicly supported microarray service facilities have been organized so far. One is the Arabidopsis Functional Genomics Consortium (AFGC: http://afgc.stanford.edu/) in the United States (Wisman and Ohlrogge 2000), funded by National Science Foundation; the other is the Genomic Arabidopsis Resources Network (GARNet: http://www.york.ac.uk/res/garnet/garnet.htm) in the United Kingdom. The microarray service facility of GARNet is part of the Biotechnology and Biological Sciences Research Counsel (BBSRC) Investigating Gene Function initiative. The AFGC provided the year 2000 array which contains 11,500 nonredundant cDNAs derived from the Michigan State University EST collections. It was announced that the year 2001 array was going to consist of 14,700 nonredundant cDNAs which include new sets of EST clones

Table 2. Transcription profiling by DNA arrays

Conditions	Array type	Number of clones	Reference
Root, leaf, flower bud, and mature flower	cDNA	1,443	Ruan et al. (1998)
Nitrate response	cDNA	5,524	Wang et al. (2000)
Defense response	cDNA	2,375	Schenk et al. (2000)
Circadian clock	Oligonucleotide	8,200 genes represented	Harmer et al. (2000)
Systemic acquired resistance	cDNA	10,000	Maleck et al. (2000)
Developing seeds	cDNA	2,715	Girke et al. (2000)
Systemic acquired resistance	cDNA	9,861	Petersen et al. (2000)
Diurnal and circadian-regulated genes	cDNA	11,521	Schaffer et al. (2001)
Drought and cold stresses	cDNA	1,300	Seki et al. (2001)
Salt stress tolerance	cDNA	9,212	Bohnert et al. (2001)
Oxidative stress	cDNA	11,000	Desikan et al. (2001)
Iron deficiency stress	cDNA (nylon membrane)	16,128	Thimm et al. (2001)
Light control of development	cDNA	9,216	Ma et al. (2001)
Dst mutants	cDNA	11,521	Perez-Amador et al. (2001)

and DNA fragments amplified from genomic DNA. If investigators send RNA samples to AFGC, the AFGC performs labeling of the probes, hybridization, and data collection. The data is made available to the public within 3 months through the Stanford Microarray Database (SMD: http://genome-www5.stanford.edu/MicroArray/SMD/).

The GARNet recently began a transcriptomics service using the Affymetrix Arabidopsis genechip. This chip holds approximately 8,000 genes represented by multiple oligonucleotides derived from portions of each gene sequence. As a part of GARNet consortium, Nottingham Arabidopsis Stock Centre (NASC: http://nasc.nott.ac.uk/) will establish an *A. thaliana* DNA array covering the entire gene components. This array is made from Gene Specific Tag (GST) clones produced by a Complete Arabidopsis Transcriptome MicroArray (CATMA). NASC will provide both the array itself in both micro (slide) and macro (filter) format to the community, as well as a complete "labeling/ hybridization/ analysis/data-mining" service. The array data will be released through integration with the Arabidopsis Genome Resource (AGR) immediately or within 3 to 6 months.

5.2 Analysis of Gene Function by Systematic Gene Disruption

Gene disruption is an effective method for obtaining information on gene functions in organisms for which gene sequences and the DNA transformation technology are available. In *A. thaliana*, a reverse genetic approach is especially useful because the complete genomic sequence is available. However, whereas this approach has been commonly used in bacteria and animals, disruption of genes by homologous recombination (gene targeting) is not feasible in flowering plants up to present. Instead, insertional mutagenesis has become a popular and effective approach in analyzing gene function by creating knockout plants. A large number of insertional lines have been produced all over the world (Table 3) by introduction of T-DNAs or transposons (gene knockout). Many of the lines are available through Arabidopsis Biological Resource Center (ABRC: http://www.biosci.ohio-state.edu/~plantbio/Facilities/abrc/abrchome.htm) or NASC.

Knockout lines can be selected from a bulk of mutant populations by PCR-based screening with a border primer of inserted DNA and a gene specific primer using DNA pools derived from the insertional lines as a template (Krysan et al. 1999). Isolation and characterization of insertional lines for unique and redundant genes have been reported (Winkler et al. 1998; Meissner et al. 1999). In the United States, the Arabidopsis Knockout Facility [http://www.biotech.wisc.edu/Arabidopsis/] was established at the University of Wisconsin (Madison) in 1999 as a part of the AFGC (Sussman et al. 2000). In this facility, 60,480 T-DNA insertion lines and 72,960 activation-Tag lines are available for screening.

Information of the flanking sequences of the inserted DNAs allows us to find knockout lines computationally. Sequencing of the flanking regions of the insertion points has been carried out in a small scale and the data have been

Table 3. The type and number of insertional lines generated

Name of the lines	Effect by the insertion	Type of the inserted elements	Number of the lines	Reference
Feldmann	Disruption	T-DNA	8,000	Feldmann (1991)
INRA-Versailles	Promoter trap	T-DNA	4,000	Bechtold et al. (1993)
Cold Spring Harbor	Enhancer and gene traps	*Ds*	491	Sundaresan et al. (1995)
Jack	Enhancer trap	T-DNA	11,370	Campisi et al. (1999)
SLAT	Disruption	*dSpm*	48,000	Tissier et al. (1999)
ITS	Disruption	*En/I*	2,592	Speulman et al. (1999)
Weigel	Activation	T-DNA	25,000	Weigel et al. (2000)
Dellaporta	Disruption	T-DNA	38,000	Galbiati et al. (2000)
AFGC	Enhancer trap	T-DNA	60,480	Sussman et al. (2000)

reported (Parinov et al. 1999). A program that creates a sequence-indexed library of insertion mutations in the *A. thaliana* genome was initiated at the Salk Institute Genomic Analysis Laboratory (SIGnAL: http://signal.salk.edu/) in 2001. In this program, over 140,000 insertion sites will be determined within 2 years. The data of insertion sites are available via their website, T-DNA Express (http://signal.salk.edu/cgi-bin/tdnaexpress), and a BLAST search of the insertion sites can be performed in web site of the Arabidopsis Information Resource (TAIR: http://www.arabidopsis.org/). About 6,000 new lines will be sequenced and subsequently released by the ABRC and the NASC each month.

Activation Tagging is an effective method for analyzing gene function. In this method, as transcriptional enhancer sequences are introduced into the genome randomly, the dominant effects generated by activated transcription near the insertion sites are analyzed. Hence, the effects that do not appear in the insertional loss-of-function mutation may be observed. The genes related to cytokinin signal transduction (Kakimoto 1996) or flowering time (Kardailsky et al. 1999) were isolated by this method.

Gene and enhancer traps are further attractive approaches for obtaining clues to gene function by looking at gene expression patterns. Reporter genes such as *uidA* [β-glucuronidase (GUS)], luciferase, or green fluorescent protein (GFP) are introduced randomly into the genome, followed by monitoring the reporter gene expression in plant tissues or cells (Springer 2000). This method also make it possible to analyze functions of essential genes, because expression patterns can be analyzed in the heterozygous plants even in early lethal mutants. The *PROLIFERA* (*PRL*) gene that may be required for the initiation of DNA replication (Springer et al. 1995) and the *FRUITFULL* (*FUL*) gene required for elongation of siliques (Gu et al. 1998) were identified using this method. Another use of gene and enhancer traps is the generation of tissue- or cell-specific markers that are useful for analyzing plant development. Changes in expression patterns of the markers would suggest that the developmental pattern may be altered in tissues or cells of the mutants. These markers are reported in some tissues or cells during the development of lateral root, leaf, flower development and embryogenesis (Topping et al. 1994; Malamy and Benfey 1997). Insertional mutagenesis resources in *A. thaliana* are listed in a website, Tagging Gateway at http://www.jic.bbsrc.ac.uk/staff/michael-bevan/atis/Resources1.htm.

5.3 Analysis of Gene Function by RNA-Interference

Post-transcriptional gene silencing (PTGS) by anti-sense or co-suppression constructs has long been utilized as a technology for repressing gene expression in plant genetics. A problem with this technology is that degree of transcriptional repression varies among individual transgenic plants, therefore a number of transgenic lines are required for each gene. Furthermore, silencing could affect other genes among the gene families, which makes it difficult to connect the phenotype and its causative gene. Recently, a related technology named RNA-interference (RNAi) has become available in *A. thaliana*. RNAi utilizes specific RNA degradation by introduction of double-stranded RNA (dsRNA), and this method has been commonly used for repression of specific gene expression in *Caenorhabditis elegans* and *Drosophila*. In *A. thaliana*, there were several reports on RNAi and gene specific interferences (Chuang and Meyerowitz 2000). Recently, it has been shown that self-complementary "hairpin" RNA (hpRNA) had the ability to cause gene silencing efficiently. Intron-containing hairpin RNA constructs (ihpRNA) are especially effective in silencing genes (Wesley et al. 2001).

5.4 Map-Based Cloning Using SNP and INDEL Markers

Map-based cloning has been a commonly used procedure for gene isolation in *A. thaliana*, but in a practical sense it is a time-consuming process to narrow down the genes responsible for mutations. One of the reasons for the slowness is the paucity of PCR-based molecular markers available for fine mapping. The complete sequence of the *A. thaliana* genome provides the opportunity to create a large number of new molecular markers. A total of 56,670 polymorphisms, including single-nucleotide polymorphisms (SNPs) and small insertion/deletion (INDELs) between the Columbia and Landsberg *erecta* accessions, were released by Cereon Genomics (http://www.arabidopsis.org/Cereon/index.html). These polymorphisms will be useful sources for generating new molecular markers to dramatically accelerate map-based cloning.

5.5 TILLING Project

Characterization of mutants caused by a point mutation is important because even mutations in essential genes may be amenable to isolation and analysis as weak alleles or temperature sensitive mutants. The targeting-induced local lesions in genomes (TILLING) reverse genetic strategy is an attractive system for large-scale screening of chemically induced mutations in *A. thaliana* (McCallum et al. 2000; Colbert et al. 2001). In the TILLING system, total genomic DNAs are prepared from a large population of the M2 generation of ethylmethanesulfonate (EMS) mutagenized plants, which are used as templates for PCR amplification with gene-specific primers. When PCR products are incubated with an endonuclease that preferentially cleaves mismatches in heteroduplexes between wild type and mutants, point mutations can then be detected by separating the cleavage products by gel electrophoresis in a state of denaturation. The *A. thaliana* TILLING Project (ATP) has been set up for the community to deliver EMS-induced mutations in genes of interest (http://tilling.fhcrc.org:9366/).

5.6 Proteome Analysis

Proteome analysis is an approach for monitoring gene expression at the translational level. Protein mixtures prepared from different tissues, plants exposed to various environmental conditions, plants of different mutant backgrounds, etc., are separated by two-dimensional (2-D) gel electrophoresis, followed by extraction of proteins from individual spots. Each protein is characterized by partial amino acid sequencing or mass spectrometry and subsequently connected to corresponding genes based on the structural information. This approach is of course effective especially when the structures of the entire gene components are known. In *A. thaliana*, the proteome analyses have been reported for the plasma membrane (Santoni et al. 1998), chloroplast (Vener et al. 2001), mitochondria (Kruft et al. 2001; Millar et al. 2001), pollen coat (Mayfield et al. 2001), and germinating seed (Gallardo et al. 2001).

6 Conclusions and Perspectives

The nucleotide sequence of the entire genome of *A. thaliana* provides both the information on the genome and gene structures carried by a flowering plant and a basis for the subsequent analyses of functions of individual genes and of the network of regulation. As described in this chapter, the complexity of the genes and the gene organization in the *A. thaliana* genome is greater than was expected when the genome sequencing project began, implying that complicated events have taken place in connection with the genome of this plant and of other plant species during their evolution. Intensive analysis of the genome structure holds promise of revealing the mechanisms of generation and evolution of the plant genes and the genomes. Another direction that can be taken after having the entire genome structure in hand is to utilize it to look at the functional aspect of the genome. Comprehensive approaches for studying gene functions at the genome-wide level, summarized in this chapter, together with the approaches of conventional genetics and physiology, likewise promise to accelerate acquisition of information on gene functions in *A. thaliana*. It is natural to expect that a significant amount of knowledge on the genetic system in this plant will be accumulated in a short period of time.

Improvement of technologies has made it possible to expand the targets for genome analysis from rather limited model plants to a wide variety of plant species that have their own characteristics and advantages in studying individual phenomena. A large number of DNA markers, ESTs, and genomic libraries are drastically changing the strategy of plant genetics, and one can simply transfer knowledge on gene structures and functions obtained from one plant species to another by using the nucleotide sequences as a common language. Application of accumulated knowledge from *A. thaliana* to other plant species, such as those of agronomical importance, through this common language has every appearance of being one of the best ways to obtain universal knowledge of the genetic systems in all the plant species on this planet.

References

Altschul SF, Madden TL, Schaffer AA, Zhang J, Zhang Z, Miller W, Lipman DJ (1997) Gapped BLAST and PSI-BLAST: a new generation of protein database search programs. Nucleic Acids Res 25:3389–3402

Asamizu E, Nakamura Y, Sato S, Tabata S (2000) A large scale analysis of cDNA in *Arabidopsis thaliana*: generation of 12,028 non-redundant expressed sequence tags from normalized and size-selected cDNA libraries. DNA Res 7:175–180

Bechtold N, Ellis J, Pelletier G (1993) *In planta Agrobacterium* mediated gene transfer by infiltration of adult *Arabidopsis thaliana* plants. C R Acad Sci Paris Life Sci 316:1194–1199

Bohnert HJ, Ayoubi P, Borchert C, Bressan RA, Burnap RL, Cushman JC, Cushman MA, Deyholos M, Fischer R, Galbraith DW, Hasegawa PM, Jenks M, Kawasaki S, Koiwa H, Kore-eda S, Lee BH, Michalowski CB, Misawa E, Nomura M, Ozturk N, Postier B, Prade R, Song CP, Tanaka Y, Wang H, Zhu JK (2001) A genomics approach towards salt stress tolerance. Plant Physiol Biochem 39:295–311

Brendel V, Kleffe J (1998) Prediction of locally optimal splice sites in plant pre-mRNA with applications to gene identification in *Arabidopsis thaliana* genomic DNA. Nucleic Acids Res 26:4748–4757

Burge C, Karlin S (1997) Prediction of complete gene structures in human genomic DNA. J Mol Biol 268:78–94

Camilleri C, Lafleuriel J, Macadre C, Varoquaux F, Parmentier Y, Picard G, Caboche M, Bouchez D (1998) A YAC contig map of *Arabidopsis thaliana* chromosome 3. Plant J 14:633–642

Campisi L, Yang Y, Yi Y, Heilig E, Herman B, Cassista AJ, Allen DW, Xiang H, Jack T (1999) Generation of enhancer trap lines in *Arabidopsis* and characterization of expression patterns in the inflorescence. Plant J 17:699–707

Choi S, Creelman RA, Mullet JE, Wing RA (1995) Construction and characterization of a bacterial artificial chromosome library of *Arabidopsis thaliana*. Weeds World 2:17–20

Chuang C-F, Meyerowitz EM (2000) Specific and heritable genetic interference by double-stranded RNA in *Arabidopsis thaliana*. Proc Natl Acad Sci USA 97:4985–4990

Colbert T, Till BJ, Tompa R, Reynolds S, Steine MN, Yeung AT, McCallum CM, Comai L, Henikoff S (2001) High-throughput screening for induced point mutations. Plant Physiol 126:480–484

Cooke R, Raynal M, Laudié M, Grellet F, Delseny M, Morris PC, Guerrier D, Giraudat J, Quigley F, Clabault G, Li YF, Mache R, Krivitzky M, Gy IJ, Kreis M, Lecharny A, Parmentier Y, Marbach J, Fleck J, Clément B, Philipps G, Hervé C, Bardet C, Tremousaygue D, Lescure B, Lacomme C, Roby D, Jourjon MF, Chabrier P, Charpenteau JL, Desprez T, Amselem J, Chiapello H, Höfte H (1996) Further progress towards a catalogue of all *Arabidopsis* genes: analysis of a set of 5000 non-redundant ESTs. Plant J 9:101–124

Creusot F, Fouilloux E, Dron M, Lafleuriel J, Picard G, Billault A, Paslier DL, Cohen D, Chaboute M-E, Durr A, Fleck J, Gigot C, Camilleri C, Bellini C, Caboche M, Bouchez D (1995) The CIC library: a large insert YAC library for genome mapping in *Arabidopsis thaliana*. Plant J 8:763–770

Desikan R, A-H-Mackerness S, Hancock JT, Neill SJ (2001) Regulation of the *Arabidopsis* transcriptome by oxidative stress. Plant Physiol 127:159–172

Feldmann KA (1991) T-DNA insertion mutagenesis in *Arabidopsis*: mutational spectrum. Plant J 1:71–82

Frishman D, Albermann K, Hani J, Heumann K, Metanomski A, Zollner A, Mewes HW (2001) Functional and structural genomics using PEDANT. Bioinformatics 17:44–57

Galbiati M, Moreno MA, Nadzan G, Zourelidou M, Dellaporta SL (2000) Large-scale T-DNA mutagenesis in *Arabidopsis* for functional genomic analysis. Funct Integr Genomics 1:25–34

Gallardo K, Job C, Groot SP, Puype M, Demol H, Vandekerckhove J, Job D (2001) Proteomic analysis of *Arabidopsis* seed germination and priming. Plant Physiol 126:835–848

Girke T, Todd J, Ruuska S, White J, Benning C, Ohlrogge J (2000) Microarray analysis of developing *Arabidopsis* seeds. Plant Physiol 124:1570–1581

Goodman HM, Ecker JR, Dean C (1995) The genome of *Arabidopsis thaliana*. Proc Natl Acad Sci USA 92:10831–10835

Gu Q, Ferrandiz C, Yanofsky MF, Martienssen R (1998) The *FRUITFULL* MADS-box gene mediates cell differentiation during *Arabidopsis* fruit development. Development 125:1509–1517

Harmer SL, Hogenesch JB, Straume M, Chang H-S, Han B, Zhu T, Wang X, Kreps JA, Kay SA (2000) Orchestrated transcription of key pathways in *Arabidopsis* by the circadian clock. Science 290:2110–2113

Hebsgaard SM, Korning PG, Tolstrup N, Engelbrecht J, Rouze P, Brunak S (1996) Splice site prediction in *Arabidopsis thaliana* pre-mRNA by combining local and global sequence information. Nucleic Acids Res 24:3439–3452

Höfte H, Desprez T, Amselem J, Chiapello H, Caboche M, Moisan A, Jourjon MF, Charpenteau JL, Berthomieu P, Guerrier D, Giraudat J, Quigley F, Thomas F, Yu DY, Mache R, Raynal M, Cooke R, Grellet F, Delseny M, Parmentier Y, Marcillac G, Gigot C, Fleck J, Philipps G, Axelos M, Bardet C, Tremousaygue D, Leacure B (1993) An inventory of 1152 expressed sequence tags obtained by partial sequencing of cDNAs from *Arabidopsis thaliana*. Plant J 4:1051–1061

Kakimoto T (1996) CKI1, a histidine kinase homolog implicated in cytokinin signal transduction. Science 274:982–985

Kardailsky I, Shukla VK, Ahn JH, Dagenais N, Christensen SK, Nguyen JT, Chory J, Harrison MJ, Weigel D (1999) Activation tagging of the floral inducer *FT*. Science 286:1962–1965

Kotani H, Sato S, Fukami M, Hosouchi T, Nakazaki N, Okumura S, Wada T, Liu Y-G, Shibata D, Tabata S (1997) A fine physical map of *Arabidopsis thaliana* chromosome 5: construction of a sequence-ready contig map. DNA Res 4:371–378

Kruft V, Eubel H, Jansch L, Werhahn W, Braun HP (2001) Proteomic approach to identify novel mitochondrial proteins in *Arabidopsis*. Plant Physiol 127:1694–1710

Krysan PJ, Young JC, Sussman MR (1999) T-DNA as an insertional mutagen in *Arabidopsis*. Plant Cell 11:2283–2290

Liu Y-G, Mitsukawa N, Vazquez-Tello A, Whittier RF (1995) Generation of a high-quality P1 library of *Arabidopsis* suitable for chromosome walking. Plant J 7:351–358

Liu Y-G, Shirano Y, Fukaki H, Yanai Y, Tasaka M, Tabata S, Shibata D (1999) Complementation of plant mutants with large genomic DNA fragments by a transformation-competent artificial chromosome vector accelerates positional cloning. Proc Natl Acad Sci USA 96:6535–6540

Lowe TM, Eddy SR (1997) tRNAscan-SE: a program for improved detection of transfer RNA genes in genomic sequence. Nucleic Acids Res 25:955–964

Ma L, Li J, Qu L, Hager J, Chen Z, Zhao H, Deng XW (2001) Light control of *Arabidopsis* development entails coordinated regulation of genome expression and cellular pathway. Plant Cell 13:2589–2607

Malamy JE, Benfey PN (1997) Organization and cell differentiation in lateral roots of *Arabidopsis thaliana*. Development 124:33–44

Maleck K, Levine A, Eulgem T, Morgan A, Schmid J, Lawton KA, Dangl JL, Dietrich RA (2000) The transcriptome of *Arabidopsis thaliana* during systemic acquired resistance. Nature Genet 26:403–410

Mayfield JA, Fiebig A, Johnstone SE, Preuss D (2001) Gene families from the *Arabidopsis thaliana* pollen coat proteome. Science 292:2482–2485

McCallum CM, Comai L, Greene EA, Henikoff S (2000) Targeting induced local lesions in genomes (TILLING) for plant functional genomics. Plant Physiol 123:439–442

Meinke DW, Cherry JM, Dean C, Rounsley SD, Koornneef M (1998) *Arabidopsis thaliana*: a model plant for genome analysis. Science 282:662–682

Meissner RC, Jin H, Cominelli E, Denekamp M, Fuertes A, Greco R, Kranz HD, Penfield S, Petroni K, Urzainqui A, Martin C, Paz-Ares J, Smeekens S, Tonelli C, Weisshaar B, Baumann E, Klimyuk V, Marillonnet S, Patel K, Speulman E, Tissier AF, Bouchez D, Jones JJD, Pereira A, Wisman E, Bevan M (1999) Function search in a large transcription factor gene family in *Arabidopsis*: assessing the potential of reverse genetics to identify insertional mutations in R2R3 *MYB* genes. Plant Cell 11:1827-1840

Millar AH, Sweetlove LJ, Giege P, Leaver CJ (2001) Analysis of the *Arabidopsis* mitochondrial proteome. Plant Physiol 127:1711–1727

Mozo T, Fischer S, Shizuya H, Altmann T (1998) Construction and characterization of IGF *Arabidopsis* BAC library. Mol Gen Genet 258:562–570

Newman T, de Bruijn FJ, Green P, Keegstra K, Kende H, McIntosh L, Ohlrogge J, Raikhel N, Somerville S, Thomashow M, Retzel E, Somerville C (1994) Genes galore: a summary of methods for accessing results from large-scale partial sequencing of anonymous *Arabidopsis* cDNA clones. Plant Physiol 106:1241–1255

Parinov S, Sevugan M, Ye D, Yang W-C, Kumaran M, Sundaresan V (1999) Analysis of flanking sequences from *Dissociation* insertion lines: a database for reverse genetics in *Arabidopsis*. Plant Cell 11:2263–2270

Perez-Amador MA, Lidder P, Johnson MA, Landgraf J, Wisman E, Green PJ (2001) New molecular phenotypes in the *dst* mutants of *Arabidopsis* revealed by DNA microarray analysis. Plant Cell 13:2703–2717

Petersen M, Brodersen P, Naested H, Andreasson E, Lindhart U, Johansen B, Nielsen HB, Lacy M, Austin MJ, Parker JE, Sharma SB, Klessig DF, Martienssen R, Mattsson O, Jensen AB, Mundy J (2000) *Arabidopsis* map kinase 4 negatively regulates systemic acquired resistance. Cell 103:1111–1120

Ruan Y, Gilmore J, Conner T (1998) Towards *Arabidopsis* genome analysis: monitoring expression profiles of 1400 genes using cDNA microarrays. Plant J 15:821–833

Santoni V, Rouquie D, Doumas P, Mansion M, Boutry M, Degand H, Dupree P, Packman L, Sherrier J, Prime T, Bauw G, Posada E, Rouze P, Dehais P, Sahnoun I, Barlier I, Rossignol M (1998) Use of a proteome strategy for tagging proteins present at the plasma membrane. Plant J 16:633–641

Sato S, Kotani H, Hayashi R, Liu Y-G, Shibata D, Tabata S (1998) A physical map of *Arabidopsis thaliana* chromosome 3 represented by two contigs of CIC YAC, P1, TAC and BAC clones. DNA Res 5:163–168

Sato S, Kaneko T, Kotani H, Hayashi R, Liu Y-G, Shibata D, Tabata S (1999) A sequence-ready contig map of the top arm of *Arabidopsis thaliana* chromosome 3. DNA Res 6:117–121

Schaffer R, Landgraf J, Accerbi M, Simon V, Larson M, Wisman E (2001) Microarray analysis of diurnal and circadian-regulated genes in *Arabidopsis*. Plant Cell 13:113–123

Schenk PM, Kazan K, Wilson I, Anderson JP, Richmond T, Somerville SC, Manners JM (2000) Coordinated plant defense responses in *Arabidopsis* revealed by microarray analysis. Proc Natl Acad Sci USA 97:11655–11660

Schmidt R, West J, Cnops G, Love K, Balestrazzi A, Dean C (1996) Detailed description of four YAC contigs representing 17 Mb of chromosome 4 of *Arabidopsis thaliana* ecotype Columbia. Plant J 9:755–765

Schmidt R, Love K, West J, Lenehan Z, Dean C (1997) Description of 31 YAC contigs spanning the majority of *Arabidopsis thaliana* chromosome 5. Plant J 11:563–572

Seki M, Narusaka M, Abe H, Kasuga M, Yamaguchi-Shinozaki K, Carninci P, Hayashizaki Y, Shinozaki K (2001) Monitoring the expression pattern of 1300 *Arabidopsis* genes under drought and cold stresses by using a full-length cDNA microarray. Plant Cell 13:61–72

Speulman E, Metz PLJ, van Arkel G, te Lintel Hekkert B, Stiekema WJ, Pereira A (1999) A two-component *enhancer-inhibitor* transposon mutagenesis system for functional analysis of the *Arabidopsis* genome. Plant Cell 11:1853–1866

Springer PS (2000) Gene traps: tools for plant development and genomics. Plant Cell 12:1007–1020

Springer PS, McCombie WR, Sundaresan V, Martienssen RA (1995) Gene trap tagging of *PROLIFERA*, an essential *MCM2-3-5*-like gene in *Arabidopsis*. Science 268:877–880

Sundaresan V, Springer P, Volpe T, Haward S, Jones JD, Dean C, Ma H, Martienssen R (1995) Patterns of gene action in plant development revealed by enhancer trap and gene trap transposable elements. Genes Dev 9:1797–1810

Sussman MR, Amasino RM, Young JC, Krysan PJ, Austin-Phillips S (2000) The *Arabidopsis* knockout facility at the University of Wisconsin-Madison. Plant Physiol 124:1465–1467

The Arabidopsis Genome Initiative (2000) Analysis of the genome sequence of the flowering plant *Arabidopsis thaliana*. Nature 408:796–815

Thimm O, Essigmann B, Kloska S, Altmann T, Buckhout TJ (2001) Response of *Arabidopsis* to iron deficiency stress as revealed by microarray analysis. Plant Physiol 127:1030–1043

Tissier AF, Marillonnet S, Klimyuk V, Patel K, Torres MA, Murphy G, Jones JDG (1999) Multiple independent defective *suppressor-mutator* transposon insertions in *Arabidopsis*: a tool for functional genomics. Plant Cell 11:1841–1852

Topping JF, Agyeman F, Henricot B, Lindsey K (1994) Identification of molecular markers of embryogenesis in *Arabidopsis thaliana* by promoter trapping. Plant J 5:895–903

Uberbacher EC, Mural RJ (1991) Locating protein-coding regions in human DNA sequences by a multiple sensor-neural network approach. Proc Natl Acad Sci USA 88:11261–11265

Vener AV, Harms A, Sussman MR, Vierstra RD (2001) Mass spectrometric resolution of reversible protein phosphorylation in photosynthetic membranes of *Arabidopsis thaliana*. J Biol Chem 276:6959–6966

Wang R, Guegler K, LaBrie ST, Crawford NM (2000) Genomic analysis of a nutrient response in *Arabidopsis* reveals diverse expression patterns and novel metabolic and potential regulatory genes induced by nitrate. Plant Cell 12:1491–1510

Weigel D, Ahn JH, Blazquez MA, Borevitz JO, Christensen SK, Fankhauser C, Ferrandiz C, Kardailsky I, Malancharuvil EJ, Neff MM, Nguyen JT, Sato S, Wang ZY, Xia Y, Dixon RA, Harrison MJ, Lamb CJ, Yanofsky MF, Chory J (2000) Activation tagging in *Arabidopsis*. Plant Physiol 122:1003–1014

Wesley SV, Helliwell CA, Smith NA, Wang M-B, Rouse DT, Liu Q, Gooding PS, Singh SP, Abbott D, Stoutjesdijk PA, Robinson SP, Gleave AP, Green AG, Waterhouse PM (2001) Construct design for efficient, effective and high-throughput gene silencing in plants. Plant J 27:581–590

Winkler RG, Frank MR, Galbraith DW, Feyereisen R, Feldmann KA (1998) Systematic reverse genetics of Transfer-DNA-tagged lines of *Arabidopsis*. Isolation of mutations in the cytochrome P450 gene superfamily. Plant Physiol 118:743–750

Wisman E, Ohlrogge J (2000) *Arabidopsis* microarray service facilities. Plant Physiol 124: 1468–1471

Zachgo EA, Wang ML, Dewdney J, Bouchez D, Camilleri C, Belmonte S, Huang L, Dolan M, Goodman HM (1996) A physical map of Chromosome 2 of *Arabidopsis thaliana*. Genome Res 6:19–25

3 The Sequence of the *Arabidopsis* Genome as a Tool for Comparative Structural Genomics in Brassicaceae

R. SCHMIDT[1], A. ACARKAN[2], K. BOIVIN[3], O. CLARENZ[4], and M. ROSSBERG[5]

1 Introduction

Comparative structural genomics aims to describe the differences in genome organization seen among diverse organisms and ultimately to unravel the molecular and genetic mechanisms accounting for these changes. Comparisons can be undertaken at different levels of resolution.

Karyotype analysis has revealed a considerable variability in chromosome number for angiosperms. Species with as few as 2n = 4 chromosomes (e.g. *Haplopappus gracilis*) and as many as 2n≈600 chromosomes (*Voanioala gerardii*) have been identified (Bennett 1998). A similarly high variability has been observed for genome sizes. A large compilation of genome size estimates for more than 2,500 angiosperm species has shown that approximately half of the flowering plants analyzed have genome sizes between 0.1 and 3.5 pg and values as high as 125 pg have been reported (Leitch et al. 1998). Repetitive sequence abundance correlates well with genome size (Flavell et al. 1974; Flavell 1980); however, a small portion of the variability can be attributed to changes of ploidy. Polyploidy is a widespread phenomenon in flowering plants, between 50 and 70% of which are believed to have experienced chromosome doubling at least once in their evolutionary history (Wendel 2000).

Comparative analyses of genome structure rely on sequences that have been conserved during evolution. This criterion is met by protein coding sequences, which generally show a high degree of DNA similarity when derived from closely related species. It is possible to determine the proportion of protein coding sequences of a given species that cross-hybridizes with DNA sequences derived from a related species. Such studies are particularly suitable for assessing the conservation of gene repertoires in related species (Schmidt 2002).

Max-Delbrück-Laboratorium in der Max-Planck-Gesellschaft, Carl-von-Linné-Weg 10, 50829 Köln, Germany
[1] Max-Planck-Institut für Molekulare Pflanzenphysiologie, 14424 Potsdam, Germany
[2] Bayer AG, Landwirtschaftszentrum PF-F-MWF, Geb. 6240, Alfred-Nobel-Str. 50, 40789 Monheim, Germany
[3] INRA–URGV, 2, rue Gaston Crémieux - CP 5708, 91057 Évry cedex, France
[4] Max-Planck-Institut für Züchtungsforschung, Carl-von-Linné-Weg 10, 50829 Köln, Germany
[5] AMGEN GmbH, Riesstr. 25, 80992 Munich, Germany

Biotechnology in Agriculture and Forestry, Vol. 52
Nagata/Tabata (Eds.) Brassicas and Legumes

Insight into the gross organization of a plant genome can be obtained by using molecular markers to assemble a genetic linkage map. Genetic mapping not only places a molecular marker on a particular linkage group, but makes it possible to determine the order of markers relative to each other. Restriction fragment length polymorphism (RFLP) analysis and PCR-based approaches are used to generate molecular markers and genetic linkage maps have been assembled for many different species (Schwarzacher 1994). RFLP markers often correspond to protein coding sequences, consequently they cross-hybridize to DNA of related species. Thus, the linkage arrangements of markers in closely related species can be directly compared if the same set of RFLP markers is employed for genetic mapping (Schmidt 2000, 2002).

Comparative genetic mapping requires only a limited set of markers, thus it is an efficient way to compare the overall organization of two or more species. A drawback of this approach is the limited resolution of such studies. Very often, the order of closely, physically-linked markers is not unequivocally determined using genetic mapping. A more detailed comparison of orthologous regions in respect to local gene order requires comparative physical mapping and sequencing studies. Libraries of cloned genomic DNA fragments are a prerequisite for these analyses. The gene content of a given cloned DNA fragment is determined and then compared to the orthologous region in the other species to be studied. Comparisons that are even more detailed can be achieved using sequence analysis. Not only can the gene repertoire of orthologous regions be established, but also the spacing of genes and their orientation relative to each other. Furthermore, the exon/intron structures of orthologous genes can be compared and conserved sequences identified (Bennetzen 2000; Schmidt 2002).

Members of Brassicaceae offer excellent opportunities for comparative genome analysis, especially since the genome of one of the crucifer species, *Arabidopsis thaliana*, has been completely sequenced. At 125 kbp this genome is one of the smallest known for higher plants (The Arabidopsis Genome Initiative 2000). *Arabidopsis* is exceptionally well suited for genetic and molecular analyses. This species has thus become the most important model for plant biology, and all aspects of development, biochemistry, physiology, and plant environment interactions are being investigated.

Approximately 3,350 species make up the Brassicaceae, with far-flung distributions (Price et al. 1994). The genus *Brassica* largely accounts for the agricultural importance of the family as a whole. *Brassica* species are important as seed oil, vegetable, and fodder crops. The close phylogenetic relationship of *Arabidopsis* to these crops presents a unique chance to test whether (and to what extent) information and resources derived from the Arabidopsis Genome Project can be transferred to crop plants.

2 The *Arabidopsis* Genome

Comprehensive genetic and physical maps have been assembled for the *Arabidopsis* genome. Initially, RFLP maps were developed for the five *Arabidopsis* chromosomes (Chang et al. 1988; Nam et al. 1989; Liu et al. 1996); recently marker development has shifted more and more toward PCR-based approaches (Reiter et al. 1992; Konieczny and Ausubel 1993; Bell and Ecker 1994; Alonso-Blanco et al. 1998; Cho et al. 1999).

The small genome size of 125 Mbp facilitated the establishment of clone contig maps for the *Arabidopsis thaliana* genome. Libraries of clones containing large inserts of *Arabidopsis* DNA were developed in yeast and bacterial artificial chromosome (BAC) vectors (Ecker 1990; Creusot et al. 1995; Choi et al. 1995; Mozo et al. 1998) and used to assemble clone contig maps for all five chromosomes (Schmidt et al. 1995; Zachgo et al. 1996; Kotani et al. 1997; Camilleri et al. 1998; Sato et al. 1998; Marra et al. 1999; Mozo et al. 1999). Most importantly, the resulting physical maps are extensively cross-referenced with the molecular marker maps. The contig maps were essential resources for the large-scale genomic sequencing project. By December 2000, the nucleotide sequence of the majority of the genome had been determined and annotated (The Arabidopsis Genome Initiative 2000).

For the annotation of genes, comparisons with coding sequences from *Arabidopsis* and other organisms provide important information. Especially valuable are large collections of expressed sequence tag (EST) and cDNA sequences. Currently more than 110,000 *Arabidopsis* EST sequences can be accessed in public databases (Höfte et al. 1993; Newman et al. 1994; Rounsley et al. 1996; Quackenbush et al. 2000, 2001). They provide a reliable and rapid means to determine whether a transcribed sequence for a particular genomic DNA sequence is already known. Since EST sequences are single-run sequences they are often of low quality and contain ambiguities. An EST sequence with an identity of 90% or higher with the genomic DNA most likely represents a cognate cDNA for this locus, whereas EST sequences showing lower sequence identity values most probably correspond to other loci in the *Arabidopsis thaliana* genome. Approximately 60% of the annotated *Arabidopsis* genes are supported by EST sequences. Thus, for a comprehensive genome annotation one also has to rely on predictions with appropriate computer algorithms (The Arabidopsis Genome Initiative 2000).

All five chromosomes are characterized by a high gene density. In the 115 Mbp of the genome that has been sequenced, a gene is found on average every 4.5 kbp. In heterochromatic regions, however, the density of genes is much lower and certain classes of repeated sequences are particularly abundant. A lot of *Arabidopsis* genes are present in more than one copy, and clusters of related genes are frequently observed. More than 1,500 tandem arrays consisting of 4,140 individual genes have been found (The Arabidopsis Genome Initiative 2000).

Analysis of the genome sequence has revealed evidence of ancient duplications (Terryn et al. 1999; Blanc et al. 2000; Paterson et al. 2000; The

Arabidopsis Genome Initiative 2000). These are characterized by a set of genes in common to each of the duplicated segments. Genes unique to any one of the duplications are found interspersed with the duplicated genes. This pattern can be explained by assuming a duplication followed by loss of genes. Approximately 60% of the genome corresponds to these large segmental duplications, suggesting the polyploid ancestry of *Arabidopsis*.

3 The *Arabidopsis* Genome Sequence as a Tool for Comparative Genetic Mapping

Using the same set of RFLP markers for genetic mapping in related species, linkage maps for both species are constructed that can be directly compared. Such experiments require the identification of conserved sequences. Moreover, it is essential that polymorphisms be detected for the markers in the mapping populations of both species analyzed. Such experiments have generally revealed collinear chromosomal segments if species belonging to the same family are analyzed, however, their size varies (Schmidt 2000). In some studies entire chromosomes show collinearity, whereas in other cases collinear segments span only few cMs. A comparison of the *Arabidopsis* chromosome maps with those of different *Brassica* species have revealed genome collinearity over short genetic distances (Kowalski et al. 1994; Lagercrantz 1998; Paterson et al. 2000).

The conclusions drawn from comparative genetic mapping are limited by a number of technical shortcomings. Firstly, conservation of marker repertoires is assessed using cross-hybridization studies. With this type of analysis it is not possible to distinguish if a particular sequence is completely absent from a genome or if it has diverged to an extent that it cannot be detected in hybridization experiments. Secondly, it is essential that only orthologous loci are taken into account for collinearity studies. Very often markers cross-hybridize to several sequences in the species analyzed, thus it cannot be determined whether the positions of orthologus or paralogous loci are compared. Consequently, it is often not possible to distinguish a deviation from collinearity from the mapping of a paralogous sequence. Thirdly, the resolution of genetic mapping is rather limited. Often small mapping populations are used, therefore it is not always feasible to determine the order of markers in both species in an unambiguous way. This makes it impossible to judge if a segment shows a truly collinear arrangement of markers or not.

By using the sequence map of the *Arabidopsis* genome as a tool for comparative genetic mapping in the Brassicaceae, these shortcomings can be largely overcome. Several hundred *Arabidopsis* RFLP markers have been placed on genetic linkage maps and are available for comparative mapping (Chang et al. 1988; Nam et al. 1989; Liu et al. 1996). This resource is even more valuable if the sequences of markers are determined. Due to the availability of the *Arabidopsis* genome sequence it is possible to determine unambiguously the chromosomal position and copy number for each marker if its sequence is known.

A comparison of a marker sequence with *Arabidopsis* EST sequences and/or the annotated gene sequences will reveal whether this marker harbors protein coding sequences or not. By identifying markers corresponding to genes, a comparative analysis can exclusively be focused on protein coding sequences. This is advantageous, since it is established that the vast majority of genes is present in genomes of related plants and not unique to a single species. Of 86 *Arabidopsis* markers analyzed (Fabri and Schäffner 1994; Liu et al. 1996), 80 (93.0%) hybridized to genomic DNA of *Capsella*. For 78 markers (90.7%) corresponding EST sequences and/or annotated gene sequences have been found. These results reveal the conservation of gene repertoires in *Arabidopsis* and *Capsella*.

It is essential to focus on the evaluation of orthologous loci in comparative studies. This can be easily achieved if only single-copy sequences are taken as markers for the comparative analysis. For markers for which several homologues exist, physical proximity to each other can be studied. Genes in the *Arabidopsis* genome are often arranged in tandem arrays (The Arabidopsis Genome Initiative 2000). When using small genetic mapping populations, it is normally impossible to detect recombination events between such closely-linked genes; such arrangements can thus be treated as a single genetic locus. Those markers sharing low nucleotide sequence identity values (<75%) with one or more homologues in the *Arabidopsis* genome can also be regarded as single-copy genes with respect to genetic mapping because the hybridization conditions can be chosen such that poorly conserved sequences will not be detected in cross-hybridization studies. Even members of a multigene family may be incorporated into comparative mapping studies if all loci corresponding to the marker sequence are determined in *Arabidopsis*. Analysis can then determine whether any of these loci are situated in a collinear arrangement with flanking single-copy sequences in the species of interest.

Twenty-four markers located on chromosome IV of *Arabidopsis thaliana* were used for genetic mapping in *Capsella*. Ten of the markers correspond to single-copy sequences in the *Arabidopsis* genome, whereas sequence comparisons of the marker sequences to the *Arabidopsis* genome established two or more loci for the remainder. For 20 markers a single locus was mapped in *Capsella* and 4 markers revealed two loci. Two linkage groups were established in *Capsella*, encompassing 8 and 16 loci, respectively. The positions of these 24 loci showed correspondence to the locations of the marker sequences on *Arabidopsis thaliana* chromosome IV (Fig. 1A). For all but two markers a collinear arrangement was established in *Arabidopsis* and *Capsella*. Markers 22023 and m518A mark an inversion (Fig. 1A).

For the four markers for which two loci have been mapped in *Capsella* (ATTS2212, ATTS3374, ATTS3804, mi330) all loci which share at least 80% sequence identity over a range of minimally 100 bp when compared to the marker sequence were located on the sequence maps of the *Arabidopsis* chromosomes. Comparing the positions of the resulting loci with those on the *Capsella* linkage groups it is evident that for three of the markers both

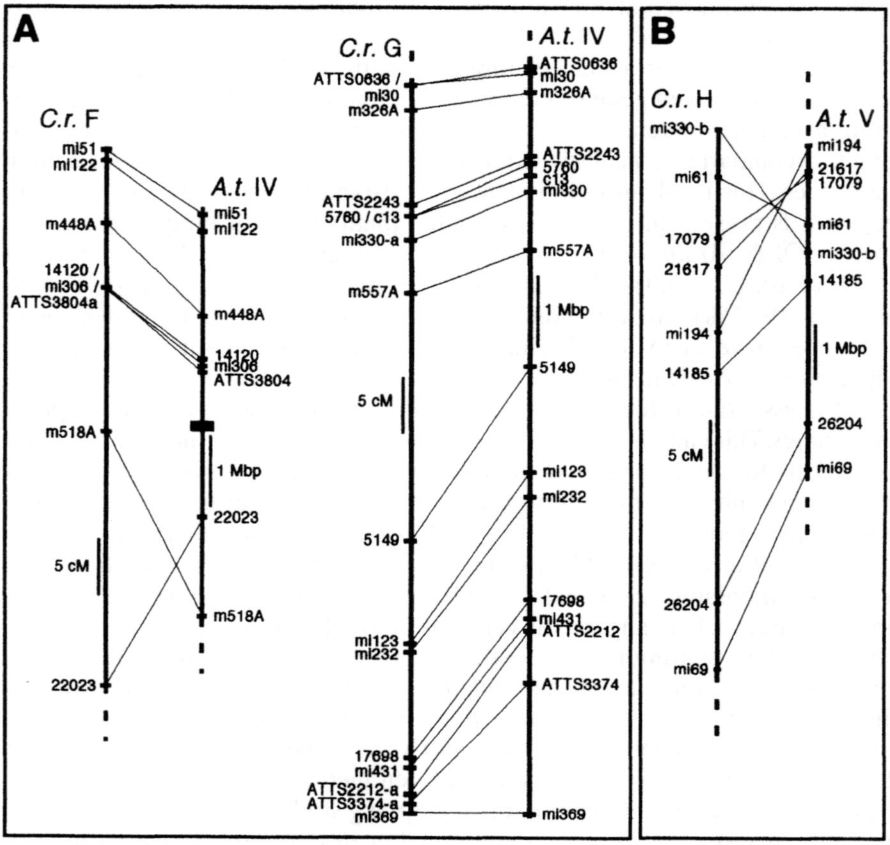

Fig. 1. Comparative genetic mapping reveals collinear segments in the *Arabidopsis* and *Capsella* genomes. *Arabidopsis* RFLP markers and EST sequences were used for genetic mapping in *Capsella*. The order of markers on the resulting *Capsella* linkage groups (*C.r.* . . .) is compared to their arrangement on the sequence maps of the *A. thaliana* chromosomes (*A.t.* . . .). Corresponding loci on the different maps are connected by *thin lines*. A *thick black bar* marks a centromeric region. *Dashed lines* represent parts of the chromosome maps which are not shown. **A** Markers located on chromosome IV of *Arabidopsis thaliana* were mapped in *Capsella*. For markers ATTS3804, ATTS2212, ATTS3374 and mi330 two loci have been mapped in *Capsella*. Only the loci residing in collinear arrangement with the marker sequences on *Arabidopsis* chromosome IV are shown. **B** Markers located on *Arabidopsis* chromosome V were used for genetic mapping in *Capsella*. The order of markers mi330-b, mi61, 17079, 21617 and mi194 is inverted in *Capsella* when compared to the arrangement on the *Arabidopsis* sequence map

Capsella loci are residing in collinear arrangement with loci in *Arabidopsis* (cf. mi330-a and -b, Fig. 1). Marker ATTS3374, however, is a single-copy sequence in the *Arabidopsis* genome whereas in *Capsella* two loci mapping on two different linkage groups are present. Thus, a change in locus number causes a deviation from collinearity.

Rather than relying on existing RFLP marker collections, it is also possible to analyze the *Arabidopsis* genome for the presence of single-copy gene sequences and take only these for comparative mapping. Since every single-copy gene can be utilized, a very large repertoire of markers is available for comparative studies in any given genomic region. This is particularly useful if a marker has been identified which indicates a deviation from collinearity. Then additional genes flanking the marker of interest can be chosen for genetic mapping. Such an analysis will show if a break of collinearity is only seen for a single gene or whether a small region has been rearranged. For example, the order of markers mi330-b and mi194 in *Capsella* is inverted when compared to their arrangement on the *Arabidopsis* sequence map. Mapping of additional markers located in this interval confirmed an inversion (Fig. 1B).

For many species of the Brassicaceae molecular marker maps have been developed (Paterson et al. 2000). It is no longer necessary to carry out laborious genetic mapping in *Arabidopsis* to establish a comparative map since the sequences of the molecular markers can be determined and aligned to the *Arabidopsis* sequence map. The average identity of exon sequences at the nucleotide level has been estimated to be 87% for the closely related genera *Arabidopsis* and *Brassica* (Cavell et al. 1998), a value of >90% is found for *Arabidopsis* and *Capsella* (Acarkan et al. 2000; Rossberg et al. 2001). Thus it will be possible to identify one or more homologues in the *Arabidopsis* genome for the vast majority of markers which harbor exon sequences. This is exemplified in a comparison carried out between *Capsella* and *Arabidopsis*. Sequences of 137 randomly cloned *Capsella* genomic DNA fragments with an average size of 434 bp were compared to the sequence of the *Arabidopsis* genome. Significant sequence identity was found for 82.5% of the sequences. Forty sequences were homologous to rDNA or organellar sequences, whereas 73 *Capsella* sequences corresponded to single- or low-copy regions in the *Arabidopsis* genome.

4 The *Arabidopsis* Genome Sequence as a Tool for Comparative Physical Mapping

Comparative physical mapping of orthologous regions allows detailed comparisons in respect to gene repertoire and local gene order. For such an analysis it is necessary to determine the gene content of a given cloned DNA fragment.

Screening of cDNA libraries with fragments of clones spanning the area of interest is an efficient way to identify protein coding sequences located in a particular genomic region. A drawback of this strategy is that very lowly or specifically expressed genes will often not be detected. In screening approximately 200,000 clones of the PRL2 library, cDNAs corresponding to seven different genes were identified in a 60 kbp region of *Arabidopsis thaliana* chromosome IV. However, the comparison of the genomic sequence to the EST sequence collections revealed the presence of several other genes in this area (Acarkan et al. 2000).

The availability of the *Arabidopsis* genome sequence offers a unique opportunity for comparative physical mapping experiments in the Brassicaceae since the annotation provides information about the gene content in any genomic segment of interest (The Arabidopsis Genome Initiative 2000). It needs to be considered, however, that annotation of a gene sequence may often rely solely on predictions by suitable computer algorithms, thus no experimental evidence has been obtained that this sequence indeed represents an expressed gene. A comparison of the annotated gene sequences with the *Arabidopsis* EST and cDNA sequence collections will show if a transcribed sequence for a particular genomic DNA sequence has already been described.

The identified genes can be used as probes to detect homologous sequences in a library of cloned genomic DNA fragments derived from the other species of interest. Libraries containing large inserts of genomic DNA are particularly suitable for such an analysis. The genome size of the species to be analyzed determines the choice of library. *Capsella rubella* is characterized by a high gene density, similar to that of *Arabidopsis thaliana*, thus cosmid clones which on average contain 20–40 kbp of plant DNA are readily suitable for comparative physical mapping, since most clones will contain several genes (Fig. 2A; Acarkan et al. 2000; Rossberg et al. 2001). For *Brassica* species moderately-sized genomes have been reported. Estimates range from 470 to 520 Mbp and 600 to 670 Mbp for different *B. rapa* and *B. oleracea* subspecies, respectively. A value of 470 Mbp has been established for *B. nigra*. Consistent with the amphidiploid nature of *B. napus* and *B. juncea,* larger genome sizes of 1,105–1,235 Mbp have been reported for these (Arumuganathan and Earle 1991). Thus, BAC libraries with an average insert size of 100 kbp or larger are well suited for physical mapping in *Brassica* species (O'Neill and Bancroft 2000; Ryder et al. 2001).

For the screening by colony hybridizations the clones of a genomic DNA library are gridded onto nylon membranes. PCR fragments of genomic DNA or cDNA clones are suitable probes, several of which can be combined in a single experiment. It is advisable, however, to exploit only single- and low-copy sequences as probes. The copy number of a particular gene sequence in the *Arabidopsis* genome can be easily established by a comparison of its coding sequence with that of the complete genome.

All clones identified in the colony hybridization analysis are then subjected to Southern blot analysis using single genes as probes. This type of analysis establishes whether for all different *Arabidopsis* genes tested a corresponding copy can be found in the clone library. In those cases where no counterpart is found in the clone library, a Southern blot analysis of genomic DNA of the species of interest should be undertaken to verify that homologous sequences are not detectable by cross-hybridization.

Based on the results of the iterative Southern blot hybridizations using single genes as probes, the large-insert clones are arranged into contigs. If microcollinearity is completely maintained in the two genomes it is expected that for all coding sequences which are residing in close proximity in *Arabidopsis* the homologues are also found in close neighborhood of each

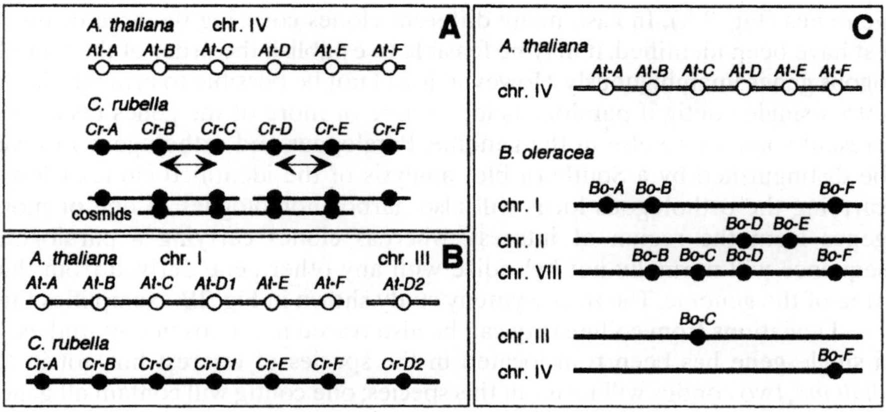

Fig. 2. Comparison of gene repertoire and arrangement in genomic regions of *Arabidopsis thaliana*, *Capsella rubella* and *Brassica oleracea*. **A** *Circles* indicate the presence of six different genes (*At-A – At-F*) in a region of *Arabidopsis* chromosome IV shown as a *line*. Colony hybridizations using the *Arabidopsis* cDNAs as probes were used to identify *Capsella rubella* cosmid clones carrying homologous sequences. *Black lines* depict three different cosmids. *Filled circles* designate the presence of *Capsella* homologues of the *Arabidopsis* genes on the individual cosmids. The cosmids were arranged into a single contig indicated as a *solid black line*, sizes of cosmid clones and the contig are not drawn to scale. *Arrows* point out that the exact order of some genes has not been determined in *Capsella* (*Cr-B – Cr-C, Cr-D – Cr-E*). **B** Six different *Arabidopsis* genes (*At-A – At-F;* At1g55580, At1g55590, At1g55600, At1g55610, At1g55620, At1g55630) were used as probes in colony hybridizations to reveal *Capsella* clones carrying homologous sequences. *Arabidopsis* gene *At-D* is present in two copies in the *Arabidopsis* genome, located on chromosomes I (At1g55610) and III (At3g13380), respectively. A collinear region carrying homologues of all five different *Arabidopsis* genes is found in *Capsella rubella*, furthermore a second contig which harbors a paralogous sequence of gene *D* is detected. **C** Comparison of a region of *A. thaliana* chromosome IV and corresponding segments in *B. oleracea*. Six single-copy *Arabidopsis* gene sequences (*At-A – At-F*; X68145, AJ288958, AJ288959, AF110227, AJ006043, AY034902), shown as *circles*, have been used as probes in colony hybridization experiments to identify *B. oleracea* BAC clones carrying homologous sequences. *Black circles* indicate which homologues of the *Arabidopsis* genes are present on the assembled BAC contigs shown as *solid black lines*. Genetic mapping was used to assign the contigs to linkage groups in *B. oleracea*. The contigs on chromosomes I, II and VIII carry homologues for two or more of the six *Arabidopsis* genes. *Brassica* homologues of the single-copy *Arabidopsis* genes *At-C* and *At-F* are also found unlinked to any other of the genes on chromosomes III and IV, respectively

other in the species of interest, and it will often be possible to arrange the large-insert clones into a single contig. Figure 2A shows the results of a microsynteny analysis between the *Arabidopsis thaliana* and *Capsella rubella* genomes. Probes representing six different single-copy genes (AF057044, AJ288956, AJ288957, X68145, AJ288958, AJ288959) located in a 35 kbp region on the long arm of *Arabidopsis* chromosome IV were used to detect homologous sequences in a cosmid library made from *Capsella rubella* genomic DNA. Cosmid clones harboring three or four different genes were identified and arranged into a single contig. Although the exact order of some genes was not determined, it is evident that the arrangement of the six genes is very similar if not identical in the *Arabidopsis thaliana* and *Capsella rubella* genomes (Fig. 2A). In case many different clones covering the area of interest have been identified, it may be feasible to establish the order of the homologous genes unambiguously. However, it will not be possible to arrange clones into a single contig if paralogous loci of one or more of the genes tested are present somewhere else in the genome. Paralogous and orthologous loci can be distinguished by a Southern blot analysis of the identified clones. Clones carrying the orthologous locus will also harbor homologues of one or more genes from the region of interest, whereas clones carrying a paralogous sequence will normally not hybridize with any other gene derived from that area of the genome. The microsynteny study shown in Fig. 2B exemplifies this.

Deviations from collinearity can be also traced in microsynteny studies. If a single gene has been translocated in the species of interest but not in *A. thaliana*, two contigs will result in this species; one contig will contain all genes of the region of interest but the translocated one, and a second set of clones will only hybridize to the gene involved in the translocation. Genetic mapping can then be used to establish whether the two contigs are still found in close vicinity of each other or in entirely different areas of the genome in the species of interest.

Such studies are also useful for highlighting duplicated areas of a genome. This is readily seen in comparisons of areas of the *Arabidopsis* genome with corresponding regions derived from *Brassica*. Genetic mapping using molecular markers has revealed the multiplicated nature of the *Brassica oleracea*, *Brassica nigra,* and *Brassica rapa* genomes. In these species each chromosomal segment is present in three copies (Lagercrantz and Lydiate 1996). Hence, it is expected that even those genes which are present as single-copy sequences in *Arabidopsis thaliana* will often correspond to multiple loci in *Brassica*. This is exemplified in a comparison of a region located on *Arabidopsis thaliana* chromosome IV and the corresponding areas of the *Brassica oleracea* genome. In any one of the three segments mapping to *Brassica oleracea* chromosomes I, II, and VIII one or several homologues of the *Arabidopsis* genes are missing, but the order of the remaining genes is the same as in the *Arabidopsis* region. The gene content of all three regions taken together equals that of *Arabidopsis*. Additionally, homologues of two *Arabidopsis* single-copy genes are found on *B. oleracea* chromosomes III and IV, indicative of two translocations (Fig. 2C). Similar collinearity relationships have been described in an independent study comparing regions of *Arabidopsis thaliana* chromosomes IV and V and the corresponding segments in *Brassica oleracea* (O'Neill and Bancroft 2000).

5 The *Arabidopsis* Genome Sequence as a Tool for Comparative Sequence Analysis

Comparative sequence analysis can give even more detailed information than physical mapping studies. Most importantly, the nature of differences in gene arrangements is identified by comparative sequence analysis. The sequence repertoire of orthologous regions is established and conserved sequences are identified.

Almost complete microcollinearity was observed by comparing two different *A. thaliana* regions harboring eleven different protein coding genes to their counterparts in *C. rubella*. In both cases the gene repertoire was identical and the genomic regions were similarly sized in the two species. Gene order and orientation of genes relative to each other were also maintained (Acarkan et al. 2000; Rossberg et al. 2001). However, one of the genes has been duplicated in *C. rubella* but not in *A. thaliana* (Fig. 3A; Acarkan et al. 2000). Another deviation from microcollinearity was found in the immediate vicinity of one of these regions. The orthologue of *A. thaliana* gene At4g16690 is interrupted by two large insertions in *C. rubella* (Fig. 3B) which share homology with mobile elements. Comparing a region of *A. thaliana* chromosome IV ecotype Columbia to the corresponding region of the ecotype Landsberg *erecta* revealed similar results. The region contains two retroelement-like sequences in the Columbia ecotype, whereas in Landsberg *erecta* three are found in different positions (Noël et al. 1999). These data show that small differences can be found in the arrangement of the *Arabidopsis* and *Capsella* genomes and that a comparison of ecotypes might show similar deviations. Nevertheless, the *Arabidopsis* genome can serve as a blueprint for the study of closely related genomes.

Aligning *A. thaliana* cDNA sequences with genomic DNA sequences of *A. thaliana* and *C. rubella* reveals the conservation of number and position of intron sequences in both species (Fig. 3; Acarkan et al. 2000; Rossberg et al. 2001). A high degree of sequence identity is confined to exon sequences, whereas in introns and intergenic regions overall sequence identity is generally not found. Moreover, intergenic regions and introns are differently sized in the two species (Fig. 3; Acarkan et al. 2000; Rossberg et al. 2001).

The exon/intron structure deduced from alignments of *Arabidopsis* cDNA and genomic DNA sequences differs from that of a predicted gene in four out of nine cases analyzed (Acarkan et al. 2000; The Arabidopsis Genome Initiative 2000; Rossberg et al. 2001). Most notably, no conserved gene structure is revealed if such (most likely incorrectly) predicted *Arabidopsis* coding sequences are compared to *Capsella* genomic DNA sequences. In contrast, aligning *Arabidopsis* cDNA sequences with *Arabidopsis* and *Capsella* genomic DNA sequences has yielded a conserved gene structure in all cases analyzed to date. These results show that comparative sequence analysis can be exploited for improvement of gene structure predictions if conservation of exon length and sequences in different species are taken into account. Many annotated genes lack corresponding EST or cDNA sequences; in these cases the presence of a conserved gene structure in another species indicates that this model may represent an expressed gene.

For one out of nine genes analyzed in *Arabidopsis* and *Capsella*, all exons differed in size. Interestingly, identity of exon sequences for this particular gene sequence was approximately 80% at the nucleotide level, whereas for the other eight genes values of >90% were determined (Fig. 3C; Rossberg et

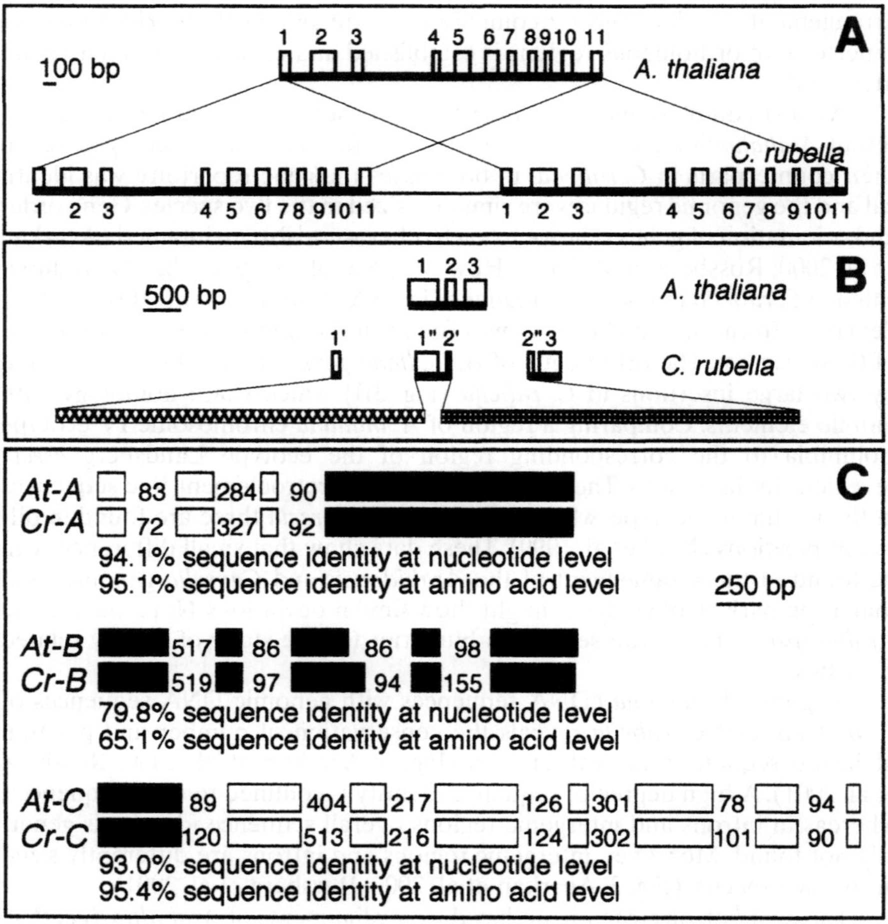

Fig. 3. Comparative sequence analysis reveals similarities and differences in orthologous genes of *A. thaliana* and *C. rubella*. **A** An *A. thaliana* single-copy gene is found duplicated in tandem in *C. rubella*. The three gene copies consist of 11 exons, depicted as *boxes*. All three copies have identically sized exons, whereas most intron sequences differ in size. **B** A *Capsella* gene is distinguished by two insertions from its orthologue in *A. thaliana*. The insertions show homology to mobile elements. **C** Three different protein coding sequences (*A*, *B*, *C*) which are present in collinear segments of the *Arabidopsis* (*At*) and *Capsella* (*Cr*) genomes (Rossberg et al. 2001) are shown. The gene structures encompass the regions between start and stop codons. Exon sequences are represented as *boxes* and drawn to scale, the sizes of intron sequences are given in bp. Exons of identical length in both species are shown as *white boxes*, whereas *black shading* indicates those exons differing in size in *Arabidopsis* and *Capsella*. Sequence identity values for exon sequences are given for the three pairs of orthologous genes at nucleotide and amino acid level

al. 2001). This case illustrates that comparative sequence analysis is especially valuable if fast-evolving sequences are to be compared.

6 Conclusions and Perspectives

Arabidopsis and *Capsella* have base chromosome numbers of five and eight, respectively. Thus, large chromosomal rearrangements distinguishing these genomes are expected. Apart from translocations, evidence for inversions was found in comparative mapping studies. In addition to these changes involving large chromosome segments, the genome arrangement of the two species is distinguished by small rearrangements, such as deletions/insertions, duplications and/or translocations of gene sequences.

Comparative genetic mapping between *A. thaliana*, *B. oleracea*, and *B. nigra* has also revealed collinear chromosome segments, although on average they are much shorter than the ones which are found for *Arabidopsis* and *Capsella* (Kowalski et al. 1996; Lagercrantz 1998, Paterson et al. 2000). Moreover, genetic mapping has provided evidence that most *Arabidopsis* chromosome segments are represented three times in *Brassica* (Lagercrantz et al. 1996; Lagercrantz 1998). The view that these *Brassica* species are derived from hexaploid ancestors is corroborated by results of microsynteny studies, since comparative physical mapping has shown that three different *B. oleracea* chromosome segments correspond to one particular *Arabidopsis* region. However, the three *Brassica* regions are characterized by a different gene content, only all three copies taken together representing the full complement of genes found in *Arabidopsis* (O'Neill and Bancroft 2000). The resulting complex collinearity relationships can be explained by assuming that loss of gene sequences is very frequent in duplicated segments. Segmental duplications in the *Arabidopsis* genome also differ in their gene content (Terryn et al. 1999; Blanc et al. 2000; The Arabidopsis Genome Initiative 2000), this suggests the same mode of genome evolution as seen in the *Brassica* genome.

From the results of comparative genome analysis studies to date it is clear that the availability of the complete sequence of the *Arabidopsis* genome is a unique resource for analyzing genomes of other Brassicaceae species, although complicated collinearity relationships may result from duplications. If a locus or trait of interest in a given species has been mapped to an interval of markers, the sequences of these molecular markers can be determined and aligned with the sequence of the *Arabidopsis* genome. For markers representing exon sequences there is a high likelihood of placing them directly on the *Arabidopsis* chromosome maps, thus identifying the corresponding region in the *Arabidopsis* genome. For any given interval of the completely annotated *Arabidopsis* genome a large population of gene sequences can be selected as markers and used for genetic mapping in the species of interest. This may ultimately pinpoint the gene coding for the locus or trait of interest.

7 Protocols

7.1 Standard Molecular Biology Techniques

Standard molecular biology techniques were carried out according to Sambrook et al. (1989). Plant genomic DNA isolation and Southern blot analysis were performed as described (Schmidt et al. 1999). Random primer extension was used for labeling reactions (Feinberg and Vogelstein 1983, 1984).

7.2 Genetic Analyses

Fifty self-compatible F_2 plants derived from an interspecific cross of *C. grandiflora* × *C. rubella* make up the population used for genetic mapping in *Capsella* (Acarkan et al. 2000). *Arabidopsis* RFLP markers (Fabri and Schäffner 1994; Liu et al. 1996) or ESTs (Höfte et al. 1993; Newman et al. 1994) were used as probes on restricted genomic DNA of *C. rubella* and a pool of F_2 plants to reveal polymorphisms. Genetic distances between linked markers were calculated using the MapMaker program with the Kosambi cM function (Lander et al. 1987).

The population used for the genetic mapping in *Brassica oleracea* consists of doubled haploid lines established from an F1 of an "alboglabra" × "italica" cross (Bohuon et al. 1996). A subset of 40 lines was chosen for genetic linkage analysis. *A. thaliana* cDNA sequences or DNA fragments derived from BAC clones carrying inserts of *B. oleracea* DNA were used as probes in RFLP mapping experiments. Sequence information derived from *Brassica oleracea* BAC subclone fragments was exploited to develop SSCP markers (Slabaugh et al. 1997). The resulting PCR products were separated on gels prepared with the MDE gel solution as recommended by the manufacturer (FMC Bioproducts, Vallenbak Strand, Denmark). Evaluation of the gels for polymorphic fragments was carried out after staining with silver (Sanguinetti et al. 1994).

7.3 *C. rubella* Library of *Mbo*I Fragments

Total genomic *C. rubella* DNA was digested with *Mbo*I, and fragments ranging in size from 0.5 to 1.5 kbp were cloned into vector pGEM 7Zf+ (Promega, Mannheim, Germany). The sequences of 137 different cloned fragments were determined. A BLAST analysis (Altschul et al. 1990) identified corresponding sequences derived from the *Arabidopsis* nuclear and organellar genomes. The following parameters were used for the alignments: nucleotide match 1, nucleotide mismatch –2, gap open penalty 5 and gap extension penalty 1. A match was reported if an expect value of $<E^{-9}$ was found.

7.4 Mapping Sequences on the *Arabidopsis* Chromosome Maps

SeqViewer of TAIR (http://www.arabidopsis.org/) was used to establish map positions of marker sequences on the sequence maps of the *Arabidopsis* chromosomes.

7.5 Identification of Cognate cDNAs

The genomic DNA sequences of *A. thaliana* regions of interest were used for a BLAST analysis (Altschul et al. 1990) to search for corresponding EST sequences. Putative cognate cDNA clones (>90% identity to the genomic sequence at the nucleotide level) were ordered from the *Arabidopsis* Biological Resource Center (ABRC, The Ohio State University, Columbus, Ohio, USA) and sequenced.

7.6 Isolation of *C. rubella* Cosmid Clones and *B. oleracea* BAC Clones

Colony hybridizations using *Arabidopsis* genomic DNA fragments or cDNA clones as probes were used to screen clone libraries made from *C. rubella* (Schmidt et al. 1999) and *B. oleracea* genomic DNA (O'Neill and Bancroft 2000; Ryder et al. 2001). The protocols for the preparation of cosmid library filters and the conditions for hybridizations have been described elsewhere (Schmidt et al. 1999). Screening of BAC libraries was carried out according to O'Neill and Bancroft (2000).

7.7 Subcloning and Sequencing

Genomic DNA inserts of cosmid and BAC clones were restricted with appropriate enzymes and cloned into pGEM vectors (Promega). DNA sequences of subclones and PCR products were determined on PE/Applied Biosystems 377 and 3700 sequencers using BigDye terminator chemistry (Perkin Elmer, Überlingen, Germany) by the ADIS unit at the Max-Planck-Institut für Züchtungsforschung (Köln, Germany). Oligonucleotides were purchased from Metabion (Martinsried, Germany) or MWG Biotech (Ebersberg, Germany). Gene sequences were sequenced on both strands. The Wisconsin Package (Version 10.0-UNIX, Genetics Computer Group (GCG), Madison, Wis., USA) and BLAST (Altschul et al. 1990) were used for sequence analysis. For comparisons of orthologous gene sequences the regions from start to stop codons were aligned.

Acknowledgements. The research was funded by a grant from the Bundesministerium für Bildung und Forschung (BMBF, grant no. 0311107). *A. thaliana* cDNA clones were provided by the *Arabidopsis* Biological Resource Center (ABRC, The Ohio State University, Columbus, Ohio, USA). Dr. G. King (Horticultural Research Institute Wellesbourne, United Kingdom) and Dr. Ian

Bancroft (John Innes Centre, Norwich, United Kingdom) donated filters of *Brassica oleracea* BAC libraries and BAC clones. Dr. Isobel Parkin and Dr. Derek Lydiate (Agriculture and Agri-Food Canada, Saskatoon Research Centre, Saskatoon) kindly provided the *B. oleracea* mapping population and genetic linkage data for this population. Prof. Dr. H. Hurka donated seed of *C. rubella* and *C. grandiflora* (Universität Osnabrück, Germany). We thank Dr. Katharina Schneider (Max-Planck-Institut für Züchtungsforschung, Köln, Germany) for discussions about SSCP analysis.

References

Acarkan A, Rossberg M, Koch M, Schmidt R (2000) Comparative genome analysis reveals extensive conservation of genome organisation for *Arabidopsis thaliana* and *Capsella rubella*. Plant J 23:55–62

Alonso-Blanco C, Peeters AJ, Koornneef M, Lister C, Dean C, van den Bosch N, Pot J, Kuiper MT (1998) Development of an AFLP based linkage map of Ler, Col and Cvi *Arabidopsis thaliana* ecotypes and construction of a Ler/Cvi recombinant inbred line population. Plant J 14:259–271

Altschul SF, Gish W, Miller W, Myers EW, Lipman D (1990) Basic local alignment search tool. J Mol Biol 215:403–410

Arumuganathan K, Earle ED (1991) Nuclear DNA content of some important plant species. Plant Mol Biol Rep 9:208–218

Bell CJ, Ecker JR (1994) Assignment of 30 microsatellite loci to the linkage map of *Arabidopsis*. Genomics 19:137–144

Bennett MD (1998) Plant genome values: how much do we know? Proc Natl Acad Sci USA 95:2011–2016

Bennetzen JL (2000) Comparative sequence analysis of plant nuclear genomes: microcollinearity and its many exceptions. Plant Cell 12:1021–1029

Blanc G, Barakat A, Guyot R, Cooke R, Delseny M (2000) Extensive duplication and reshuffling in the *Arabidopsis* genome. Plant Cell 12:1093–1101

Bohuon EJR, Keith DJ, Parkin IAP, Sharpe AG, Lydiate DJ (1996) Alignment of the conserved C genome of *Brassica oleracea* and *Brassica napus*. Theor Appl Genet 93:833–839

Camilleri C, Lafleuriel J, Macadre C, Varoquaux F, Parmentier Y, Picard G, Caboche M, Bouchez D (1998) A YAC contig map of *Arabidopsis thaliana* chromosome 3. Plant J 14:633–642

Cavell AC, Lydiate DJ, Parkin IAP, Dean C, Trick M (1998) Collinearity between a 30-centimorgan segment of *Arabidopsis thaliana* chromosome 4 and duplicated regions within the *Brassica napus* genome. Genome 41:62–69

Chang C, Bowman JL, DeJohn AW, Lander ES, Meyerowitz EM (1988) Restriction fragment length polymorphism linkage map for *Arabidopsis thaliana*. Proc Natl Acad Sci USA 85: 6856–6860

Cho RJ, Mindrinos M, Richards DR, Sapolsky RJ, Anderson M, Drenkard E, Dewdney J, Reuber TL, Stammers M, Federspiel N, Theologis A, Yang WH, Hubbell E, Au M, Chung EY, Lashkari D, Lemieux B, Dean C, Lipshutz RJ, Ausubel FM, Davis RW, Oefner PJ (1999) Genome-wide mapping with biallelic markers in *Arabidopsis thaliana*. Nat Genet 23:203–207

Choi S, Creelman RA, Mullet JE, Wing RA (1995) Construction and characterization of a bacterial artificial chromosome library of *Arabidopsis thaliana*. Plant Mol Biol Rep 13:124–128

Creusot F, Fouilloux E, Dron M, Lafleuriel J, Picard G, Billault A, Le Paslier D, Cohen D, Chaboute M-E, Durr A, Fleck J, Gigot C, Camilleri C, Bellini C, Caboche M, Bouchez D (1995) The CIC library: a large insert YAC library for genome mapping in *Arabidopsis thaliana*. Plant J 8: 763–770

Ecker JR (1990) PFGE and YAC analysis of the *Arabidopsis* genome. Methods 1:186–194

Fabri C, Schäffner A (1994) An *Arabidopsis thaliana* RFLP mapping set to localize mutations to chromosomal regions. Plant J 5:149–156

Feinberg AP, Vogelstein B (1983) A technique for radiolabelling DNA restriction endonuclease fragments to high specific activity. Anal Biochem 132:6–13

Feinberg AP, Vogelstein B (1984) A technique for radiolabelling DNA restriction endonuclease fragments to high specific activity. Addendum. Anal Biochem 137:266–267

Flavell R (1980) The molecular characterization and organization of plant chromosomal DNA sequences. Annu Rev Plant Phys 31:569–596

Flavell RB, Bennett MD, Smith JB, Smith DB (1974) Genome size and the proportion of repeated nucleotide sequence DNA in plants. Biochem Genet 12:257–269

Höfte H, Desprez T, Amselem J, Chiapello H, Rouzé P, Caboche M, Moison A, Jourjon M-F, Charpenteau J-L, Berthomieu P, Guerrier D, Giraudat J, Quigley F, Thomas F, Yu D-Y, Mache R, Raynal M, Cooke R, Grellet F, Delseny M, Parmentier Y, de Marcillac G, Gigot C, Fleck J, Philipps G, Axelos M, Bardet C, Tremousaygue D, Lescure B (1993) An inventory of 1152 expressed sequence tags obtained by partial sequencing of cDNAs from *Arabidopsis thaliana*. Plant J 4:1051–1061

Konieczny A, Ausubel F (1993) A procedure for quick mapping of *Arabidopsis* mutants using ecotype specific markers. Plant J 4:403–410

Kotani H, Sato S, Fukami M, Hosouchi T, Nakazaki N, Okumura S, Wada T, Liu YG, Shibata D, Tabata S (1997) A fine physical map of *Arabidopsis thaliana* chromosome 5: construction of a sequence-ready contig map. DNA Res 4:371–378

Kowalski SP, Lan T-H, Feldmann KA, Paterson AH (1994) Comparative mapping of *Arabidopsis thaliana* and *Brassica oleracea* chromosomes reveals islands of conserved organization. Genetics 138:499–510

Lagercrantz U (1998) Comparative mapping between *Arabidopsis thaliana* and *Brassica nigra* indicates that *Brassica* genomes have evolved through extensive genome replication accompanied by chromosome fusions and frequent rearrangements. Genetics 150:1217–1228

Lagercrantz U, Lydiate D (1996) Comparative genome mapping in *Brassica*. Genetics 144:1903–1910

Lagercrantz U, Putterill J, Coupland G, Lydiate D (1996) Comparative mapping in *Arabidopsis* and *Brassica*, fine scale genome collinearity and congruence of genes controlling flowering time. Plant J 9:13–20

Lander ES, Green P, Abrahamson J, Barlow A, Daly MJ, Lincoln SE, Newburg L (1987) MapMaker: an interactive computer package for constructing primary genetic linkage maps of experimental and natural populations. Genomics 1:174–181

Leitch IJ, Chase MW, Bennett MD (1998) Phylogenetic analysis of DNA C-values provides evidence for a small ancestral genome size in flowering plants. Ann Bot 82(Suppl A):85–94

Liu Y-G, Mitsukawa N, Lister C, Dean C, Whittier RF (1996) Isolation and mapping of a new set of 129 RFLP markers in *Arabidopsis thaliana* recombinant inbred lines. Plant J 10:733–736

Marra M, Kucaba T, Sekhon M, Hillier L, Martienssen R, Chinwalla A, Crockett J, Fedele J, Grover H, Gund C, McCombie WR, McDonald K, McPherson J, Mudd N, Parnell L, Schein J, Seim R, Shelby P, Waterston R, Wilson R (1999) A map for sequence analysis of the *Arabidopsis thaliana* genome. Nat Genet 22:265–270

Mozo T, Fischer S, Shizuya H, Altmann T (1998) Construction and characterization of the IGF *Arabidopsis* BAC library. Mol Gen Genet 258:562–570

Mozo T, Dewar K, Dunn P, Ecker JR, Fischer S, Kloska S, Lehrach H, Marra M, Martienssen R, Meier-Ewert S, Altmann T (1999) A complete BAC-based physical map of the *Arabidopsis thaliana* genome. Nat Genet 22:271–275

Nam H-G, Giraudat J, den Boer B, Moonan F, Loos WDB, Hauge BM, Goodman H (1989) Restriction fragment length polymorphism linkage map of *Arabidopsis thaliana*. Plant Cell 1:699–705

Newman T, de Bruijn FJ, Green P, Keegstra K, Kende H, McIntosh L, Ohlrogge J, Raikhel N, Somerville S, Thomashow M, Retzel E, Somerville C (1994) Genes galore: a summary of methods for accessing results from large-scale partial sequencing of anonymous *Arabidopsis* cDNA clones. Plant Physiol 106:1241–1255

Noël L, Moores TL, van der Biezen EA, Parniske M, Daniels MJ, Parker JE, Jones JDG (1999) Pronounced intraspecific haplotype divergence at the *RPP5* complex disease resistance locus of *Arabidopsis*. Plant Cell 11:2099–2111

O'Neill CM, Bancroft I (2000) Comparative physical mapping of segments of the genome of *Brassica oleracea* var. *alboglabra* that are homoeologous to sequenced regions of chromosomes 4 and 5 of *Arabidopsis thaliana*. Plant J 23:233–243

Paterson AH, Bowers JE, Burow MD, Draye X, Elsik CG, Jiang C-X, Katsar CS, Lan T-H, Lin Y-R, Ming R, Wright RJ (2000) Comparative genomics of plant chromosomes. Plant Cell 12:1523–1539

Price RA, Palmer JD, Al-Shebaz IA (1994) Systematic relationships of *Arabidopsis*: a molecular and morphological perspective. In: Meyerowitz EM, Somerville CR (eds) *Arabidopsis*. Cold Spring Harbor Lab Press, Cold Spring Harbor, NY, pp 7–19

Quackenbush J, Liang F, Holt I, Pertea G, Upton J (2000) The TIGR gene indices: reconstruction and representation of expressed gene sequences. Nucleic Acids Res 28:141–145

Quackenbush J, Cho J, Lee D, Liang F, Holt I, Karamycheva S, Parvizi B, Pertea G, Sultana R, White J (2001) The TIGR gene indices: analysis of gene transcript sequences in highly sampled eukaryotic species. Nucleic Acids Res 29:159–164

Reiter RS, Williams JGK, Feldmann KA, Rafalski JA, Tingey SV, Scolnik PA (1992) Global and local genome mapping in *Arabidopsis thaliana* by using recombinant inbred lines and random amplified polymorphic DNAs. Proc Natl Acad Sci USA 89:1477–1481

Rossberg M, Theres K, Acarkan A, Herrero R, Schmitt T, Schumacher K, Schmitz G, Schmidt R (2001) Comparative sequence analysis reveals extensive microcollinearity in the *Lateral suppressor* regions of the tomato, Arabidopsis and Capsella genomes. Plant Cell 13:979–988

Rounsley SD, Glodek A, Sutton G, Adams MD, Somerville CR, Venter JC, Kerlavage AR (1996) The construction of *Arabidopsis* expressed sequence tag assemblies. A new resource to facilitate gene identification. Plant Physiol 112:1177–1183

Ryder CD, Smith LB, Teakle GR, King GJ (2001) Contrasting genome organisation: two regions of the *Brassica oleracea* genome compared with collinear regions of the *Arabidopsis thaliana* genome. Genome 44:808–817

Sambrook J, Fritsch EF, Maniatis T (1989) Molecular cloning: a laboratory manual, 2nd edn. Cold Spring Harbor Lab Press, Cold Spring Harbor, NY

Sanguinetti CJ, Neto ED, Simpson AJG (1994) Rapid silver staining and recovery of PCR products separated on polyacrylamide gels. BioTechniques 17:915–919

Sato S, Kotani H, Hayashi R, Liu YG, Shibata D, Tabata S (1998) A physical map of *Arabidopsis thaliana* chromosome 3 represented by two contigs of CIC YAC, P1, TAC and BAC clones. DNA Res 5:163–168

Schmidt R (2000) Synteny: recent advances and future prospects. Curr Opin Plant Biol 3:97–102

Schmidt R (2002) Plant genome evolution: lessons from comparative genomics at the DNA level. Plant Mol Biol 48:21–37

Schmidt R, West J, Love K, Lenehan Z, Lister C, Thompson H, Bouchez D, Dean C (1995) Physical map and organization of *Arabidopsis thaliana* chromosome 4. Science 270:480–483

Schmidt R, Acarkan A, Koch M, Rossberg M (1999) A strategy for comparative physical mapping in cruciferous plants. In: van Raamsdonk LWD, den Nijs JCM (eds) Plant evolution in man-made habitats. Proceedings of the VIIth Symposium of the International Organization of Plant Biosystematics. Hugo de Vries Lab, University of Amsterdam, Amsterdam, pp183–196

Schwarzacher T (1994) Mapping in plants: progress and prospects. Curr Opin Genet Dev 4:868–874

Slabaugh MB, Huestis GM, Leonard J, Holloway JL, Rosato C, Hongtrakul V, Martini N, Toepfer R, Voetz M, Schell J, Knapp SJ (1997) Sequence-based genetic markers for genes and gene families: single-strand conformational polymorphisms for the fatty acid synthesis genes of *Cuphea*. Theor Appl Genet 94:400–408

Terryn N, Heijnen L, De Keyser A, Van Asseldonck M, De Clercq R, Verbakel H, Gielen J, Zabeau M, Villarroel R, Jesse T, Neyt P, Hogers R, Van Den Daele H, Ardiles W, Schueller C, Mayer K, Déhais P, Rombauts S, Van Montagu M, Rouzé P, Vos P (1999) Evidence for an ancient chromosomal duplication in *Arabidopsis thaliana* by sequencing and analyzing a 400-kb contig at the *Apetala2* locus on chromosome 4. FEBS Lett 445:237–245

The Arabidopsis Genome Initiative (2000) Analysis of the genome sequence of the flowering plant *Arabidopsis thaliana*. Nature 408:796–815

Wendel JF (2000) Genome evolution in polyploids. Plant Mol Biol 42:225–249

Zachgo EA, Wang ML, Dewdney J, Bouchez D, Camilleri C, Belmonte S, Huang L, Dolan M, Goodman HM (1996) A physical map of chromosome 2 of *Arabidopsis thaliana*. Genome Res 6:19–25

4 Application of Molecular Markers in *Brassica* Coenospecies: Comparative Mapping and Tagging

M. LAKSHMIKUMARAN[1], S. DAS[2], and P.S. SRIVASTAVA[2]

1 Introduction

The great bulk of genetic variation at the nucleotide level may not be visible at the phenotypic level. It is this variation that is exploited to generate molecular markers. The study of molecular genetics and plant breeding has been aided variously and to an unprecedented degree and accuracy with the developments in the field of molecular markers. Techniques such as restriction fragment length polymorphism (RFLP), randomly amplified polymorphic DNA (RAPD), amplified fragment length polymorphism (AFLP), simple sequence repeats (SSR), expressed sequence tags (EST), cleaved amplified polymorphic sequence (CAPS), sequence characterized amplified region (SCAR) and single nucleotide polymorphisms (SNP) have permitted rapid and precise analysis of germplasm, trait mapping, and marker-assisted breeding and selection. These markers have also been used to further our insight into genome evolution, and to study gene order in terms of conservation (synteny) and rearrangement across taxa. Molecular or DNA-based markers offer many advantages over conventional phenotypic markers. They are heritable, easy to score, free from developmental and environmental influences, detectable in all tissues, insensitive to epistatic or pleiotropic interactions, and they provide a choice of codominant (simultaneous detection of various alleles) or dominant (no allelic information conveyed) markers.

DNA markers can be classified broadly into hybridization- and PCR-based. RFLP and minisatellite/microsatellite markers comprise the major hybridization-based marker system, whereas PCR-based markers include RAPD, AFLP, CAPS, EST, SCAR, and SSR. Besides these, high-throughput techniques of micro-arrays, which employ both PCR and hybridization and can be used to simultaneously analyze a large number of loci, have been developed (Drenkard et al. 2000; Wisman and Ohlrogge 2000; Zhu and Wang 2000; Seki et al. 2001; Zhu et al. 2001). Genetic maps are constructed using markers that survey both coding and noncoding sequences. The maps are prepared by analyzing the segregating populations derived from crosses using genetically

[1] Bioresources and Biotechnology Division, TERI, Habitat Place, Lodhi Road, New Delhi 110 003, India
[2] Centre for Biotechnology, Faculty of Science, Hamdard University, Hamdard Nagar, New Delhi 110 062, India

Biotechnology in Agriculture and Forestry, Vol. 52
Nagata/Tabata (Eds.) Brassicas and Legumes

diverse parents and estimating the recombination frequency among genetic loci. The individuals of the population may be either derived by selfing, backcrossing, or are double-haploid plants from microspores. A trait when it exhibits cosegregation with a genetic locus, defined by a molecular marker, is said to be tagged or linked to the particular molecular marker. The principle of marker-assisted selection is based on following the linkage of an easily screenable or identifiable molecular marker to a trait of interest, and consequently to the gene responsible for the trait. The linked marker can therefore be employed for following the segregation of the character (marker-assisted selection) and for map-based cloning wherein the gene responsible for the trait is isolated and cloned after the map position of the marker/locus is known.

Brassica species are valued for their use as oilseeds, vegetables, and condiments the world over. The genus *Brassica* comprises 35 species (Warwick 1993) of which 6 are agronomically important. These are three cultivated diploid species, namely, *B. campestris/B. rapa* (2n = 20, AA), *B. nigra* (2n = 16, BB) and *B. oleracea* (2n = 18, CC), and three amphidiploid species namely *B. juncea* (2n = 36, AABB), *B. carinata* (2n = 34, BBCC), and *B. napus* (2n = 38, AACC). The three amphidiploid species are derived from the diploid genomes through natural hybridization, and the relationship among the six species can be explained in the form of U's triangle (U 1935). Apart from the economic importance, *Brassica* is also a close relative of *Arabidopsis thaliana*, the model dicot plant. The relatedness of *Brassica* and *Arabidopsis* gives us a unique opportunity to present these as a case study where genomic information derived from a model species can be applied to an economically important crop species. Often the advances made in *Brassica* genome analysis have been amply demonstrated to be a direct result of the advances made in *Arabidopsis* genome analysis.

2 Molecular Markers

A brief survey of the molecular markers reveals that these fall into two categories – those based on hybridization and those based on amplification.

2.1 Hybridization-Based Markers

This class of markers detects polymorphism using probes that are hybridized to restricted and electrophoresed DNA samples. The probes may simultaneously bind to and detect one locus or a few loci, as do single-copy RFLP probes, or bind and detect multiple loci, as do probes such as variable number of tandem repeats (VNTR) and simple sequence repeat (SSR) oligonucleotides that target minisatellite and microsatellite sequences. The single-copy probes are codominant in nature and are best suited for investigations such as mapping (Slocum et al. 1990; Cheung et al. 1997) and evolution (Song et al. 1988a,b, 1990, 1995a; Miller and Tanksley 1990; Zheng et al. 1994). Probes targeting minisatellite and microsatellite regions are dominant in nature and are applied for genetic diversity studies (Jeffrey et al. 1985; Bhatia et al. 1995; Prasad et al. 2000).

2.2 Amplification-Based Markers

Amplification markers are based on polymerase chain reactions (PCR) and are further classified as target PCR and arbitrarily primed PCR.

2.2.1 *Target PCR*

A prerequisite for target PCR is the prior sequence information of the locus to be amplified. Primers are designed based on the unique sequences and used to amplify a single locus or a few loci. The resultant products are codominant in nature and thus applied for mapping. These marker systems include CAPS (Konieckzny and Ausubel 1993), EST (Adams et al. 1991), SCAR (Paran and Michelmore 1993), and STS (Thomas and Scott 1994). A combination of these PCR-based markers and sequence information has expedited the process of map construction. Sequencing of cDNA clones from both ends generate expressed sequence tags (ESTs), whereas, any part of the genome flanked by known sequence would be considered as sequence tagged sites (STSs). Primers are designed based on the prior sequence data and used to amplify corresponding regions of segregating individuals from a mapping population. Amplified products generating polymorphism among segregating individuals can be scored as codominant markers. If no polymorphism is obtained, then the products are restricted with enzymes and analyzed to generate cleaved amplified polymorphic sequences (CAPS). Both CAPS and STS markers have been successfully employed in mapping and tagging of genes of agronomic importance in *Brassica* species (Plieske and Struss 2001a; Vicente and King 2001), and recently an EST enriched map of *B. oleracea* has been developed (Lan et al. 2000). Among other target PCR tools, primer pairs targeted towards microsatellite regions have been used for genetic diversity (Plieske and Struss 2001b) and in mapping studies in *Brassica* (Saal et al. 2001).

2.2.2 *Arbitrarily Primed PCR*

Arbitrarily primed PCR circumvents the need to have any prior sequence information about a particular genome and therefore can be applied across a number of taxa. Since the primers are random, they amplify multiple alleles and are dominant. The technique of RAPD was one of the first modifications of PCR for genome scanning and analysis (Welsh and McClelland 1990; Williams et al. 1990). It involves amplification of genomic DNA with a single arbitrary, short primer (usually ten nucleotides) at a low annealing temperature (37–40 °C). The technique of RAPD has been used for genetic diversity analysis (Demeke et al. 1992; Jain et al. 1994; Thormann et al. 1994; Divaret et al. 1999) and mapping (Hu et al. 1995; Camarago et al. 1997; Chen et al. 1997; Costa et al. 2000).

Another arbitrary primed amplification technique, AFLP, was developed by Vos et al. (1995). Based on selective amplification of a subset of restricted DNA fragments using PCR. it detects polymorphism due to changes in restriction site distribution and flanking nucleotide sequences that affect primer binding and extension capability. Like RFLP, AFLP is reproducible, heritable, and robust, and does not need prior genome information; it is amenable to automation, as is the case with RAPD. AFLP markers have been used for assessment of genetic diversity studies in a number of plant species, including *Arabidopsis thaliana* (Erschadi et al. 2000), *B. campestris* (Das et al. 1999), *B. juncea* (Srivastava et al. 2001), sunflower (Hongtrakul et al. 1997), *Mangifera* species (Eiadthong et al. 2000); in construction and saturation of linkage maps, as in barley (Becker et al. 1995); and finally in the tagging of genes, as in *Brassica* species (Voorrips et al. 1997; Negi et al. 2000).

The products of arbitrary primed PCR (RAPD and AFLP) being dominant in nature can used for mapping purposes after conversion to sequence characterized amplified regions (SCARs), which are codominant (Paran and Michelmore 1993).

Very recently, a novel semi-arbitrary primed PCR-based method, termed sequence-related amplified polymorphism (SRAP), has been developed. Its utility was demonstrated for both genetic diversity (Riaz et al. 2001) and for mapping and tagging of genes responsible for glucosinolate desaturation (Li and Quiros 2001).

3 Applications

Molecular markers have made it possible to dissect the genome and identify genetic differences which are revealed in the form of DNA polymorphism. These genetic markers have been variously applied for linkage analysis, map construction, comparative mapping to study synteny and genome evolution, gene tagging and map-based cloning, and analysis of genetic diversity. These applications with special emphasis on *Brassica* species are discussed.

3.1 Development of Linkage Maps

Cosegregation analysis of genetic markers in progenies derived from single or two parents are used to construct linkage maps. The individual progenies are a result of fusion of meiotic products (the gametes from the two parents) and therefore their genetic analysis reflects the recombination events and frequencies. Markers that are present on different chromosomes or are physically at a distant location on the same chromosome will exhibit independent assortment. In comparison, markers derived from the same chromosome and closely placed will exhibit linkage (except when separated by recombination, since recombination is a function of the physical distance between two loci/markers; therefore the shorter the distance the tighter the linkage, and vice versa). A measure of recombination events among different markers will thus reveal their relative order of arrangement and distance from one another. Molecular maps are thus seen to be a linear arrangement of the genetic markers, the distance between each representing the recombination frequency in terms of centimorgan (cM) units. The number of linkage groups in maps equals the haploid chromosome number. Molecular maps have been developed for a number of plant species using either one type of marker or a combination of different markers (Quiros et al. 1994; Mazur and Tingey 1995; Mohan et al. 1997; Gale and Devos 1998; Gupta et al. 1999; Botha and Venter 2000; Costa et al. 2000; Heusden et al 2000; Ulloa et al. 2002). Initially, molecular maps in *Brassica* species utilized isozyme markers (Arus and Orton 1983; Chen et al. 1989); later they used DNA markers as listed in Table 1.

Mapping in *Brassica* species has made use of mapping population and alien addition lines. The mapping populations are derived by repeated selfing (Sharma et al. 1994; Lan and Paterson 2000), backcrossing (Lagercrantz and Lydiate 1995), and through diploidization of haploid lines, i.e., doubled daploid (DH) lines (Cheung et al. 1997; Lombard and Delourme 2001).

A number of workers have also employed mapping populations obtained by various combinations of the three strategies outlined above. For example, a population of backcrossed individuals of DH lines (Ramsay et al. 1996; Kearsey et al. 1996) and inbred backcrossed lines (IBLs) have been used (Buttruille et al. 1999).

Table 1. Some RFLP maps constructed in *Brassica* species

Species	Marker system	Reference
B. oleracea	RFLP	Slocum et al. (1990)
B. napus	RFLP + EST	Landry et al. (1991)
B. campestris	RFLP + Isozyme	McGrath and Quiros (1991)
B. campestris	RFLP	Song et al. (1991)
B. oleracea	RFLP	Kianian and Quiros (1991)
B. oleracea	RFLP	Landry et al. (1992)
B. campestris	RFLP	Chyi et al. (1992)
B. nigra	RFLP + RAPD	Truco and Quiros (1994)
B. napus	RFLP	Ferreira et al. (1994)
B. campestris	RFLP	Teutonico and Osborn (1994)
B. nigra	RFLP	Lagercrantz and Lydiate (1995)
B. napus	RFLP	Uzuonova et al. (1995)
B. oleracea	RFLP	Bohuon et al. (1996)
B. oleracea	RFLP + AFLP	Voorrips et al. (1997)
B. juncea	RFLP	Cheung et al. (1997)
B. napus	RFLP	Butruille et al. (1999)
B. juncea	RFLP	Axelson et al. (2000)
B. oleracea	AFLP + RFLP	Sebastian et al. (2000)
B. oleracea	EST	Lan et al. (2000)
B. napus	Isozyme + RFLP + RAPD + AFLP	Lombard and Delourme (2001)

Alien addition lines have been used to investigate and characterize B and C genome species of *Brassica* (Quiros et al. 1987; McGrath et al. 1990; This et al. 1990; Chevre et al. 1991; Kaneko et al. 2000, 2001). Maps for the five agronomically important U-triangle-member species have been published, the exception being *B. carinata*. Salient features of these maps are discussed.

3.1.1 **B. campestris** *and* **B. rapa** *Map*

An F2 population derived from a cross between "Michihili" × "Spring Broccoli" was used to construct and report the first map for *B. rapa* (Song et al. 1991). A total of 188 genomic clones identified 280 RFLP loci that were spread over ten linkage groups (LG) and covered a total map length of 1,850 cM. Segregation analysis of 220 RFLP loci and 28 morphological traits including leaf, stem, and flowering features led to a further refinement of the same map. The total map length was estimated at 1,593 cM spread over 10 LG. A variable number of QTL ranging between 0–5 were associated with the different traits studied (Song et al. 1995b). Another map developed using an F2 mapping population of "R500" × "Horizon" was reported by Chyi et al. (1992). The map consisted of 360 marker loci spread over 10 LG identified by 269 genomic clones. The physical distance as measured by recombinational units for the 10 LG ranged from 89–300 cM, and the total map length was 1,876 cM (Chyi et al. 1992).

Maps have also been reported by McGrath and Quiros (1991) employing 3 genomic, 28 cDNA, and 4 isozyme markers spread over 8 LG and a total of 262 cM, and by Schilling (1991) using an F2 mapping population with 58 markers.

3.1.2 **B. nigra** *Map*

Alien addition lines of *B. nigra* in various genetic backgrounds like *D. erucoides-B. nigra* (This et al. 1990), *B. napus-B. nigra* (Chevre et al. 1991) were used to develop the first maps on which RFLP, RAPD, and isozyme markers were placed. Further, using these three types of markers Truco and Quiros (1994) developed genetic map covering 8 LG and two unlinked groups. Maps with higher marker density have been developed by Lagercrantz and Lydiate (1995) and later expanded to include additional probes from *A. thaliana* (Lagercrantz 1998). A number of markers have been physically assigned to their respective chromosomes using the *B. nigra* addition lines (Struss et al. 1991, 1996).

3.1.3 **B. oleracea** *Map*

A map of *B. oleracea* spread over 9 LG was generated by Slocum et al. (1990). Another map was generated using nearly 200 RFLP markers spanning 9 LG, covering about 1,112 cM (Landry et al. 1992). Recently, Camarago et al. (1997) have reported a composite map having 112 RFLP and 47 RAPD loci covering 9 LG, using an F2 mapping population. Another map was developed based on a combination of RFLP and AFLP markers in a doubled-haploid mapping population by Voorrips et al. (1997). An EST-enriched map using sequence information from *A. thaliana* has been reported by Lan et al. (2000). An integrated AFLP and RFLP map of *B. oleracea* having 547 markers covering 893 cM on 9 LG with an average spacing of 2.6 cM has been recently reported by Sebastian et al. (2000). Alien-addition lines of *B. oleracea* in a *B. rapa* background were employed by McGrath et al. (1990) to assemble a C-genome-specific map. The chromosomes carrying the genes responsible for flower and seed color in *B. oleracea* were identified and characterized using *B. oleracea* addition lines in a *B. rapa* background (Chen et al. 1992; Cheng et al. 1994a,b, 1995; Jorgensen et al. 1996).

3.1.4 **B. napus** *Map*

The first map of *B. napus* was developed using an F2 mapping population from "Westar" × "Topas." It employed 92 cDNA probes that detected 120 loci on 19 LG covering approximately 1,400 cM (Landry et al. 1991). A number of the clones mapped onto duplicate positions, indicating the amphidiploid nature of *B. napus* (Landry et al. 1991). A genetic map based on doubled-haploid lines from F1 of a cross between "Stellar" (annual) × "Major" (biennial) was constructed; in it 101 RFLP probes detected 132 loci spread over 22 linkage groups. That the number of LG generated (22) exceeds the haploid number (19) is explained by the fact that only few probes were analyzed, which led to a number of unlinked loci. The map covered a total of 1,016 cM with an average spacing of 7.7 cM between loci (Ferreira et al. 1994). Among other maps are those developed by Uzunova et al. (1995) and Foisset et al. (1996), both using DH lines and composite in nature, utilizing RFLP, RAPD, and isozyme markers. The map developed by Uzunova et al. (1995) used a DH population from F1 of "Mansholt Hamburger Raps" × "Samouri" in combination with 204 RFLP, 2 RAPD and 1 phenotypic marker. The resultant map contained 19 LG over a total length of 1,441 cM. Probes for acyl carrier protein (ACP) and β-keto-acyl-ACP-synthase (KAS I) were able to localize, respectively, 3 and 2 loci responsible for these genes. The map was also used to identify and localize QTL for seed glucosinolate content (Uzunova et al. 1995).

A doubled-haploid population derived from a single F1 plant of "Darmor-*bzh*" × "Yudal" cross was employed by Foisset and coworkers (1996) to generate a map that predominantly used RAPD (68%) followed by RFLP (26.5%) and isozyme (4%) markers. The mapping population segregated for oligogenic and polygenic traits including plant development (dwarfism, earliness of flowering), seed quality (erucic acid and glucosinolate content), and disease resistance (*Phoma lingam*, *Cylindrosporium concentricum*). The probes mapped onto 19 LG and covering 1,765 cM, corresponding to nearly 71% of the *B. napus* genome (Foisset et al. 1996).

One of the most recent maps of *B. napus* is a consensus map obtained by integration of three maps from three independent DH mapping populations. The consensus map is composed of 540 different isozyme, RFLP, RAPD, and AFLP markers spread over 19 linkage groups covering a total genetic distance of 2,429 cM (Lombard and Delourme 2001).

3.1.5 **B. juncea** ***Map***

Among various maps of this species reported thus far, the first was by Sharma et al. (1994) based on an F2 population. It had only 25 markers on 9 linkage groups. The second map was more extensive, based on doubled-haploid lines with 343 loci in 18 linkage groups covering 2,073 cM (Cheung et al. 1997). Recently, two more maps of *B. juncea* have been developed using RFLP markers (Axelson et al. 2000). One of these maps used a resynthesized *B. juncea* (*B. rapa* × *B. nigra*) crossed to a natural *B. juncea* as parents, whereas the other map employed natural *B. juncea* cultivars as parents to generate a mapping population. The two maps exhibited an identical distribution of RFLP markers (Axelson et al. 2000).

Precision mapping and physical localization of markers have been achieved by developing chromosome-specific markers, use of substitution lines and chromosomes addition lines in various *Brassica* species (Hosaka et al. 1990; Quiros et al. 1991; Ramsay et al. 1996; Jorgensen et al. 1996; Kaneko et al. 2000, 2001).

3.2 Investigation of Genetic Relationships and Evolution of Species

Evolutionary studies have been found to be carried out more efficiently and effectively using molecular tools such as genome/gene sequences, RFLP, RAPD, and AFLP markers. These have been used to analyze phylogenetic relations in a number of crop species (Hosaka et al.1990; Miller and Tanksley 1990; Zheng et al. 1994; Kellog 1998; Moore 1995, 2000; Francisco-Ortega 1999; Yang et al. 1999). In *Brassica* species, Song et al. (1988a,b, 1990, 1995a) employed nuclear and chloroplast RFLP to study taxonomic relations. Their analysis indicated different pathways for evolution of *B. nigra,* with *Sinapis arvensis* or a wild relative being the possible progenitor. *B. campestris* and *B. oleracea* were found to have evolved from a common ancestral wild *B. oleracea* or another wild relative with nine chromosomes. However, earlier studies had otherwise presumed a monophyletic origin of all the three basic diploid *Brassica* species from an ancestral species with $n = 6$ (Prakash and Hinata 1980). Molecular analysis revealed a biphyletic origin of the three, and the two lineages arising out of the diploid species are the *Nigra/Sinapis* lineage and the *Rapa/Oleracea* lineage. The natural amphidiploids (*B. napus* and *B. juncea*) may have had a polyphyletic origin with the involvement of and recombination between different diploids (Song et al 1988a,b). Similar grouping of *Brassica* species reflecting their evolutionary relatedness was observed when analyzed using RAPD markers (Demeke et al. 1992). RFLP analysis has also been used to study the relationship among *Brassica* species with 2n = 18. Though all the species were 2n = 18, they formed complexes with a varying degree of genetic similarity. The clusterings were concordant with the geographical distribution of the species examined; for example, *B. oleracea*, *B. bourgeaui*, and *B. montana* formed an Atlantic-western Mediterranean cluster; *B. incana*, *B. rupestris,* and *B. villosa* formed an Italian cluster; and *B. insularis* and *B. macrocarpa* formed a Central Mediterranean group (Lanner et al. 1997).

3.3 Comparative Mapping and Synteny

The availability of genetic maps of the three diploid genomes allowed for their comparative analysis which has thrown light on the evolution of the "A", "B," and "C" genome (Slocum 1989; Hosaka et al. 1990; McGrath and Quiros 1991; Lagercrantz and Lydiate 1996; Truco et al. 1996; Schmidt et al. 2001).

Comparative analysis revealed that the linear order in which genes are arranged is conserved over a large evolutionary time scale (Heslop-Harrison 2000). The phenomenon of synteny or collinearity has been observed not only among the three monogenomic species but also between the amphidiploid *B. juncea* (AB) and *B. napus* (AC) genomes. The QTL from a RFLP map of *B. rapa* exhibits synteny when compared with linkage maps from *B. napus*, *B. oleracea* and *A. thaliana* (Teutonico and Osborn 1994).

A comparative mapping of the "A" genome from *B. campestris* and "derived A genome" from *B. napus* indicated conserved loci in both the genomes (Hoencke and Chyi 1991). Similar conservation of linkage groups has been observed when the "A" and "C" genomes from *B. rapa* and *B. oleracea,* respectively. were compared with the "AC" genome from *B. napus* (Teutonico and Osborn 1994; Parkin et al. 1995; Bohuon et al. 1996). In a recent study, two RFLP maps of *B. juncea* were developed and compared. One of the maps was generated by using a synthetic *B. juncea* (a chromosome-doubled interspecific hybrid of *B. rapa* × *B. nigra*) crossed to a natural *B. juncea*; the other generated using two natural *B. juncea* cultivars. The comparison revealed that the genomic segments derived from "A" (*B. rapa*) and "B" (*B. nigra*) genomes were perfectly conserved in the "AB" genome (*B. juncea*) and the two maps were highly collinear, suggesting that synteny can extend throughout the entire genome (Axelson et al. 2000).

Synteny or collinearity has also been observed across different genera, as observed for genes controlling flowering time between *A. thaliana* and *Brassica* species (Lagercrantz et al. 1996; Osborn et al. 1997), and in a cluster of five genes (Sadowski et al. 1996). A comparative analysis of *A. thaliana* and *B. oleracea* maps revealed "islands of conserved organization" (Kowalaski et al. 1994). The analysis of the S locus (controlling the incompatibility response) from *B. campestris* and *A. thaliana* revealed a high degree of synteny at the submegabase scale with the homologous region in *A. thaliana* (Conner et al. 1998). Conservation of gene order or synteny can extend for a few hundred kb, as has been observed in a 222-kb region between *A. thaliana* and *B. oleracea* ("microsynteny"; O'Neill and Bancroft 2000); it can incorporate several megabase pairs, as in a 7.5-Mbp region (corresponding to approx. 30 cM) between *A. thaliana* and *B. napus* ("macrosynteny"; Cavell et al. 1998), and it can involve an entire genomic level, as observed between "AB" genome (from natural *B. juncea*) and "A" and "B" genome derived from *B. rapa* and *B. nigra* (Axelson et al. 2000).

Comparisons of genetic maps also reveal changes in the *Brassica* genome such as rearrangements, large-scale duplications, deletions, inversions, and translocations of genetic segments (Hu and Quiros 1991; Sharpe et al. 1995). Some of the regions of one chromosome or a complete chromosome from one genome may share homology with multiple regions in another genome (Slocum 1989). Nearly 40% of the RFLP markers examined localized at two to three loci in *B. nigra*, a diploid species (Truco and Quiros 1994; Lagercrantz and Lydiate 1995). In *B. napus* more than 90% of the RFLP loci analyzed mapped to as many as four loci present on different linkage groups (Landry et al. 1991). Similar observations were reported in *B. napus* by Ferreira et al. (1994) and Uzunova et al. (1995). Comparative physical mapping between *Arabidopsis thaliana* and *Brassica rapa* in the genomic segments corresponding to chromosome II of *Arabidopsis* (Jackson et al 2000), and comparative genetic mapping between *Arabidopsis thaliana* and *Brassica* species (*B. nigra, B. oleracea, B. rapa* and *B. juncea*) in the genomic region controlling flowering time have revealed extensive duplication in the *Brassica* genome (Axelsson et al. 2001). These reports arising from observations in different species support the hypothesis that the present day diploid *Brassica* species are secondary polyploids (Prakash and Hinata 1980; Quiros et al. 1994). Deletions of genomic segments leading to null alleles have been observed by Quiros et al. (1988), and Kianian and Quiros (1992) in their mapping studies. This phenomenon of deletions in the C genome has been confirmed through cytology (Hu and Quiros 1991). The evidence gathered from molecular mapping studies indicates that the ancestral genome has undergone extensive intra- and intergenomic recombination through hybridization events while simultaneously maintaining overall arrangements (synteny) to give rise to the present day *Brassica* genome.

Collinearity of mapped markers such as those observed between *Arabidopsis* and *Brassica* (Meinke et al. 1998; Bevan and Murphy 1999), *Arabidopsis* and *Capsella rubella* (Acarkan et al. 2000), and a number of cereals such as rice, maize, wheat, and oats (Gale and Devos 1998; Kellogg 1998; Moore, 1995, 2000; Keller and Feulliet 2000) has widespread implications (McCouch 1998). The *Arabidopsis* genome may act as an anchor genome such that markers positioned on it can be utilized for reciprocal localization of markers on *Brassica* species (Kowalaski et al. 1994; Kurata et al. 1994; Sadowski et al. 1996; Sillito et al. 2000). Knowledge of position of genes controlling qualitative traits as well as quantitative traits (QTL) in one species can be used to predict the location of homologous genes in other species. This will be of immense help in plant breeding. Comparative map analysis to date has been most extensive in cereal crops. Comparisons of the maps of various members of Gramineae have revealed large-scale synteny. The maps of rice, maize, wheat, foxtail millet, sorghum, and sugarcane can in fact be aligned on the basis of conserved blocks which can usually be demonstrated to be derived from the rice genome, the maps of the other cereal crops simply being constructed with some rearrangement of the blocks (Gale and Devos 1998; Moore 2000). As more and more saturated genetic maps continue to be generated, synteny is being observed across species over a wide evolutionary distribution, providing fresh information for taxonomic studies and construction of evolutionary trees. A high level of synteny has been maintained over an evolutionary time scale estimated to be between 10 and 20 million years when *Arabidopsis* and *Brassica* diverged (Gale and Devos 1998; Schmidt et al. 2001), and up to 50–60 million years in cereals (Moore 1995, 2000; Bennetzen 2000).

Highly saturated maps would also allow the analysis of genomic rearrangements such as inversions and duplications (of less than a few centimorgans, the level of resolution achieved presently with recombinational mapping) (Bennetzen 2000). However, it is important to note that these observed examples of synteny are based only on single-copy and cDNA sequences, which contribute only small fraction of the genome. For a thorough understanding, studies on high-copy sequences are urgently required.

3.4 Mapping of Traits, Marker-Assisted Breeding, and Map-Based Cloning

Once a framework genetic map for any species is constructed, the next logical task is to link the various traits, qualitative and quantitative, to the markers. These molecular tags are a prerequisite for any marker-assisted breeding and selection programs for crop improvement. One of the most successful approaches for tagging has been the application of various molecular markers on bulked samples, termed as bulked segregant analysis (BSA). A number of traits in *Brassica* have been tagged (Table 2), and some of these are discussed with special reference to the seed coat color trait.

3.4.1 Bulked Segregant Analysis (BSA)

Bulked segregant analysis was developed by Michelmore and his coworkers (Michelmore et al. 1991) to rapidly identify markers linked to any specific gene or trait. BSA employs two bulked DNA samples collected from a number of individuals segregating in a population. Each bulked or pooled sample contains individuals that are similar for a particular trait or genomic region, but arbitrary for other unselected loci. These bulks can be based on phenotypic segregation of character or genotypic segregation of DNA markers. Bulks, after selection, are screened for polymorphism using an RAPD, RFLP, or AFLP technique. Any polymorphism thus detected is potentially due to differences in the trait/genomic region/phenotype based on which the

Table 2. Some of the traits mapped in *Brassica* species

Species	Trait	Mapping / tagging information	Reference
B. rapa	Seed coat color	Yls locus mapped to LG 5	Teutonico and Osborn (1994)
B. rapa	Resistance to *Albugo candida* race2	Resistance locus *ACA1* mapped to LG 4 flanked by markers ec2b3a and wg6c1a	Kole et al. (1996)
B. rapa	Resistance to race 4 of *Xanthomonas campestris* pv. *campestris*	RAPD marker WE(22)980 associated with resistance	Ignatov et al. (2000)
B. rapa ssp. *oleifera*	Oleic acid concentration	RAPD marker OPH-17 linked to QTL for oleic acid concentration	Tanhuanpää et al. (1996)
B. rapa	Erucic acid content	Eru locus mapped to LG 1	Teutonico and Osborn (1994)
B. oleracea	Glucosinolate desaturation	Tagging of glucosinolate desaturation gene BoGLS-ALK with semi-arbitrarily primed SRAP marker	Li and Quiros (2001)
B. oleracea	Glucosinolate biosynthesis	ALK locus in *B. oleracea* co-segregating with OHP gene (from *A. thaliana*), known to be involved in glucosinolate biosynthesis	Hall et al. (2001)
B. oleracea	Curd-related traits	86 QTL controlling 8 curd-related traits	Lan and Paterson (2000)
B. oleracea	Plant size	47 QTL reflecting 35 different genetic loci for such morphological traits as lamina length, lamina width, petiole length, stem length, stem width, and node number detected	Lan and Paterson (2001)
B. oleracea	Early bolting trait	RAPD markers specific for monosomic addition lines linked to early bolting trait	Kaneko et al. (2000)
B. oleracea	Disease resistance genes	CAPS marker developed for disease resistance gene like sequences (RGLs) and mapped onto four different linkage groups	Vicente and King (2001)
B. oleracea	Resistance to *Xanthomonas campestris*	QTL controlling to *X. campestris* mapped on LG1 and LG 9	Camarago et al. (1995)
B. oleracea	Resistance to *Plasmodiophora brassicae* Wor. Race 7	Marker 14a on LG 1C, 48 on LG 4C and 177b on LG 9C linked to clubroot resistance	Figdore et al. (1993)
B. oleracea	Resistance to *Plasmodiophora brassicae*	*pb-3* and *pb-4* gene linked to RFLP and AFLP markers	Voorrips et al. (1997)

B. juncea	Resistance to white rust caused by *Albugo candida*	Two RAPD markers WR2 and WR3 flanking the resistance locus *$Ac2_1$* identified in a DH population;	Prabhu et al. (1998)
		Eight AFLP markers linked to white rust resistance	Somers et al. (2002)
B. juncea	Seed Coat color	AFLP marker (AFLP8) converted to SCAR (SCM-08) for seed coat color	Negi et al. (2000)
B. napus	Loci controlling vernalization requirement and flowering time	Marker on linkage group 9 linked to both the traits; FLC co-segregating with a QTL region corresponding to VFR-2	Ferreira et al. (1995b) Kole et al. (2001)
B. napus	Fertility restorer gene for n*ap* CMS	RFLP probe 3NF2 linked to the *Rfn/Mmt* gene responsible for restoration of fertility in the *nap* CMS system	Li et al. (1998)
B. napus	Fertility restorer gene for *Ogu* CMS	Four RAPD marker $OPC02_{1150}$, $OPD02_{1000}$, $OPF06_{1200}$ and $OPG02_{700}$ linked to fertility restorer gene *Rfo* for Ogura radish CMS; 13 RAPD and 4–10 RFLP probes (new) identified for *Rfo* gene and used to characterize the introgressed genome from radish	Delourme et al. (1994, 1998)
B. napus	Fertility restorer gene for *Pol* CMS	Fertility restorer locus *Rfp1*, for polima CMS, linked to 10 RFLP and one RAPD loci; cRF1b showed perfect linkage to *Rfp1* and *Rfp2*	Jean et al. (1997)
B. napus	Oleic acid level	Three AFLP markers linked to oleic acid variation; the markers mapped close to the *fad*2 gene	Schierholt et al. (2000)
B. napus	Linolenic acid concentration	RAPD marker $K01\text{-}_{1100}$ linked to locus controlling linolenic acid concentration	Hu et al. (1995)
B. napus	Erucic acid level	Two QTL *E1* and *E2* identified; *E1* group has 6 RAPD markers and *E2* has 8 RAPD markers	Jourden et al. (1996c)

Table 2. *Continued*

Species	Trait	Mapping / tagging information	Reference
B. napus	Erucic acid synthesis	Loci controlling erucic acid synthesis linked to markers RP841.H1 (LG 6) and RP1544. E1(LG 12)	Ecke et al. (1995)
B. napus	Linolenic acid concentration	RAPD marker 25a linked to locus controlling linolenic acid concentration	Tanhuanpää et al. (1995)
B. napus	Erucic acid and Linolenic acid concentration	Two QTL for loci controlling erucic acid and Linolenic acid mapped on LG 7 and LG 15	Thormann et al. (1996)
B. napus	Seed aliphatic glucosinolate content	Five loci GSL1, GSL2, GSL3, GSL4, and GSL5 mapped on LG 20, LG 1, LG 18, LG 4, and LG 13, respectively	Toroser et al. (1995)
B. napus	Seed color genes	RFLP markers 4NF6a and pLF15 linked to seed color genes	Van Deyzne et al. (1995)
B. napus	Resistance to Turnip yellow virus (TuYV)	A major QTL on LG 17 responsible for resistance to turnip yellow virus identified through AFLP	Dreyer et al. (2001)
B. napus	Resistance to *Leptosphaeria maculans*	*LmFr$_1$* locus linked to marker cDNA011 and cDNA110	Dion et al. (1995)
B. napus	Resistance to turnip mosaic virus	Resistance gene *TuRB01* mapped on linkage group N6	Walsh et al. (1999)
B. napus	Resistance to light leaf spot (*Pyrenopeziza brassicae*) and blackleg resistance (*Leptosphaeria maculans*)	Five to seven QTL for light leaf spot resistance on leaf and 3–5 QTL for light leaf spot on stems; Ten QTL for blackleg resistance	Pilet et al. (1998a,b)
B. napus	Clubroot resistance	One major gene, Pb-Bn1, and two QTL mapped on LG4 and LG15	Manzanares-Daulex et al. (2000)

bulks were constructed. As the unselected loci in the two bulks are random, they are likely to be monomorphic. The BSA approach has been used to identify the marker-linked restorer gene *Rfo* in the Ogura radish, CMS lines of *B. napus* (Delourme et al. 1994), seed coat color in *B. napus* (Van Deynze et al. 1995) and *B. juncea* (Negi et al. 2000), genes for erucic acid content in *B. napus* (Ecke et al. 1995; Jourden et al. 1996c), and QTL for oleic acid content in both *B. rapa* ssp. *oleifera* (Tanhuanpää et al. 1996) and *B. napus* (Schierholt et al. 2000).

3.4.2 Seed Coat Color Trait

The seed coat in rapeseed and mustard varies from yellow to brown, with intermediary shades. The yellow-seeded varieties have inherent advantage over the dark-seeded varieties in both qualitative and quantitative terms, with lower crude fiber content in the seed hull, high protein content in the seed meal, and higher oil content in the seed (Stringam et al. 1974; Daun and DeClercq 1988). Such characteristics are of benefit to the oil industry. However, most varieties grown for commercial cultivation are dark-seeded with very few yellow/light-seeded varieties. An example in *B. campestris* is the variety "Candle" which has been released for cultivation in Canada (Stringam 1980), and YsPb (yellow sarson) and Y1D which have been released for commercial cultivation in India. In *B. juncea,* all varieties grown for commercial cultivation purposes in India (BIO902, Pusa bold, Varuna, Kranti, Pusa barani, and RLM 198) are brown-seeded. Attempts have been made to develop artificially synthesized yellow-seeded varieties in *B. campestris*, *B. napus,* and *B. juncea* (Jönsson 1975, 1977; Abraham and Bhatia 1986; Chen et al. 1988; Henderson and Pauls 1992; Rashid et al. 1993).

The seed coat coloration results from deposition of condensed polyphenols or polymers of leukocyanidin pigments in the palisade layer and partially in the parenchymal layer of the testa in *Brassica* species (Vierhover et al. 1920; Vaughan 1956; Vaughan et al. 1963; Leung et al. 1979; Van Caseele et al. 1982). The difference in the colors, i.e., brown, light brown, reddish brown, and yellow has been ascribed to variable amounts of polyphenols in the seed coat, the lowest amount being detected in yellow seeds (Theander et al. 1977). Leukocyanidins are derivatives of the flavonoid biosynthesis pathway, which has been shown to affect seed coat color in different plant species. (Van Tunen et al. 1988; Wienand et al. 1986; Jende-Strid 1991; Sparvoli et al. 1994; Shirley et al 1992, 1995; Wang et al. 1994).

3.4.2.1 Genetics

The inheritance pattern of seed coat color has been studied in order to understand the genetics of control, which in turn helps in localizing regions of the genome that control this trait and in breeding varieties with a desired seed coat color. In a pioneering study, Mohammed et al. (1942) reported the involvement of three independent genes (Br_1, Br_2, and Br_3) in the control of seed coat color in oleiferous *Brassica* members.

The seed coat color in *Brassica campestris* has been variously shown to be controlled by one gene (Ahmed and Zuberi 1971; Chen and Heneen 1992; Teutonico and Osborn 1994), two (Stringam 1980), three (Mohammed et al. 1942; Jönsson 1975), and multiple genes (Schwetka 1982). In *B. napus*, Shirzadegan (1986) observed a complete maternal inheritance of seed coat color controlled by three genes. Independent investigations by Van Deyzne and Pauls (1994) also led to similar conclusions, and a three-gene model with maternal genotypic control of seed coat color in *B. napus* was proposed by these authors. The inheritance of seed coat color in mustard (*B. juncea*) has been reported by Vera et al. (1979), Vera and Woods (1982), and Anand et al. (1985). An exclusive maternal inheritance was observed and brown seed coat was dominant over yellow. A segregation ratio of 15 brown:1 yellow indicated a digenic control of the trait. (Vera and Woods 1982; Anand et al. 1985).

Environmental influence on seed coat color has been recorded by various researchers, and the observed variation that has defied explanation by laws of genetics has been ascribed to environmental factors (Jönsson 1977; Schwetka 1982; Shirzadegan 1986). A genotype-dependent reduction in seed coat pigmentation for light-seeded doubled haploid in rapeseed (*B. napus*) lines growing at an increasing day/night temperature of 16/12 °C to 24/16 °C was reported by Van Deyzne et al. (1993).

3.4.2.2 Tagging of Seed Coat Color

Since yellow/light-seeded character is recessive in nature, maternally inherited, and influenced by environmental factors, it is desirable that the locus responsible for seed coat color be linked to molecular markers. The seed coat color gene has been tagged using RFLP, RAPD, and AFLP markers. RFLP markers linked to seed coat color in *B. napus* were identified using the BSA approach where individual plants homozygous for seed coat (black vs. yellow) were bulked and their DNA screened with RFLP markers to identify linkage to seed coat color trait. RFLP markers 4NF6a and pLF15 were found to be associated with the seed coat color trait in *B. napus*. RFLP markers associated with this trait were then mapped onto *B. napus* linkage maps (Van Deynze et al. 1995). In *B. napus*, one of the mapped markers corresponded to a clone for chalcone synthase (CHS) an enzyme from the flavonoid biosynthesis pathway (Van Deyzne et al. 1995). Direct biochemical evidence to prove that the clone indeed is involved in seed coat coloration is yet to be shown in *Brassica* species, but has been demonstrated in soybean (Wang et al. 1994).

RAPD markers linked to the seed coat color in *B. campestris* using *B. campestris-oleracea* addition lines were developed by Chen et al. (1997). In another study, Teutonico and Osborn (1994) reported a 3:1 ratio of segregation of brown:yellow seed in *B. rapa* and the locus controlling brown versus yellow seed color was mapped to linkage group 5 (LG 5). A combination of bulked segregant analysis (BSA) and AFLP was used to identify the markers linked to seed coat color in *B. juncea* (Negi et al. 2000). Plants homozygous for yellow and brown seed coat were generated via selfing. A 15:1 segregation of brown:yellow coat was observed indicating a control by two genes, as reported earlier (Anand et al. 1985; Chauhan and Kumar 1987). Using the AFLP technique on pooled samples of homozygous yellow and brown-coated samples, a tightly linked marker AFLP8, among several others, was identified. Sequences adjacent to this marker were isolated and characterized by the PCR walking technique (Siebert et al. 1995; Devic et al. 1997) and converted into a simple PCR-based codominant SCAR marker (SCM-08). This AFLP/SCAR marker could be linked to one of the two genes responsible for the seed coat color in *B. juncea*. This marker could distinguish between yellow and brown lines and also between homozygous and heterozygous brown seeded lines (Negi et al. 2000) and was also effective across different accessions of *B. juncea*.

3.4.3 Mapping of Other Important Traits in Brassica

3.4.3.1 Disease Resistance

In two independent studies, molecular markers have been generated for the genes conferring resistance to *Leptosphaeria maculans* in *B. napus*. Ferreira and coworkers (1995a) generated a marker on LG 6, whereas Dion et al. (1995) showed that the resistance locus $LmFr_1$ was linked to markers cDNA 011 and cDNA 110. Resistance to clubroot disease caused by *Plasmodiophora brassicae* in *B. oleracea* is conferred by loci pb-3 and pb-4 and has been linked to RFLP and AFLP markers (Voorrips et al. 1997). In another study, markers 14a on LG 1, marker 48 on LG 4, and 177b on LG 9 were linked to resistance to *Plasmodiophora brassicae* wor. race 7 in *B. oleracea* (Figdore et al. 1993). Furthermore, a major gene Pb-Bn1 mapped on LG4, and two additional QTL on LG4 and LG15, were found to confer resistance to *Plasmodiophora brassicae* in *B napus* (Manzanares-Dauleux et al. 2000). A QTL controlling resistance to *Xanthomonas campestris* was mapped on LG 1 and LG 9 in *B. oleracea* (Camarago et al. 1995). A locus conferring resistance to white rust caused by *Albugo candida* has been mapped by various workers (Ferreira et al. 1995c; Cheung et al. 1998). One locus *ACA1*, conferring resistance to *Albugo candida* race 2 in *B. rapa* was mapped to LG 4 and was flanked by markers ec2b3a and wg6c1a (Kole et al. 1996). In *B. juncea* two RAPD markers, WR2 and WR3, linked to white rust resistance locus $Ac2_1$ were identified after BSA (Prabhu et al. 1998). In an independent investigation, one cosegregating RFLP marker (X140a) and two closely linked RFLP markers (X42 and X83) were linked to the *Acr* gene that confers resistance to white rust, and mapped in *B. juncea* on LG 7 (Cheung et al. 1998). Bulked segregant analysis coupled with the AFLP tool was employed to identify markers linked to white rust resistance in *B. juncea*. These markers were introgressed into the *B. juncea* genome from the *B. napus* donor genome during an interspecific breeding program (Somers et al. 2002).

3.4.3.2 Vernalization Requirement and Flowering Time

A number of markers linked to the loci controlling the vernalization requirement and flowering time have been identified (Ferreira et al. 1995b; Bohuon et al. 1998). In *B. napus*, linked markers for flowering time were identified and mapped on LG 9 (Ferreira et al. 1995b). In *B. rapa*, two genomic regions determining the plant habit (annual/biennial) were identified. These regions correspond to the genomic segment carrying genes related to the flowering time in *B. napus* (Ferreira et al. 1995b; Teutonico and Osborn 1995). In *B. oleracea*, a total of five QTL were detected for flowering time and reproductive traits. One of these QTL was associated with S-locus *slg6* and another with petiole length (Camarago and Osborn 1996; Camarago et al. 1997). In contrast, Bohuon and coworkers (1998) detected 6 QTL for flowering time in *B. oleracea* on different chromosomes, and significantly, all of these corresponded to a single tract of *A. thaliana* chromosome 5 which contains a number of flowering-related genes (Bohuon et al. 1998). In a fine-scale comparative mapping analysis between *A. thaliana* and *B. nigra*, homologues of the CONSTANS (CO) gene determining flowering time in *A. thaliana* were detected in *B. nigra* (Lagercrantz et al. 1996). The CO gene, which promotes flowering under long-daylight conditions in *A. thaliana*, was isolated by chromosome walking (Putterill et al. 1993; Robert et al. 1998). In *B. rapa*, the VFR-2 locus, identified to be responsible for vernalization-responsive flowering time, was shown to cosegregate perfectly with the flowering locus (FLC) identified from *A. thaliana* (Kole et al. 2001).

3.4.3.3 Fatty Acid Content

Brassica species are grown as a major oilseed crop, and manipulation of oil quality and quantity has been a primary objective of crop improvement programs. Towards this end, studies have been undertaken to generate markers linked to genes for various fatty acids such as linolenic, linoleic, oleic, palmitic, and erucic acids. Two RAPD markers, K-01_{1100} and 25a, were found to be linked to the linolenic acid concentration (Hu et al. 1995; Tanhuanpää et al. 1995). Later, Hu et al. (1999) reported the conversion of an RAPD marker to a codominant SCAR marker linked to linolenic acid content in *B. napus*. One of the alleles of the codominant SCAR markers was linked to low linolenic acid concentration (allele A) and another allele was linked to the high linolenic acid concentration (allele B). This SCAR marker is also probably linked to the omega-3-desaturase gene in *B. napus* (Hu et al. 1999). The same study also reports identification and linkage of RAPD markers to oleic and linoleic acid levels in *B. napus* (Hu et al. 1999). Markers linked to genomic regions controlling linolenic acid concentration in *B. napus* that corresponded to *fad3* (omega-3-desaturase) gene in *A. thaliana* (Arondel et al. 1992) were generated by Jourden et al. (1996a,b) and Thormann et al. (1996). One QTL containing six markers was found to be associated with the contents of oleic, palmitic, and linoleic acids in *B. rapa* (Tanhuanpää et al. 1996). Three AFLP markers linked to oleic acid content in *B. napus* were identified using BSA. On mapping, these markers localized near one copy of the *fad*2 gene (Schierholt et al. 2000).

Different workers have reported the involvement of two QTL in erucic acid synthesis. Two loci controlling synthesis of erucic acid were linked to markers RP841.H1 on LG 6 and marker RP1544.E1 on LG 12 in *B. napus* (Ecke et al. 1995). Two QTL controlling erucic acid and linolenic acid were detected in *B. napus* which mapped onto LG 7 and LG 15 (Thormann et al. 1996). In another study, two QTL, E1, and E2 were identified as being associated with the erucic acid level in *B. napus* (Jourden et al. 1996c). QTL E1 has six RAPD markers and QTL E2 has eight RAPD markers. Later, it was shown that QTL E1 and E2 corresponds to the *β*-ketoacyl-synthase (KCS) gene, a candidate gene for control of the erucic acid level in *B. napus*. The two regions correspond to the *Fae*1 gene derived from *B. campestris* and *B. oleracea*, the two parental species of *B. napus* (Fourmann et al. 1998). Erucic acid loci in *B. napus* has been linked to molecular markers by various workers (Ecke et al. 1995; Thormann et al. 1996; Jourden et al. 1996c; Barret et al. 1998b) using BSA and RFLP analysis. Ecke et al. (1995) have placed the RAPD markers on LG 6 and LG 12, whereas Thormann et al. (1996) have localized the RAPD markers onto LG 7 and LG 15 of *B. napus*. Jourden et al. (1996c) have mapped 14 RAPD markers linked to erucic acid content to two different loci which were termed E1 and E2 loci. Among other *Brassica* species, erucic acid loci have been linked to molecular markers in *B. rapa* (syn *campestris*) using RFLP markers (Teutonico and Osborn 1994).

3.4.3.4 Seed Glucosinolate Content

A number of QTL that varies between four (Uzunova et al. 1995; S.S. Banga, pers. comm.) and five (Toroser et al. 1995) has been detected for seed glucosinolate content in *B. napus*. These QTL, GSL-1, GSL-2, GSL-3, GSL-4, and GSL-5 mapped on LG 20, LG 1, LG 18, LG 4, and LG 13, respectively. Of these, GSL-1 and GSL-2 were associated with low and high glucosinolate content, respectively (De Quiroz and Mithen 1996).

3.4.3.5 Fertility Restorer Gene

Molecular markers have been generated for the two endogenous (*nap* and *pol*) and one introgressed (*ogu* from radish) nuclear restorer genes of the cytoplasmic male sterile (CMS) system in *B. napus*. BSA was employed by Delourme and his coworkers (1994) to identify four RAPD markers OPC-02_{1150}, OPD-02_{1000}, OPF-06_{1200}, and OPG-02_{700} linked to the fertility restorer gene for the Ogura radish CMS system (Delourme et al. 1994). These RAPD fragments linked to restorer gene *Rfo*, in Ogura radish CMS lines of *B. napus* were converted into SCAR markers (Delourme et al. 1994). Further analysis led to the identification of 13 new RAPD and 4–10 new RFLP markers linked to the *Rfo* gene (depending on the mapping population analyzed). These markers proved effective in identifying *B. napus* linkage group carrying the radish introgression (Delourme et al. 1998).

The *Rfp1* and *Rfp2* loci responsible for fertility restoration in polima CMS were analyzed in three different mapping populations. – CMS "Karat" (rfprfp:pol) × pol "restorer Westar" (RfpRfp:pol), in population KW; "Westar" (rfprfp:nap) × "Westar Rf" (RfpRfp:pol), in population WW; and CMS "Karat" (rfprfp:pol) × pol "restorer UM2353" (Rfp2Rfp2:pol), in population KU. The *Rfp1* locus was linked to ten RFLP and one RAPD marker. Of the 11 markers, 5 were exclusive to one population (KW), 2 to another (WW), and 4 markers were common to both the populations (KW and WW). For *Rfp2* locus, three linked markers were identified in the KU population (Jean et al. 1997). One probe, cRF1b, showed perfect linkage to *Rfp1* and *Rfp2* loci (Jean et al. 1997). The identical map position of the two loci, *Rfp1* and *Rfp2*, obtained in different mapping populations indicates that these could be allelic.

The restorer for *nap* CMS system in *B. napus* is encoded by the *Rfn* locus identical to the *Mmt* gene that modifies the transcripts from several mitochondrial regions. The *Rfn* locus was linked to an RFLP probe 3NF2 that segregates with male fertility. The RFLP mapping data indicate that the *Rfn*/*Mmt* and the *Rfp* loci are located at similar positions on the *B. napus* map, and thus could be different alleles of a single nuclear locus (Li et al. 1998). Mapping studies indeed revealed that different forms of the endogenous *Brassica* CMS system are restored by different alleles of same nuclear gene. The restorer gene product could be associated with the modification of the CMS-associated mitochondrial transcripts (Li et al. 1998).

*3.4.3.6 Dwarf BREIZH Gene (*bzh*)*

The dwarf BREIZH (bzh) gene in *B. napus* was tagged using RAPD markers OPM-07_{730} (Foisset et al. 1995). The dominant RAPD marker was converted into a codominant SCAR marker which amplified two loci, a and b. Locus a was localized on linkage group DY 6 of *B. napus* at 0.8 cM from the *bzh* locus and at 8.1 cM from the RFLP locus 1NG3a. Locus b has been mapped on linkage group DY12 at 8.1 cM from the RFLP locus 1NG3b. The two loci, a and b, correspond to the genomic segment derived from the *B. campestris* and *B .oleracea* genomes, respectively (Barret et al. 1998a).

3.4.4 Map-Based Cloning

The tagged markers can be used to clone important genes, a procedure called map-based cloning or positional cloning. The process of target locus/loci identification has been hastened by the combined approach of forward (phenotype to genotype) and reverse (genotype to phenotype) genetics and high throughput analysis (microarrays and SNP). The high-density maps of *Arabidopsis* that are being developed and the availability of a vast range of mutants are likely to speed up the process of positional cloning (as aided by the molecular markers tagged to the locus/loci controlling various traits). There are now a variety of strategies available to create, identify, and analyze the affected locus/loci (Michaels and Amasino 1998; Mayer et al. 1999; Lin et al. 1999; Liu et al. 1999; Ponce et al. 1999; Lukowitz et al. 2000; Springer 2000). A number of genes from *Arabidopsis,* such as SUP (Sakai et al. 1995), NIM1 (Ryals et al. 1997), WUS (Mayer et al. 1998), plus others, have been the target of detailed investigations and map-based or positional cloning strategies. Similarly, based on the conserved peptide sequence from *A. thaliana* corresponding to six loci, namely, *AGAMOUS* (*AG*; floral meristem control), *APETALAE* (*AP*; floral meristem control), *LEAFY* (*LFY3*; floral meristem control), *FAD7* (chloroplastic linoleate desaturase), *FAD3* (cytosolic linoleate desaturase), and *ADH* (alcohol dehydrogenase) were amplified from *B. napus*,

B. oleracea, and *B. rapa* based on *A. thaliana* genome information (Brunel et al. 1999). This method demonstrated the feasibility of isolating homologues from the *Brassica* crops (Brunel et al. 1999). It is expected that similar approaches of identification of important genes and their map-based cloning will become a reality in *Brassica* rather sooner than later.

4 Conclusions

The usefulness of molecular markers and their application in advancing plant breeding and genetics is enormous. It is now certain that the modern practices of breeding to expedite and increase productivity will be dependent on molecular markers and techniques. These markers have been applied at all stages of plant breeding – from trait identification and introgression, through marker-assisted breeding, and finally selection. Molecular markers play a significant role in selecting parental genotypes in breeding programs, elimination of linkage drag in backcross breeding, and selection for traits which are otherwise difficult to score using morphological markers. Molecular markers thus have made the process of plant breeding a very precise approach as opposed to a methodology followed till date wherein visual or phenotypic scoring was relied upon. Among the *Brassica* species, DNA-based markers (RFLP, RAPD, AFLP, CAPS, SCAR) have been used to construct genetic maps after initial attempts with morphology and isozymes were unsuccessful. Work is now underway to saturate these maps with a high density positioning of markers in order to achieve fine-scale mapping. A comparison of maps between species has provided information about the genomic relationships and genome evolution. Such investigations have revealed genome collinearity or synteny not only among *Brassica* species but also between *Brassica* and *Arabidopsis*. An analysis of the genetic maps also indicated that extensive inter- and intragenomic hybridization and recombination events, duplication, deletions, inversions, and translocations in the ancestral genome were responsible for the evolution of the present day *Brassica* species, which are secondary polyploids.

With the advancements in generating highly saturated maps covering parts of the chromosomes, molecular markers have been linked to a wide range of traits. However, there are still large tracts of the chromosomal region that need to be mapped. This is because of the presence of a vast amount of repetitive DNA elements in the *Brassica* genome which render mapping of these regions nearly impossible. The traits in *Brassica* that have been linked to the markers include those controlled by single or few genes, i.e., mono- or digenomic traits following Mendelian segregation (seed coat color) and those which have a complex inheritance pattern and are governed by multiple genes, termed quantitative trait loci or QTL (fatty acid content, vernalization and flowering-time requirement). Markers tightly linked to a trait have provided means for devising strategies for chromosomal landing and map-based cloning in which the linked markers are used to identify and isolate genes from large-fragment DNA libraries. Such strategies are and will continue to be extensively used for isolation and cloning of genes of importance. The relationship between *A. thaliana* and *Brassica* species presents an opportunity to evaluate comparative mapping and genomics, to understand evolution and the functioning of biological systems, and to extrapolate information gathered in model plant systems to important crop plants.

Acknowledgements. Research in our laboratories has received financial support from Department Biotechnology, Government of India, and Indo-French Centre for the Promotion of Advanced Research (Centre Franco Indien Pour La Promotion de la Recherche Avancee, IFCPAR-CEFIPERA).

References

Abraham V, Bhatia CR (1986) Development of strains with yellow seed coat in Indian mustard (*Brassica juncea* Czern. and Coss.). Plant Breed 97:86–88

Acarkan A, Rossberg M, Koch M, Schmidt R (2000) Comparative genome analysis reveals extensive conservation of genome organization for *Arabidopsis thaliana* and *Capsella rubella*. Plant J 23:55–62

Adams MD, Kelley JM, Gocayne JD, Dubnick M, Polymeropoluos M H, Xiao H, Merril CR, Wu A, Olde B, Moreno RF, Kerlavage AR, McCombie W R, Venter JC (1991) Complimentary DNA sequencing: expressed sequence tags and human genome project. Science 252:1651–1656

Ahmed SU, Zuberi MI (1971) Inheritance of seed coat colour in *Brassica campestris* – variety Toria. Crop Sci 11:309–310

Anand IJ, Reddy WR, Rawat DS (1985) Inheritance of seed colour in mustard. Indian J Genet 45:34–37

Arondel V, Lemieux B, Hwang T, Gibson S, Goodman HM, Somerville CR (1992) Map-based cloning of a gene controlling omega-3 fatty acid desaturation in *Arabidopsis*. Science 258:1352–1355

Arus P, Orton TJ (1983) Inheritance and linkage relationships of isozyme loci in *Brassica oleracea*. J Hered 74:405–412

Axelson T, Bowman CM, Sharpe AG, Lydiate DJ, Lagercrantz U (2000) Amphidiploid *Brassica juncea* contains conserved progenitor genomes. Genome 43:679–688

Axelsson T, Shavorskaya O, Lagercrantz U (2001) Multiple flowering time QTLs within several *Brassica* species could be the result of duplicated copies of one ancestral gene. Genome 44:856–64

Barret P, Delourme R, Foisset N, Renard M (1998a) Development of a SCAR (sequence characterized amplified region) marker for molecular tagging of the dwarf BREIZH (*Bzh*) gene in *Brassica napus* L. Theor Appl Genet 97:828–833

Barret P, Delourme R, Renard M, Domergue F, Lessire L Delseny M, Roscoe TJ (1998b) A rapeseed *FAE1* gene is linked to the E1 locus associated with variation in the content of erucic acid. Theor Appl Genet 96:177–186

Becker J, Vos P, Kuiper M, Salamini F, Heun M (1995) Combined mapping of AFLP and RFLP markers in barley. Mol Gen Genet 249:65–73

Bennetzen JL (2000) Comparative sequence analysis of the plant nuclear genomes: microcollinearity and its many exceptions. Plant Cell 12:1021–1029

Bevan M, Murphy G (1999) The small, the large and the wild: the value of comparisons in plant genetics. Trends Genet 15:211–214

Bhatia S, Das S, Jain A, Lakshmikumaran M (1995) DNA fingerprinting of the *B. juncea* cultivars using the microsatellite probes. Electrophoresis 16:1750–1754

Bohuon EJR, Keith DJ, Parkin IAP, Sharpe AG, Lydiate DJ (1996) Alignment of the conserved C genomes of *Brassica oleracea* and *Brassica napus*. Theor Appl Genet 93:833–839

Bohuon EJ, Ramsay LD, Craft JA, Arthur AE, Marshall DF, Lydiate DJ, Kearsey MJ (1998) The association of flowering time quantitative trait loci with duplicated regions and candidate loci in *Brassica oleracea*. Genetics 50:393–401
Botha AM, Venter E (2000) Molecular maker technology linked to pest and pathogen resistance in wheat breeding. S Afr J Sci 96:233–240
Brunel D, Froger N, Pelletier G (1999) Development of amplified consensus genetic marker (ACGM) in *Brassica napus* from *Arabidopsis thaliana* sequences of known biological function. Genome 42:387–402
Butruille DV, Guries RP, Osborn TC (1999) Linkage analysis of molecular markers and quantitative trait loci in populations of inbred backcross lines of *Brassica napus* L. Genetics 153: 949–964
Camarago LEA, Osborn TC (1996) Mapping loci controlling flowering time in *Brassica oleracea*. Theor Appl Genet 92:610–616
Camarago LEA, Williams PH, Osborn TC (1995) Mapping of quantitative trait loci controlling resistance of *Brassica oleracea* to *Xanthomonas campestris* pv. *campestris* in the field and green house. Phytopathology 85:1296–1300
Camarago LEA, Savides L, Jung G, Nienhuis J, Osborn TC (1997) Location of self-incompatibility locus in an RFLP and RAPD map of *Brassica oleracea*. J Hered 88:57–59
Cavell AC, Lydiate DJ, Parkin IA, Dean C, Trick M (1998) Collinearity between a 30–centimorgan segment of *Arabidopsis thaliana* chromosome 4 and duplicated regions within the *Brassica napus* genome. Genome 41:62–69
Chauhan YS, Kumar K (1987) Genetics of seed colour in mustard (*Brassica juncea* Czern. and Coss.). Cruciferae Newsl 12:22–23
Chen BY, Heneen WK (1992) Inheritance of seed color in *Brassica campestris* L. and breeding for yellow-seeded *B. napus* L. Euphytica 59:157–163
Chen BY, Heneen WK, Jönsson R (1988) Resynthesis of *Brassica napus* L. through interspecific hybridization between *B. alboglabra* Bailey and *B. campestris* L. with special emphasis on seed color. Plant Breed 101:52–59
Chen BY, Heneen WK, Simonsen V (1989) Comparative and genetic studies of isozymes in resynthesized and cultivated *Brassica napus* L., *B. campestris* and *B. alboglabra* Bailey. Theor Appl Genet 77:673–679
Chen BY, Simonsen V, Lanner-Herrera C, Heneen WK (1992) A *Brassica campestris-alboglabra* addition line and its use for gene mapping, intergenomic gene transfer and generation of trisomics. Theor Appl Genet 84:592–599
Chen BY, Jorgensen RB, Cheng BF, Heneen WK (1997) Identification and chromosomal assignment of RAPD marker linked with a gene for seed coat colour in a *Brassica campestris-alboglabra* addition line. Hereditas 126:133–138
Cheng BF, Chen BY, Heneen WK (1994a) Addition of *Brassica alboglabra* Bailey chromosome to *B. campestris* with special emphasis to seed colour. Heredity 73:185–189
Cheng BF, Heneen WK, Chen BY (1994b) Meiotic studies of a *Brassica campestris-alboglabra* monosomic addition line and derived *B. campestris* primary trisomics. Genome 37:584–589
Cheng BF, Heneen WK, Chen BY (1995) Mitotic karyotypes of *Brassica campestris* and *B. alboglabra* and identification of the *B. alboglabra* chromosome in an addition line. Genome 38:313–319
Cheung WY, Friesen L, Rakow GFW, Seguin-Swartz G, Landry BS (1997) An RFLP based linkage map of mustard [*Brassica juncea* (L) Czern. and Coss.]. Theor Appl Genet 94:841–851
Cheung WY, Gugel RK, Landry BS (1998) Identification of RFLP markers linked to the white rust resistance gene (*Acr*) in mustard (*Brassica juncea* (L.) Czern. and Coss.). Genome 41: 626–628
Chevre AM, This P, Eber F, Deschamps M, Renard M, Delseny M, Quiros CF (1991) Characterization of disomic addition lines of *Brassica napus-Brassica nigra* by isozymes, fatty acids and RFLP markers. Theor Appl Genet 81:43–49
Chyi YS, Hoeneke ME, Sernyk JL (1992) A genetic linkage map of restriction fragment length polymorphism loci for *Brassica rapa* (syn. *campestris*). Genome 35:746–757
Conner JA, Conner P, Nasrallah ME, Nasrallah JB (1998) Comparative mapping of the *Brassica* S locus region and its homolog in *Arabidopsis*: implications for the evolution of mating systems in the Brassicaceae. Plant Cell 10:801–812

Costa P, Pot D, Dubos C, Frigerio JM, Pion C, Bodeness C, Bertocchi E, Cervera M-T, Remington DL, Plomion C (2000) A genetic map of maritime pine based on AFLP, RAPD and protein markers. Theor Appl Genet 100:39–48

Das S, Rajagopal J, Bhatia S, Srivastava PS, Lakshmikumaran M (1999) Assessment of genetic variation within *Brassica campestris* cultivars using AFLP and RAPD markers. J BioSci 24: 433–440

Daun JK, DeClercq DR (1988) Quality of yellow and dark seeds in *Brassica campestris* canola varieties Candle and Tobin. JAOCS 65:122–126

Delourme R, Bouchereau A, Hubert N, Renard M, Landry BS (1994) Identification of RAPD markers linked to a fertility restorer gene for the Ogura radish cytoplasmic male sterility of rapeseed (*Brassica napus* L). Theor Appl Genet 88:741–748

Delourme R, Foisset N, Horvais R, Barret P, Champagne G, Cheung WY, Landry BS, Renard M (1998) Characterization of the radish introgression carrying the *Rfo* restorer gene for the *Ogu*-INRA cytoplasmic male sterility in rapeseed (*Brassica napus* L). Theor Appl Genet 97: 129–134

Demeke T, Adams RP, Chibbar R (1992) Potential taxonomic use of random amplified polymorphic DNA (RAPD): a case study in *Brassica*. Theor Appl Genet 84:990–994

De Quiroz HC, Mithen R (1996) Molecular markers for low-glucosinolate alleles in oilseed rape (*Brassica napus* L.). Mol Breed 2:277–281

Devic M, Albert S, Delseny M, Roscoe TJ (1997) Efficient PCR-walking on plant genomic DNA. Plant Physiol Biochem 35:331–339

Dion Y, Gugel RK, Rakow GFW, Seguin-Swartz G, Landry BS (1995) RFLP mapping of resistance to the blackleg disease [casual agent, *Leptosphaeria maculans* (Desm.) Ces. et de Not.] in canola (*Brassica napus* L.). Theor Appl Genet 91:1190–1194

Divaret I, Margale E, Thomas G (1999) RAPD markers on seed bulks efficiently assess the genetic diversity of a *Brassica oleracea* L. collection. Theor Appl Genet 98:1029–1035

Drenkard E, Richter BG, Rozen S, Stutius LM, Angell NA, Mindrinos M, Cho RJ, Oefner PJ, Davis RW, Ausubel FM (2000) A simple procedure for the analysis of single nucleotide polymorphisms facilitates map-based cloning in *Arabidopsis*. Plant Physiol. 124:1483–1492

Dreyer F, Graichen K, Jung C (2001) A major quantitative trait locus for resistance to Turnip Yellow Virus (TuYV, syn. beet western yellows virus, BWYV) in rapeseed. Plant Breed 120: 457–460

Ecke W, Uzunova M, Weissleder K (1995) Mapping the genome of rapeseed (*Brassica napus* L) II. Localization of genes controlling erucic acid synthesis and oil content. Theor Appl Genet 91:972–977

Eiadthong W, Yonemori K, Kanzaki S, Sugiura A, Utsunomiya N, Subhadrabandhu S (2000) Amplified fragment length polymorphism analysis for studying genetic relationships among *Mangifera* species in Thailand. J Am Hortic Soc 125:160–164

Erschadi S, Haberer G, Schöniger M, Torres-Ruiz RA (2000) Estimating genetic diversity of *Arabidopsis thaliana* ecotypes with amplified fragment length polymorphisms (AFLP). Theor Appl Genet 100:633–640

Ferreira ME, Williams PH, Osborn TC (1994) RFLP mapping of *Brassica napus* using doubled haploid lines. Theor Appl Genet 89:615–621

Ferreira ME, Rimmer SR, Williams PH, Osborn TC (1995a) Mapping loci controlling *Brassica napus* resistance to *Leptosphaeria maculans* under different screening conditions. Genetics 85:213–217

Ferreira ME, Satagopan J, Yandell BS, Williams PH, Osborn TC (1995b) Mapping loci controlling vernalization requirement and flowering time in *Brassica napus*. Theor Appl Genet 90:727–732

Ferreira ME, Williams PH, Osborn TC (1995c) Mapping loci controlling resistance to *Albugo candida* in *Brassica napus* using molecular markers. Phytopathology 85:218–220

Figdore SS, Ferreira ME, Slocum MK, Williams PH (1993) Association of RFLP markers with trait loci affecting clubroot resistance and morphological characters in *Brassica oleracea* L. Euphytica 69:33–44

Foisset N, Delourme R, Barret P, Renard M (1995) Molecular tagging of the dwarf BREIZH (*Bzh*) gene in *Brassica napus*. Theor Appl Genet 91:756–761

Foisset N, Delourme R, Barret P, Hubert N, Landry BS, Renard M (1996) Molecular mapping analysis of *Brassica napus* using isozyme, RAPD and RFLP markers on doubled haploid progeny. Theor Appl Genet 93:1017–1025

Fourmann M, Barret P, Renard M, Pelletier G, Delourme R, Brunel D (1998) The two genes homologous to *Arabidopsis FAE1* co-segregate with the two loci governing erucic acid content in *Brassica napus*. Theor Appl Genet 96:852–858

Francisco-Ortega J, Fuertes-Aguilar J, Gomez-Campo C, Santos-Guerra A, Jansen RK (1999) Internal transcribed spacer sequence phylogeny of *Crambe* L. (Brassicaceae): molecular data reveal two Old World disjunctions. Mol Phylogenet Evol 11:361–380

Gale MD, Devos KM (1998) Comparative genetics in the grasses. Proc Natl Acad Sci USA 95:1971–1974

Gupta PK, Varshney RK, Sharma PC, Ramesh B (1999) Molecular markers and their application in wheat breeding. Plant Breed 118:369–390

Hall C, McCallum D, Prescott A, Mithen R (2001) Biochemical genetics of glucosinolate modification in *Arabidopsis* and *Brassica*. Theor Appl Genet 102:369–374

Henderson CAP, Pauls KP (1992) The use of haploidy to develop plants that express several recessive traits using light-seeded canola (*Brassica napus*) as an example. Theor Appl Genet 83:476–479

Heslop Harrison JS (2000) Comparative genome organization in plants: from sequence and markers to chromatin and chromosomes. Plant Cell 12:617–635

Heusden AW van, Ooijen JW van, Vrielink-van Ginkel R, Verbeek W HJ, Wietsma WA, Kik C (2000) A genetic map of interspecific cross in *Allium* based on amplified fragment length polymorphism (AFLP) markers. Theor Appl Genet 100:118–126

Hoenecke M, Chyi YS (1991) Comparison of *Brassica napus* and *B. rapa* genomes based on restriction fragment length polymorphism mapping. In: McGregor DI (ed) Rapeseed in a changing world. Proc 8th International Rapeseed Congr Saskatoon, Saskatchewan, Canada 4, pp 221–226

Hongtrakul V, Heustis GM, Knapp SJ (1997) Amplified fragment length polymorphisms as a tool for DNA fingerprinting sunflower germplasm: genetic diversity among oilseed inbred lines. Theor Appl Genet 95:400–407

Hosaka K, Kianian SF, McGrath JM, Quiros CF (1990) Development and chromosomal localization of genome-specific DNA markers of *Brassica* and the evolution of amphidiploids and $n = 9$ diploid species. Genome 33:131–142

Hu J, Quiros CF (1991) Molecular and cytological evidence of deletions in alien chromosomes for two monosomic addition lines of *Brassica campestris-oleracea*. Theor Appl Genet 81: 221–226

Hu J, Quiros C, Arus P, Struss D, Robbelen G (1995) Mapping of a gene determining linolenic acid concentration in rapeseed with DNA-based markers. Theor Appl Genet 90:258–262

Hu J, Li G, Struss D, Quiros CF (1999) SCAR and RAPD markers associated with 18-carbon fatty acids in rapeseed, *Brassica napus*. Plant Breed 118:145–150

Ignatov AN, Kuginuki Y, Suprunova TP, Pozmogova GE, Seitova A M, Dorokhov DB, Hirai M (2000) RAPD–markers linked to the locus for resistance to the race 4 pathogen for black rot, *Xanthomonas campestris* pv. campestris (Pamm.) Dow., in *Brassica rapa* L. Genetika 36: 357–360

Jackson SA, Cheng Z, Wang ML, Goodman HM, Jiang J (2000) Comparative fluorescence in situ hybridization mapping of a 431-kb *Arabidopsis thaliana* bacterial artificial chromosome contig reveals the role of chromosomal duplications in the expansion of the *Brassica rapa* genome. Genetics 2000 156:833–838

Jain A, Bhatia S, Banga SS, Prakash S, Lakshmikumaran M (1994) Potential use of random amplified polymorphic DNA (RAPD) technique to study the genetic diversity in Indian mustard (*Brassica juncea*) and its relationship to heterosis. Theor Appl Genet 88:116–122

Jean M, Brown GG, Landry BS (1997) Genetic mapping of nuclear fertility restorer genes for the Polima cytoplasmic male sterility in canola (*Brassica napus* L) using DNA markers. Theor Appl Genet 95:321–328

Jeffrey AJ, Wilson V, Thein SL (1985) Individual specific fingerprints of human DNA. Nature 316: 76–79

Jende-Strid B (1991) Gene-enzyme relations in the pathway of flavonoid biosynthesis in barley. Theor Appl Genet 81:668–674

Jönsson R (1975) Yellow-seeded rape and turnip rape. II. Breeding for improved quality of oil and meal in yellow-seeded materials. J Swed Seed Assoc 85:271–275

Jönsson R (1977) Breeding for improved oil and meal quality in rape (*Brassica napus* L.) and turnip rape (*Brassica campestris* L.). Hereditas 87:205–218

Jorgensen RB, Chen BY, Cheng BF, Heneen WK, Simonsen V (1996) Random amplified polymorphic DNA markers of the *Brassica alboglabra* chromosome of a *B. campestris-alboglabra* addition line. Chromosome Res 4:111–114

Jourden C, Barret P, Brunel D, Delourme R, Renard M (1996a) Specific molecular marker of genes controlling linolenic acid content in rapeseed. Theor Appl Genet 93:512–518

Jourden C, Barret P, Horvais R, Delourme R, Renard M (1996b) Identification of RAPD marker linked to linolenic acid genes in rapeseed. Euphytica 90:351–357

Jourden C, Barret P, Horvais R, Foisset N, Delourme R, Renard M (1996c) Identification of RAPD markers linked to loci controlling erucic acid level in rapeseed. Mol Breed 2:61–71

Kaneko Y, Bang SW, Matsuzawa Y (2000)Early bolting trait and RAPD markers in the specific monosomic addition line of radish carrying the e-chromosome of *Brassica oleracea*. Plant Breed 119:137–140

Kaneko Y, Yano H, Bang SW, Matsuzawa Y (2001) Production and characterization of *Raphanus sativus-Brassica rapa* monosomic addition lines. Plant Breed 120:163–168

Kearsey MJ, Ramsay LD, Jennings DE, Lydiate DJ, Bohuon EJR, Marshall DF (1996) Higher recombination frequency in female compared to male meiosis in *Brassica oleracea*. Theor Appl Genet 92:363–367

Keller B, Feulliet C (2000) Collinearity and gene density in grass genomes. Trends Plant Sci 5: 246–251

Kellog EA (1998) Relationship of cereal crops and other grasses. Proc Natl Acad Sci USA 95: 2005–2010

Kianian SF, Quiros CF (1991) Genetic analysis of major multigene families in *Brassica oleracea* and related species. Genome 35:516–527

Kianian SF, Quiros CF (1992) Generation of a *Brassica oleracea* composite RFLP map: linkage arrangements among various populations and evolutionary implications. Theor Appl Genet 84:544–554

Kole C, Teutonico R, Williams PH, Osborn TC (1996) Molecular mapping of a locus controlling resistance to *Albugo candida* in *Brassica rapa*. Genetics 86:367–369

Kole C, Quijada P, Michaels SD, Amasino RM, Osborn TC (2001) Evidence for homology of flowering-time genes VFR2 from *Brassica rapa* and FLC from *Arabidopsis thaliana*. Theor Appl Genet 102:425–430

Konieckzny A, Ausubel M (1993) A procedure for mapping *Arabidopsis* mutations using co-dominant ecotype-specific PCR-based markers. Plant J 4:403–410

Kowalaski SP, Lan T-H, Feldmann KA, Paterson AH (1994) Comparative mapping of *Arabidopsis thaliana* and *Brassica oleracea* chromosomes reveals islands of conserved organization. Genetics 138:499–510

Kurata N, Moore G, Nagamura Y, Foote T, Yano M, Minobe Y, Gale M (1994) Conservation of genome structure between rice and wheat. Bio/Technology 12:276–278

Lagercrantz U (1998) Comparative mapping between *Arabidopsis thaliana* and *Brassica nigra* indicates that *Brassica* genome have evolved through extensive genome replication accompanied by chromosome fusions and frequent rearrangements. Genetics 150:1217–1228

Lagercrantz U, Lydiate D (1995) RFLP mapping in *Brassica nigra* indicates differing recombination rates in male and female meioses. Genome 38:255–264

Lagercrantz U, Lydiate D (1996) Comparative genome mapping in *Brassica*. Genetics 144: 1903–1910

Lagercrantz U, Putterill J, Coupland G, Lydiate D (1996) Comparative mapping of *Arabidopsis* and *Brassica*, fine scale genome collinearity and congruence of genes controlling flowering time. Plant J 9:13–20

Lan T-H, Paterson AH (2000) Comparative mapping of quantitative trait loci sculpting the curd of *Brassica oleracea*. Genetics 155:1927–1954

Lan T-H, Paterson AH (2001) Comparative mapping of QTLs determining the plant size of *Brassica oleracea*. Theor Appl Genet 103:383–397

Lan T-H, DelMonte TA, Reischmann KP, Hyman J, Kowalski SP, McFerson J, Kresovich SN, Paterson AH (2000) An EST-enriched comparative map of *Brassica oleracea* and *Arabidopsis thaliana*. Genome Res 10:776–88

Landry BS, Hubert N, Etoh T, Harada JJ, Lincoln SE (1991) A genetic map for *Brassica napus* based on restriction fragment length polymorphisms detected with expressed DNA sequences. Genome 34:543–552

Landry BS, Hubert N, Crete R, Chiang MS, Lincoln SE, Etoh T (1992) A genetic map for *Brassica oleracea* based on RFLP markers detected with expressed DNA sequences and mapping of resistance genes to race 2 of *Plasmodiophora brassicae* (Woronin). Genome 35:409–420

Lanner C, Bryngelsson T, Gustafson M (1997) Relationship of wild *Brassica* species with chromosome number 2n = 18, based on RFLP studies. Genome 40:302–308

Leung J, Fenton TW, Mueller MR, Clandinin (1979) Condensed tannins of rapeseed meal. J Food Sci 44:1313–1316

Li G, Quiros CF (2001) Sequence-related amplified polymorphism (SRAP), a new marker system based on a simple PCR reaction: its application to mapping and gene tagging in *Brassica*. Theor Appl Genet 103:455–461

Li X-Q, Jean M, Landry BS, Brown GG (1998) Restorer genes for different forms of *Brassica* cytoplasmic male sterility map to a single nucleus locus that modifies transcripts of several mitochondrial genes. Proc Natl Acad Sci USA 95:10032–10037

Lin X, Kaul S, Rounsley S, Shea TP, Benito M-I, Town CD, Fujii CY, Mason T, Bowman CL, Barnstead M, Feldblyum TV, Buell CR, Ketchum KA, Lee J, Ronning CM, Koo HL, Moffat KS, Cronin LA, Shen M, Pai G, Van Aken S, Umayam L, Tallon LJ, Gill JE, Venter JC (1999) Sequence and analysis of chromosome 2 of the plant *Arabidopsis thaliana*. Nature 402:761–768

Liu YG, Shirano Y, Fukaki H, Yanai Y, Tasaka M, Tabata S, Shibata D (1999) Complementation of plant mutants with large genomic DNA fragments by a transformation competent artificial chromosome vector accelerates positional cloning. Proc Natl Acad Sci USA 96:6535–6540

Lombard V, Delourme R (2001) A consensus linkage map for rapeseed (*Brassica napus* L.): construction and integration of three individual maps from DH populations. Theor Appl Genet 103:491–507

Lukowitz W, Gillmor CS, Scheibel W-R (2000) Positional cloning in *Arabidopsis*. Why it feels good to have a genome initiative working for you. Plant Physiol 123:795–805

Manzanares-Dauleux MJ, Delourme R, Baron F, Thomas G (2000) Mapping of one major gene and of QTLs involved in resistance to clubroot in *Brassica napus*. Theor Appl Genet 101: 885–891

Mayer K, Schüller C, Wambutt R et al. (1999) Sequence and analysis of chromosome 4 of the plant *Arabidopsis thaliana*. Nature 402:769–777

Mayer KFX, Schoof H, Häcker A, Lenard M, Jürgens G, Laux T (1998) Role of WUSCHEL in regulating stem cell fate in *Arabidopsis* shoot meristem. Cell 95:805–815

Mazur BJ, Tingey SV (1995) Genetic mapping and introgression of genes of agronomic importance. Curr Opin Biotech 6:175–182

McCouch S (1998) Towards a plant genome initiative: thoughts on the value of cross genera comparisons in the grasses. Proc Natl Acad Sci USA 95:1983–1985

McGrath JM, Quiros CF (1991) Inheritance of isozyme and RFLP markers in *Brassica campestris* and comparison with *B. oleracea*. Theor Appl genet 82:668–673

McGrath JM, Quiros CF, Harada JJ, Landry BS (1990) Identification of *Brassica oleracea* monosomic alien chromosome addition lines with molecular markers reveals extensive gene duplication. Mol Gen Genet 223:198–204

Meinke DW, Cherry JM, Dean C, Rounsley SD, Koornneef M (1998) *Arabidopsis thaliana*: a model plant for genome analysis. Science 282:662–682

Michaels SD, Amasino RM (1998) A robust method for detecting single-nucleotide changes as polymorphic markers by PCR. Plant J 14:381–385

Michelmore RW, Paran I, Kesseli RV (1991) Identification of markers linked to disease-resistance genes by bulked segregant analysis: a rapid method to detect markers in specific genomic regions by using segregating populations. Proc Natl Acad Sci USA 88:9828–9832

Miller JC, Tanksley SD (1990) RFLP analysis of phylogenetic relationships and genetic variation in the genus *Lycopersicon*. Theor Appl Genet 80:437–448

Mohammed A, Sikka S M, Aziz MA (1942) Inheritance of seed colour in some oleiferous *Brassicae*. Ind J Genet Plant Breed 2:112–127

Mohan M, Nair S, Bhagwat A, Krishna TG, Yano M, Bhatia CR, Sasakai T (1997) Genome mapping, molecular markers and marker-assisted selection in crop plants. Mol Breed 3:87–103

Moore G (1995) Cereal genome evolution: pastoral pursuits with "Lego" genomes. Curr Opin Genet Dev 5:717–724

Moore G (2000) Cereal chromosome structure, evolution and pairing. Ann Rev Plant Physiol Plant Mol Biol 51:195–222

Negi MS, Devic M, Delseny M, Lakshmikumaran M (2000) Identification of AFLP fragments linked to seed coat colour in *Brassica juncea* and conversion to SCAR marker for rapid selection. Theor Appl Genet 101:146–152

O'Neill CM, Bancroft I (2000) Comparative physical mapping of segments of the genome of *Brassica oleracea* var. alboglabra that are homoeologous to sequenced regions of chromosomes 4 and 5 of *Arabidopsis thaliana*. Plant J 23:233–243

Osborn TC, Kole C, Parkin IA, Sharpe AG, Kuiper M, Lydiate DJ, Trick M (1997) Comparison of flowering time genes in *Brassica rapa*, *B. napus* and *Arabidopsis thaliana*. Genetics 146: 1123–1129

Paran I, Michelmore RW (1993) Development of reliable PCR-based markers linked to downy mildew genes in lettuce. Theor Appl Genet 85:985–993

Parkin IA, Sharpe AG, Keith DJ, Lydiate DJ (1995) Identification of the A and C genomes of amphidiploid *Brassica napus* (oilseed rape). Genome 38:1122–1131

Pilet ML, Delourme R, Foisset N, Renard M (1998a) Identification of loci controlling field resistance to blackleg disease, causal agent *Leptosphaeria maculans* (Desm.) Ces. et de Not., in winter rapeseed (*Brassica napus* L.). Theor Appl Genet 96:23–30

Pilet ML, Delourme R, Foisset N, Renard M (1998b) Identification of QTL involved in field resistance to light leaf spot (*Pyrenopeziza brassicae*) and blackleg resistance (*Leptosphaeria maculans*) in winter rapeseed (*Brassica napus* L.). Theor Appl Genet 97:398–406

Plieske J, Struss D (2001a) STS markers linked to Phoma resistance genes of the *Brassica* B-genome revealed sequence homology between *Brassica nigra* and *Brassica napus*. Theor Appl Genet 102:483–488

Plieske J, Struss D (2001b) Microsatellite markers for genome analysis in *Brassica*. I. development in *Brassica napus* and abundance in Brassicaceae species. Theor Appl Genet 102:689–694

Ponce MR, Robles P, Micol JL (1999) High throughput mapping in *Arabidopsis thaliana*. Mol Gen Genet 261:408–415

Prabhu K V, Somers DJ, Rakow G, Gugel RK (1998) Molecular markers linked to white rust resistance in mustard *Brassica juncea*. Theor Appl Genet 97:865–870

Prakash S, Hinata K (1980) Taxonomy, cytogenetics and origin of crop Brassicas, a review. Opera Bot 55:1–57

Prasad M, Varshney RK, Roy JK, Balyan HS, Gupta PK (2000) The use of microsatellites for detecting DNA polymorphism, genotype identification and genetic diversity in wheat. Theor Appl Genet 100:584–592

Putterill J, Robson F, Lee K, Coupland G (1993) Chromosome walking with YAC clones in *Arabidopsis*: isolation of 1700 kb of contiguous DNA on chromosome 5, including a 300 kb region containing the flowering time gene CO. Mol Gen Genet 239:145–157

Quiros CF, Ochoa O, Kianian SF, Douches D (1987) Analysis of *Brassica oleracea* genome by the generation of *B. campestris-oleracea* addition lines: Characterization by isozymes and rDNA genes. Theor Appl Genet 74:758–766

Quiros CF, Ochoa O, Douches DS (1988) Exploring the role of x = 7 species in *Brassica* evolution: hybridization with *B. nigra* and *B. oleracea*. J Hered 79:351–358

Quiros CF, Hu J, This P, Chevre AM, Delseny M (1991) Development and chromosomal localization of genome-specific markers by polymerase chain reaction in *Brassica*. Theor Appl Genet 88:402–406

Quiros CF, Hu J, Truco MJ (1994) DNA-based marker *Brassica* maps. In: Phillips RL, Vasil IK (eds) Advances in cellular and molecular biology of plants, vol. I: DNA based markers in plants. Kluwer, Dordrecht, pp 199–222

Ramsay LD, Jennings DE, Bohuon EJR, Arthur AE, Lydiate DJ, Kearsey MJ, Marshall DF (1996) The construction of a substitution library of recombinant backcross lines in *Brassica oleracea* for the precision mapping of quantitative loci. Genome 39:558–567
Rashid A, Rakow G, Downey RK (1993) Development of yellow seeded *Brassica napus* L. through interspecific crosses. In:8th Crucifer Genetics Workshop, Saskatoon, Saskatchewan
Riaz A, Li G, Quresh Z, Swati MS, Quiros CF (2001) Genetic diversity of oilseed *Brassica napus* inbred lines based on sequence-related amplified polymorphism and its relation to hybrid performance. Plant Breed 120:411–415
Robert LS, Robson F, Sharpe A, Lydiate D, Coupland G (1998) Conserved structure and function of the *Arabidopsis* flowering time gene CONSTANS in *Brassica napus*. Plant Mol Biol 37:763–72
Ryals J, Weymann K, Lawton K, Freidrich L, Ellis D, Steiner HY, Johnson J, Delaney TP, Jesse T, Vos P, Uknes S (1997) The *Arabidopsis* NIM1 protein shows homology to the mammalian transcription factor inhibitor I *k* B. Plant Cell 9:425–439
Saal B, Plieske J, Hu J, Quiros CF, Struss D (2001) Microsatellite markers for genome analysis in *Brassica*. II. Assignment of rapeseed microsatellites to the A and C genomes and genetic mapping in *Brassica oleracea* L. Theor Appl Genet 102:695–699
Sadowski J, Gaubier P, Delseny M, Quiros CF (1996) Genetic and physical mapping in *Brassica* diploid species of a gene cluster defined in *Arabidopsis thaliana*. Mol Gen Genet 251:298–306
Sakai H, Medrano LJ, Meyerowitz EM (1995) Role of *SUPERMAN* in maintaining *Arabidopsis* floral whorl boundaries. Nature 378:199–203
Schierholt A, Becker HC, Ecke W (2000) Mapping a high oleic acid mutation in winter oilseed rape (*Brassica napus* L.). Theor Appl Genet 101:897–901
Schilling A (1991) Development of a molecular genetic linkage map in *Brassica rapa* (syn. *Campestris* L.). MSc Thesis, University of Massachusetts, Amherst, Mass., USA
Schmidt R, Acarkan A, Boivin K (2001) Comparative structural genomics in the Brassicaceae family. Plant Physiol Biochem 39:253–262
Schwetka A (1982) Inheritance of seed color in turnip rape (*Brassica campestris* L.). Theor Appl Genet 62:161–169
Sebastian RL, Howell EC, King GJ, Marshall DF, Kearsey MJ (2000) An integrated AFLP and RFLP of *Brassica oleracea* linkage from two morphologically distinct doubled-haploid mapping populations. Theor Appl Genet 100:75–81
Seki M, Narusaka M, Yamaguchi-Shinozaki K, Carninci P, Kawal J, Hayashizaki Y, Shinozaki K (2001) *Arabidopsis* encyclopedia using full length cDNAs and its application. Plant Physiol Biochem 39:211–220
Sharma A, Mohopatra T, Sharma RP (1994) Molecular mapping and character tagging in *Brassica juncea*. I: Degree, nature and linkage relationship of RFLPs and their association with quantitative traits. J Plant Biochem Biotech 3:85–89
Sharpe AG, Parkin IA, Keith DJ, Lydiate DJ (1995) Frequent nonreciprocal translocations in the amphidiploid genome of oilseed rape (*Brassica napus*). Genome 38:1112–1121
Shirley BW, Hanley S, Goodman HM (1992) Effects of ionizing radiation on a plant genome: analysis of two *Arabidopsis transparent testa* mutations. Plant Cell 4:333–347
Shirley BW, Kubasek WL, Storz G, Bruggemann E, Koorneef M, Ausubel FM, Goodman HM (1995) Analysis of *Arabidopsis* mutants deficient in flavonoid biosynthesis. Plant J 8: 659–671
Shirzadegan M (1986) Inheritance of seed colour in *Brassica napus* L. Z Pflanzenzuchtg 96: 140–146
Siebert PD, Chenchick A, Kellogg DE, Lukyanov KA, Lukyanov SA (1995) An improved PCR method for walking in uncloned genomic DNA. Nucleic Acids Res 23:1087–1088
Sillito D, Parkin IA, Mayerhofer R, Lydiate DJ, Good AG (2000) *Arabidopsis thaliana*: a source of candidate disease-resistance genes for *Brassica napus*. Genome 43:452–460
Slocum MK (1989) Analyzing the genomic structure of *Brassica* species using RFLP analysis. In: Helentjaris T, Burr B (eds) Development and application of molecular markers to problems in plant genetics. Cold Spring Harbor Lab Press, Cold Spring Harbor, NY, pp 73–80
Slocum MK, Figdore SS, Kenard WC, Suzuki JY, Osborn TC (1990) Linkage arrangement of restriction fragment length polymorphic loci in *Brassica oleracea*. Theor Appl Genet 80:75–64

Somers DJ, Rakow G, Rimmer SR (2002) *Brassica napus* DNA markers linked to white rust resistance in *Brassica juncea*. Theor Appl Genet DOI 10.1007/s00122–001–0812–1 (online first article)
Song KM, Osborn TC, Williams PH (1988a) *Brassica* taxonomy based on nuclear restriction fragment length polymorphism (RFLPs). 1. Genome evolution of diploid and amphidiploid species. Theor Appl Genet 75:784–794
Song KM, Osborn TC, Williams PH (1988b) *Brassica* taxonomy based on nuclear restriction fragment length polymorphism (RFLPs). 2: Preliminary analysis of sub-species within *B. rapa* (syn. *campestris*) and *B. oleracea*. Theor Appl Genet 76:593–600
Song KM, Osborn TC, Williams PH (1990) *Brassica* taxonomy based on nuclear restriction fragment length polymorphism (RFLPs). 3. Genome relationship in *Brassica* and related genera and the origin of *B. oleracea* and *B. rapa* (syn. *campestris*). Theor Appl Genet 79:497–506
Song KM, Suzuki JY, Slocum MK, Williams PH, Osborn TC (1991) A linkage map of *Brassica rapa* (syn. *campestris*) based on restriction fragment length polymorphism loci. Theor Appl Genet 82:296–304
Song KM, Lu P, Tang K, Osborn TC (1995a) Rapid genome change in synthetic polyploids of *Brassica* and its implications for polyploid evolution. Proc Natl Acad Sci USA 92:7719–7723
Song KM, Slocum, MK, Osborn TC (1995b) Molecular marker analysis of genes controlling morphological variation in *Brassica rapa* (syn. *Campestris*). Theor Appl Genet 90:1–10
Sparvoli F, Martin C, Scienza A, Gavazzi G, Tonelli C (1994) Cloning and molecular analysis of structural genes involved and stilbene biosynthesis in grape (*Vitis vinifera* L.). Plant Mol Biol 24:743–755
Springer PS (2000) Gene traps: tools for plant development and genomics. Plant Cell 12:1007–1020
Srivastava A, Gupta V, Pental D, Pradhan AK (2001) AFLP-based genetic diversity assessment amongst agronomically important natural and some newly synthesized lines of *Brassica juncea*. Theor Appl Genet 102:193–199
Stringam GR (1980) Inheritance of seed color in turnip rape. Can J Plant Sci 60:331–335
Stringam GR, McGregor DI, Pawlowski SH (1974) Chemical and morphological characteristics associated with seed coat colour in rapeseed. In: Proc 4th Int Rapeseed Conf, Giessen, Germany, pp 99–108
Struss D, Bellin U, Robbelen G (1991) Development of B-genome chromosome addition lines of *B. napus* using different interspecific *Brassica* hybrids. Plant Breed 106:209–214
Struss D, Quiros CF, Plieske J, Robbelen G (1996) Construction of *Brassica* B genome synteny groups based on chromosomes extracted from three different sources by phenotypic, isozyme and molecular markers. Theor Appl Genet 93:1026–1032
Tanhuanpää PK, Vilkki JP, Vilkki HJ (1995) Association of a RAPD marker with linolenic acid concentration in the seed oil of rapeseed (*Brassica napus* L). Genome 38:414–416
Tanhuanpää PK, Vikki JP, Vikki HJ (1996) Mapping of a QTL for oleic acid concentration in spring turnip rape (*Brassica rapa* ssp. *oleifera*). Theor Appl Genet 92:952–956
Teutonico RA, Osborn TC (1995) Mapping loci controlling vernalization requirement in *Brassica rapa*. Theor Appl Genet 91:1279–1283
Teutonico RA, Osborn TC (1994) Mapping of RFLP and quantitative trait loci in *Brassica rapa* and comparison to the linkage maps of *B. napus*, *B. oleracea* and *Arabidopsis thaliana*. Theor Appl Genet 89:885–894
Theander O, Aman P, Mikscha GE, Yasuda S (1977) carbohydrates, polyphenols and lignins in seed hulls of different colours from turnip rape. J Agric Food Chem 25:270–273
This P, Ochoa O, Quiros CF (1990) Dissection of *Brassica nigra* genome by chromosome addition lines. Plant Breed 105:211–220
Thomas MR, Scott NS (1994) Sequence-tagged site markers for microsatellites: simplified technique for rapidly obtaining flanking sequences. Plant Mol Biol Rep 12:58–64
Thormann CE, Ferreira ME, Camarago LEA, Tivang JG, Osborn TC (1994) Comparison of rFLP and RAPD markers for estimating genetic relationship among and within cruciferous species. Theor Appl Genet 88:973–980
Thormann CE, Romero J, Mantet J, Osborn TC (1996) Mapping loci controlling the concentration of erucic and linolenic acids in seed oil of *Brassica napus* L. Theor Appl Genet 93: 282–286

Toroser D, Thormann CE, Osborn TC, Mithen R (1995) RFLP mapping of quantitative trait loci controlling seed aliphatic glucosinolate content in oilseed rape (*Brassica napus* L). Theor Appl Genet 91:802–808
Truco MJ, Quiros CF (1994) Structure and organization of the B genome based on a linkage map in *Brassica nigra*. Theor Appl Genet 89:590–598
Truco MJ, Hu J, Sadowski J, Quiros CF (1996) Inter- and intragenomic homology of the *Brassica* genomes: implications for their origin and evolution. Theor Appl Genet 93:1225–1233
Ulloa M, Meredith WR Jr, Shappley ZW, Kahler AL (2002) RFLP genetic linkage maps from four F2.3 populations and a joinmap of *Gossypium hirsutum* L. Theor Appl Genet 104:200–208
U N (1935) Genomic analysis in *Brassica* with special reference to the experimental formation of *B. napus* and peculiar mode of fertilization. Jpn J Bot 7:389–452
Uzunova M, Ecke W, Weissleder K, Robbelen G (1995) Mapping the genome of rapeseed (*Brassica napus* L). I. Construction of an RFLP linkage map and localization of QTLs for seed glucosinolate content. Theor Appl Genet 90:194–204
Van Caseele L, Mills JT, Summer M, Gillespie R (1982) Cytological study of the palisade development in the seed coat of Candle canola. Can J Bot 60:2469–2475
Van Deynze AE, Pauls KP (1994) The inheritance of seed colour and vernalization requirement in *Brassica napus* using doubled haploid populations. Euphytica 74:77–83
Van Deynze AE, Beversdorf WD, Pauls KP (1993) Temperature effects on seed colour in black- and yellow-seeded rapeseed. Can J Plant Sci 73:383–387
Van Deynze AE, Landry BS, Pauls KP (1995) The identification of restriction fragment length polymorphisms linked to seed colour genes in *Brassica napus*. Genome 38:534–542
Van-Tunen AJ, Koes RE, Spelt CE, van der Krol AR, Stuitje AR, Mol JNM (1988) Cloning of the two chalcone flavonone isomerase genes from *Petunia hybrida*: coordinate, light-regulated and differential expression of flavonoid genes. EMBO J 7:1253–1263
Vaughan JG (1956) The seed coat structure of *Brassica intergrifolia* (West) O.E. Schulz var. carinata (A. Br.). Phytomorphology 6:363–367
Vaughan JG, Hemingway JS, Schofield HJ (1963) Contributions to a study of variation in *Brassica juncea* Czern. and Coss. J Linn Soc (Bot) 58:435–447
Vera CL, Woods DL, Downey RK (1979) Inheritance of seed coat color in *Brassica juncea*. Can J Plant Sci 59:635–637
Vera CL, Woods DL (1982) Isolation of independent gene pairs at two loci for seed coat color in *Brassica juncea*. Can J Plant Sci. 62:47–50
Vicente JG, King GJ (2001) Characterisation of disease resistance gene-like sequences in *Brassica oleracea* L. Theor Appl Genet 102:555–563
Vierhover A, Calvenger JF, Ewing CO (1920) Studies in mustard seeds and substitutes: I. Chinese Colza (*Brassica campestris chinoleifera* Viehoever). Agric Res 20:117–140
Voorrips RE, Jongerius MC, Kanne HJ (1997) Mapping of two genes for resistance to clubroot (*Plasmodiophora brassicae*) in a population of doubled haploid lines of *Brassica oleracea* by means of RFLP and AFLP markers. Theor Appl Genet 94:75–82
Vos P, Hogers R, Bleeker M, Reijans M, van de Lee T, Hornes M, Frijters A, Pot J, Peleman J, Kuipers M, Zabeau M (1995) AFLP: a new technique for DNA fingerprinting. Nucleic Acids Res 23:4407–4414
Walsh JA, Sharpe AG, Jenner CE, Lydiate DJ (1999) Characterization of resistance to turnip mosaic virus in oilseed rape (*Brassica napus*) and genetic mapping of TuRB01. Theor Appl Genet 99:1149–1154
Wang C-S, Todd JJ, Vodkin LO (1994) Chalcone synthase mRNA and activity are reduced in yellow soybean seed coats with dominant *I* alleles. Plant Physiol 105:739–748
Warwick SI (1993) Guide to wild germplasm of *Brassica* and allied crops. Part I: Taxonomy and genome status in the tribe Brassicaceae (Cruciferae). Technical Bulletin 1993–14E. Center for Land and Biological Resources Research, Research Branch, Agriculture Canada
Welsh J, McClelland M (1990) Fingerprinting genomes using PCR with arbitrary primer. Nucleic Acid Res 18:7213–7218
Wienand U, Weydemann U, Niesbach-Kloesgen U, Peterson PA, Saedler H (1986) Molecular cloning of *C2* locus of *Zea mays*, the gene coding for chalcone synthase. Mol Gen Genet 203:202–207

Williams GK, Kubelik AR, Livak KJ, Rafalski JA, Tingey SV (1990) DNA polymorphisms amplified by arbitrary primers are useful as genetic markers. Nucleic Acids Res 18:6531–6535

Wisman E, Ohlrogge J (2000) *Arabidopsis* microarray service facilities. Plant Physiol 124: 1468–1471

Yang YW, Lai KN, Tai PY, Ma DP, Li WH (1999) Molecular phylogenetic studies of *Brassica*, *Rorippa*, *Arabidopsis* and allied genera based on the internal transcribed spacer region of 18S-25S rDNA. Mol Phylogenet Evol 13:455–462

Zheng K, Qian H, Shen B, Zhuang J, Lin H, Lu J (1994) RFLP-based phylogenetic analysis of wide compatibility varieties in *Oryza sativa* L. Theor Appl Genet 88:65–69

Zhu T, Wang X (2000) Large-scale profiling of the *Arabidopsis* transcriptome. Plant Physiol 124:1472–1476

Zhu T, Budworth P, Han B, Brown D, Chang H-C, Zou G, Wang X (2001) Toward elucidating the global gene expression patterns of developing *Arabidopsis*: parallel analysis of 8300 genes by a high density oligonucleotide probe array. Plant Physiol Biochem 39:221–242

5 The Molecular Genetic Basis of Flowering Time Variation in *Brassica* Species

T.C. Osborn and L. Lukens

1 Flowering Time Variation in *Brassica* Species

One of the most striking features of the *Brassica* genus is the wide range of morphological diversity within cultivated species. Different morphs have been selected for a variety of uses, including as vegetables, oilseeds, forages, and condiments. This variation is most apparent within *B. rapa* (syn. *campestris*; $n = 10$) and *B. oleracea* ($n = 9$), where broad, parallel diversity exists for many vegetable forms (Fig. 1). *Brassica napus* ($n = 19$), the amphidiploid species derived from hybridization of *B. rapa* and *B. oleracea* (U 1935) also includes diverse crop forms, such as winter and spring oilseed rape, rutabagas, and fodder rape. Many different components of plant growth and development contribute to this morphological variation, and each of these components may be under complex genetic control. Thus, it has been difficult to obtain a comprehensive picture of the genetic control of morphological diversity in *Brassica* crops. However, a major component of this diversity is variation in flowering time, and for this trait considerable progress in our understanding has been made.

The most extreme variants in flowering time among *Brassica* crops are biennials and annuals. Biennials do not flower during the first growing season and have been selected for their edible stems, leaves, or roots, or as overwintering seed crops [for example, Chinese cabbage, turnips, and winter turnip rapes in *B. rapa* (Fig. 1a); cabbage, kale, and kohlrabi in *B. oleracea* (Fig. 1b); and winter oilseed rape, rutabagas, and fodder rapes in *B. napus*]. They require a period at low temperatures in order to flower (vernalization), which can be achieved by over-wintering in field locations with moderate temperate climates, or by placing potted plants in cold chambers. Annuals flower during the first growing season and have been selected for their edible floral parts or as seed crops [for example, spring broccoli, spring turnip rape, and sarson in *B. rapa* (Fig. 1a); broccoli, and cauliflower in *B. oleracea* (Fig. 1b); and spring oilseed rape in *B. napus*]. Wide variation in flowering time also exists among the annual forms, including rapid cycling genotypes that were developed for research and education (Williams and Hill 1986; Fig. 1a,b). Some annual forms will respond to vernalization by flowering more quickly; and biennials also

Department of Agronomy, University of Wisconsin, 1575 Linden Drive, Madison, Wisconsin 53706, USA

Biotechnology in Agriculture and Forestry, Vol. 52
Nagata/Tabata (Eds.) Brassicas and Legumes

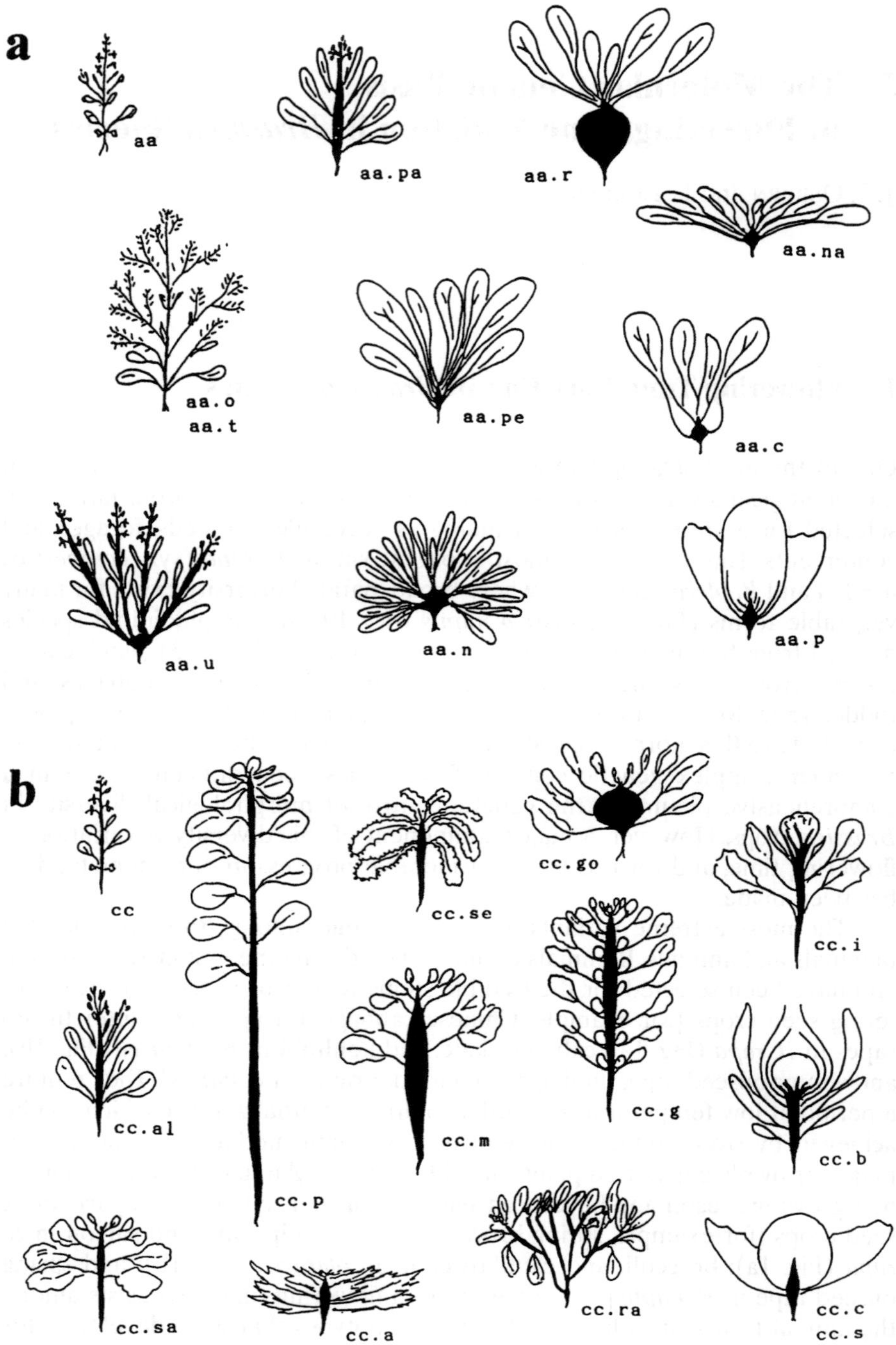

Fig. 1. Morphological diversity among the crop forms of *Brassica rapa* (**a**) and *B. oleracea* (**b**). *Brassica rapa* forms include turnip (*aa.r*), spring turnip rape and sarson (*aa.*o and *aa.t*), spring broccoli (*aa.u*), and Chinese cabbage (*aa.p*). *Brassica oleracea* forms include kohlrabi (*cc.go*), broccoli (*cc.i*), cauliflower (*cc.b*), kale (*cc.u*), and cabbage (*cc.c*). Rapid cycling forms of each species are also shown (*aa* and *cc*). (Reproduced with permission of Paul Williams from the Crucifer Genetics Cooperative Resource Book)

vary in their response to vernalization, some requiring a longer period of cold than others to induce flowering.

Flowering time is a highly heritable trait, and although it can be easily manipulated by phenotypic selection, the identification and characterization of flowering-time genes could facilitate some applications of *Brassica* improvement. For example, winter and spring oilseed rape represent two genetically distinct groups of *B. napus* (Diers and Osborn 1994), and hybridization between these groups would expand genetic diversity and might improve rapeseed germplasm (Butruille et al 1999a,b). Knowledge of the location and effects of genes controlling flowering time variation could facilitate this process by allowing breeders to select for specific genes conferring the desired growth habit while maintaining or advancing the desired genetic background in other portions of the genome. A more ambitious application of flowering time manipulation would be the temporary conversion of biennial forms to annual forms in order to speed up breeding cycles. This will require a thorough understanding of how specific genes affect flowering time and how these effects can be temporarily altered to change flowering habits. The genetic mechanisms underlying flowering-time diversity in *Brassica* crops may apply to other components of morphological diversity within *Brassica* and to variation in other crop species. Thus, the insight we gain from studying flowering time in *Brassica* could have broad applications for understanding phenotypic diversity.

2 Mapping Flowering Time Genes

The first genetic studies on flowering time in *Brassica* species relied on phenotypic analysis of segregating populations in order to estimate heritabilities, gene effects, and gene numbers (for example, Detjen 1926; Dickson 1968). While these studies provided some information on the genetic control of this trait, segregating *Brassica* populations show continuous variation for flowering time, or discontinuous variation that cannot be explained by single Mendelian loci. It was not until molecular markers became available that specific genetic effects could be identified and studied. Molecular markers allow one to construct linkage maps for segregating populations and use statistical methods to detect the positions and effects of quantitative trait loci (QTL). Molecular markers have been used to map flowering time QTL in populations of *B. rapa* (Song et al. 1995; Teutonico and Osborn 1995; Osborn et al. 1997; Axelsson et al. 2001), *B. oleracea* (Axelsson et al. 2001; Bohuon et al. 1998; Camargo and Osborn 1996; Kennard et al. 1994; Lan and Paterson 2000), *B. nigra* (Lagercrantz et al. 1996), *B. napus* (Ferreira et al. 1995; Osborn et al. 1997; Butruille et al. 1999b), and *B. juncea* (Axelsson et al. 2001).

A summary of results from these studies is shown in Table 1. For all analyses, the mapping populations were derived from parents that differed in flowering times. In some cases, the crosses were between annuals and biennials, in

Table 1. Summary of flowering time QTL mapping studies in *Brassica* species

Cross[a]	Population[c]	No. of QTL[d]	Variation explained[e] (%)	Additive effects[f]	Reference
B. rapa					
Michihili × Spr. Broccoli[b]	F_2	1	30	12	Song et al. (1995)
Per × R500[b]	F_3 families	2	75	45 to 64	Teutonico and Osborn (1995)
Per × R500[b]	RI lines	3	55	10 to 34	Osborn et al. (1997)
Per × R500[b]	RI lines	6	~80	7 to 21	Kole et al. (unpubl. data)
R-c-50a7 × rapid cycling	F_2	2	38	2.5 and 2.7	Axelsson et al. (2001)
B. oleracea					
WGA × Packman[b]	F_2	3 (A/B)	37	n.r.	Kennard et al. (1994)
WGA × Packman[b]	F_2	4 (DB)	60	n.r.	Kennard et al. (1994)
BI16 × Cr7[b]	F_3 families	3	54	10 to 16	Camargo and Osborn (1996)
Broccoli × alboglabora	DH lines	5	58	–5 to 2	Bohuon et al. (1998)
Broccoli × alboglabora	DH lines	5	93	–5 to 7	Bohuon et al. (1998)
Cantanese × rapid cycling	F_2	4 (DB)	90	0.02 to 0.05	Lan and Paterson (2000)
Pusa Katki × rapid cycling	F_2	6 (DB)	82	–0.05 to 0.15	Lan and Paterson (2000)
Bugh Kana × rapid cycling	F_2	4 (DB)	65	–0.02 to 0.07	Lan and Paterson (2000)
22.269 × rapid cycling	F_2	2	22	2.6 and 3.1	Axelsson et al. (2001)
B. napus					
Major × Stellar[b]	DH lines	3	64	7 to 22	Osborn et al. (1997)
Major × Stellar[b]	DH lines	5	~77	–5 to 21	Kole et al. (unpubl. data)
Ceres × Westar or Marnoo[b]	IB lines	7	59	–2 to 6	Butruille et al. (1999b)
B. nigra					
Cat × rapid cycling	BC to Cat	3	64	n.r.	Lagercrantz et al. (1996)
B. juncea					
J-o-3DH1 × rapid cycling	F_2	3	56	0.4–0.8	Axelsson et al. (2001)

[a] Later flowering parent in each cross is listed first.
[b] Crosses between biennials and annuals.
[c] RI, recombinant inbred; DH, doubled haploid; IB, inbred backcross; BC, backcross.
[d] Number of QTL reported for days to flowering, except A/B = annual or biennial categorical data; DB = days to bud. Various environments, analysis methods and threshold levels were used. Osborn et al. (1997) and Kole et al. (unpubl.) report results from analyzing the same *B. rapa* and *B. napus* data sets using two different statistical methods. Bohuon et al. (1998) report results from evaluations in two different environments. Axelsson et al. (2001) tested only three genomic regions in *B. rapa* and *B. oleracea* and six genomic regions in *B. juncea*, so addition QTL may have segregated.
[e] Percentage of total variation explained by models containing all QTL effects. Approximate values indicate sum of individual QTL effects.
[f] Range of additive effects for QTL alleles from the late-flowering parent in trait values. n.r., Not reported.

others they were between two annual parents. These populations were evaluated in various environments, and different statistical analyses and significance thresholds were often used to identify and report QTL. Thus, the results of these studies are not easily comparable, but a few points can be observed from the summary in Table 1. First, the number of QTL detected varied quite widely among studies (from one to seven). Some of this variation is due to differences in statistical analyses (for example, the two results shown for Per × R500 RI lines and for Major × Stellar DH lines in Table 1 were from applying different QTL detection methods to the same data sets). However, much of this variation is undoubtedly due to differences among the populations in the number of genes segregating and the magnitudes of their effects. Second, flowering time is a highly heritable trait and the percentage of total variation explained by the QTL models is generally very high. Third, populations derived from crosses of biennial and annual parents generally had QTL alleles with larger effects than populations derived from crosses between annual parents with a smaller difference in flowering time. Finally, some alleles from the late-flowering parent decreased flowering time (negative values in Table 1). These "cryptic effects" have been detected in many QTL mapping studies and provide a genetic explanation for transgressive segregation (Edwards et al. 1992).

An obvious question that arises from comparison of different QTL mapping studies is whether any of the studies detected QTL effects in the same genomic regions, suggesting that the same loci may be segregating in different populations. Here, we encounter another problem: many of these maps were created with different sets of molecular markers, and this makes it impossible to completely align all the maps. Even in cases where some of the same probes were used in different maps (these maps are based almost exclusively on RFLP markers), the replicated nature of *Brassica* genomes means that probes often hybridize to multiple loci, and we do not always know if the same loci are detected in different maps.

Fortunately, *Brassica* species are closely related to the model organism, *Arabidopsis thaliana* ($n = 5$), and many researchers have aligned their maps to the *Arabidopsis* genome. This provides two benefits. First, we can determine that two regions on two different *Brassica* maps are related if they both align to same region of the *Arabidopsis* genome (although they may be homoeologous regions and not homologous regions due to the replicated nature of *Brassica* genomes). Second, we can identify candidate genes from *Arabidopsis* for traits of interest (e.g., flowering time) many of which have been cloned and characterized in *Arabidopsis*.

3 Comparative Mapping Between *Brassica* and *Arabidopsis*

Brassica species are in the same family (Brassicaceae) as *A. thaliana*, and these genera diverged approximately 15–21 million years ago (Yang et al. 1999; Koch et al. 2000). Based on this relationship, one may predict that ancestral chromosomal segments have been conserved during the evolution of these species. The degree of conservation has been addressed in some studies by comparative linkage mapping. Since DNA sequences of homologous genes in the two taxa are quite similar (86% identical for 13 *Arabidopsis* genes on chromosome 4 and their homologues in *B. napus*; Cavell et al. 1998), it is possible to use clones from one species as RFLP probes to map loci in the other species. Results from studies using this approach have provided evidence for conserved linkage arrangements or collinearity between chromosomal segments of *A. thaliana* and *B. oleracea* (Kowalski et al. 1994; Lan et al. 2000) or *B. nigra* (Lagercrantz 1998). Evidence for micro-collinearity between *Brassica* and *Arabidopsis* genomes also has been obtained by comparisons of physical maps of small regions in these genomes (reviewed by Schmidt et al. 2001).

With the recent completion of the entire *Arabidopsis* genome sequence (www.*Arabidopsis*.org), it is now possible to compare the *Brassica* and *Arabidopsis* genomes using a bioinformatics approach. This can be achieved by sequencing DNA clones that have been used as probes to map RFLP loci in *Brassica* species, and then searching the database of the *Arabidopsis* genome sequence for close matches. These analyses are ongoing in several labs and the results are expected to give a much more comprehensive view of the genome relationships. This approach was used to identify regions of the *Arabidopsis* genome containing candidate genes for disease resistance loci in *B. rapa* (Kole et al. 2002), and it is illustrated in Fig. 2 for a region of the *Arabidopsis* genome containing candidate genes for a flowering-time locus in *B. rapa* (see below).

4 Comparative Mapping to Identify Candidate Genes for Flowering Time

To identify genes that may be allelic to *Brassica* flowering time QTL, one can align the regions of *Brassica* maps containing QTL to homoeologous regions of the *Arabidopsis* genome. The aligned *Arabidopsis* intervals can then be

Fig. 2. Genetic map of the region in linkage group R10 (equivalent to Br 8 in Kole et al. 2001) of *Brassica rapa* containing the flowering time locus VFR2 and the corresponding region in the genetic and physical maps of *Arabidopsis thaliana* chromosome V(At 5). The genetic map of *B. rapa* was based on a BC_3S_1 family that segregated in a Mendelian fashion for flowering time due to the action of VFR2, allowing exact placement of the VFR2 locus. Comparison to the genetic map of At5 was based on using RFLP probes from *A. thaliana* to help construct the *B. rapa* map. Comparison to the physical map was based on BLAST searches of the *A. thaliana* genome sequence using DNA sequences of the RFLP probes in the *B. rapa* genetic map (our unpublished data). The comparisons show strong conservation between *B. rapa* and *A. thaliana* for gene order in these regions. The alignment also provides strong evidence that VFR2 is homologous to *A. thaliana FLC*, a flowering time gene encoding a MADS domain protein. Exons 2–6 of an *A. thaliana FLC* cDNA hybridized to an RFLP locus in *B. rapa* that cosegregated exactly (resolution of <0.24 cM) with VFR2

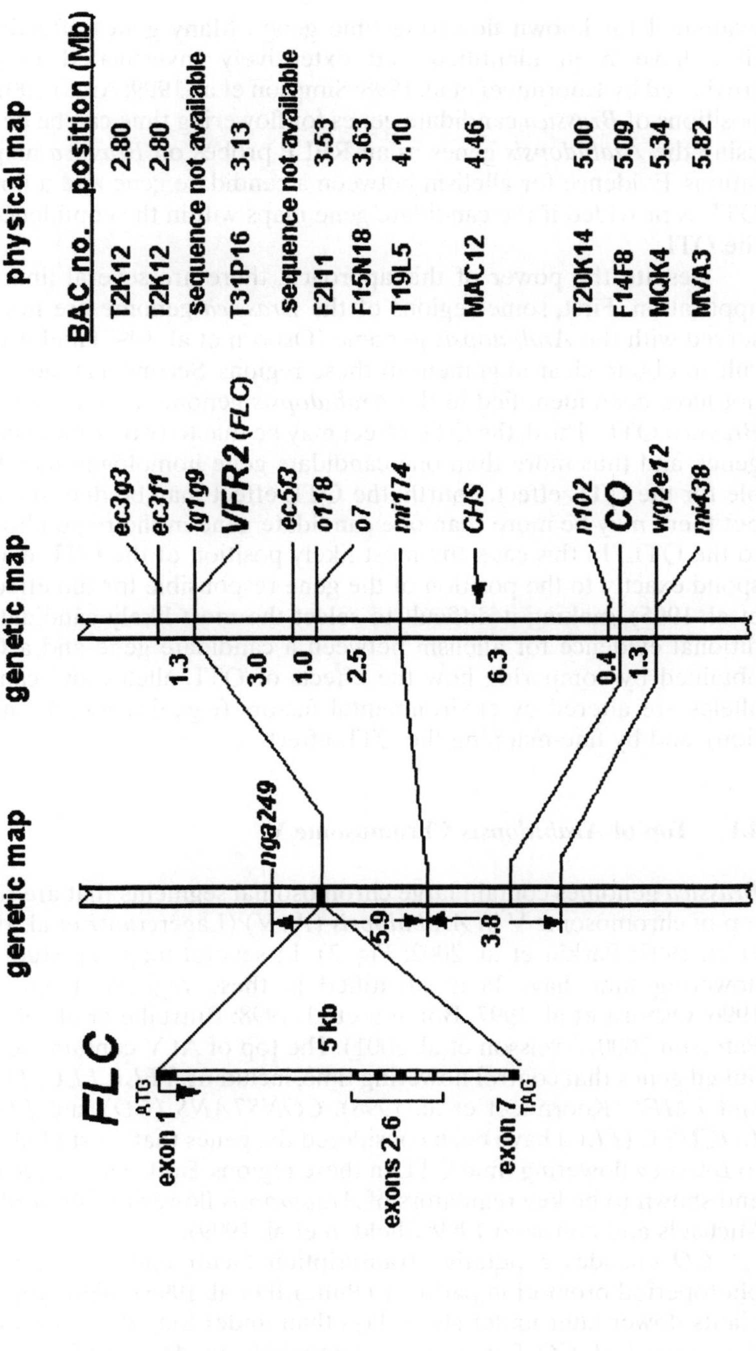
FLC
exon 1
ATG
exons 2-6
exon 7
TAG
5 kb
At 5
genetic map
nga249
5.9
3.8
R10
genetic map
1.3
3.0
1.0
2.5
6.3
0.4
1.3
ec3g3
ec3f1
tg1g9
VFR2 (FLC)
ec5f3
o118
o7
mi174
CHS
n102
CO
wg2e12
mi438
At 5
physical map
BAC no.
position (Mb)
T2K12 2.80
T2K12 2.80
sequence not available
T31P16 3.13
sequence not available
F2I11 3.53
F15N18 3.63
T19L5 4.10
MAC12 4.46
T20K14 5.00
F14F8 5.09
MQK4 5.34
MVA3 5.82

examined for known flowering time genes. Many genes affecting flowering time have been identified and extensively investigated in *Arabidopsis* (reviewed by Koornneef et al. 1998; Simpson et al. 1999; Araki 2001). The exact positions of *Brassica* candidate genes for flowering time can be determined by using the *Arabidopsis* genes as an RFLP probes on *Brassica* mapping populations. Evidence for allelism between a candidate gene and a flowering-time QTL is provided if the candidate gene maps within the confidence interval of the QTL.

Despite the power of this approach, there are several limitations in its application. First, some regions of the *Brassica* genome are not highly conserved with the *Arabidopsis* genome (Osborn et al. 1997) and it may be difficult to obtain clear alignment in these regions. Second, candidate genes may not have been identified in the *Arabidopsis* genome region that aligns to the *Brassica* QTL. Third, the QTL effect may be due to two or more closely linked genes, and thus more than one candidate gene homologue may be responsible for the QTL effect. Fourth, the QTL effect may be due to a single locus, but there may be more than one candidate gene in the region homoeologous to the QTL. In this case, the most likely position of the QTL may not correspond exactly to the position of the gene responsible for the effect. (Doebley et al. 1995), making it difficult to select the most likely candidate gene. Additional evidence for allelism between a candidate gene and a QTL can be obtained by comparing how the effects of QTL alleles and candidate-gene alleles are altered by environmental factors (e.g., day length and vernalization); and by fine-mapping the QTL effect.

4.1 Top of *Arabidopsis* Chromosome V

Brassica genomes contain large chromosomal segments that are similar to the top of chromosome V in *Arabidopsis* (At V) (Lagercrantz et al. 1996, Osborn et al. 1997; Parkin et al. 2002; Fig. 2). In several mapping studies, QTL for flowering time have been identified in these regions (Lagercrantz et al. 1996; Osborn et al. 1997; Bohuon et al. 1998; Butruille et al. 1999b; Lan and Paterson 2000; Axelsson et al. 2001). The top of At V contains several closely linked genes that control flowering time, including *TFL1*, *FLC*, *TFL2*, *CO*, *FY*, and *EMF1* (Koornneef et al. 1998). *CONSTANS* (*CO*) and *FLOWERING LOCUS C* (*FLC*) have been considered the genes that most likely correspond to *Brassica* flowering time QTL in these regions. Both genes have been cloned and shown to be key regulators of *Arabidopsis* flowering (Putterill et al. 1995; Michaels and Amasino 1999; Sheldon et al. 1999).

CO encodes a putative transcription factor and is a regulator of the photoperiod promotion pathway (Putterill et al. 1995). Wild-type *Arabidopsis* plants flower later under short days than under long days; however, mutants that have lost *CO* function are insensitive to day length and flower late under both long- and short-day conditions. Plants that are late-flowering due to a *CO* mutation have very little response to vernalization (Koornneef et al. 1991). Although *CO* alleles that cause later flowering have been generated by mutagenesis, natural allelic variants with phenotypic effects have not been identified (Koornneef et al. 1998).

FLC encodes a MADS domain containing transcription factor and is a key regulator of the autonomous flowering pathway (Michaels and Amasino 1999; Sheldon et al. 1999). Plants with late-flowering alleles have delayed flowering regardless of photoperiod, but vernalization largely negates the effects of these alleles. Levels of *FLC* expression are positively correlated with the delay in flowering, and vernalization diminishes *FLC* transcript levels. High levels of *FLC* expression require a functional allele at another flowering time locus, *FRIGIDA* (*FRI*), and loss of either *FLC* or *FRI* function causes early flowering. Allelic variation at one or both of these loci appears to be responsible for the most extreme flowering-time variants of *Arabidopsis* ecotypes (Koornneef et al. 1998; Johanson et al. 2000; Michaels and Amasino 2000).

4.1.1 CO *as a Candidate Gene*

Based on results from crosses of early- and late-flowering annual parents, several studies have focused on *CO* as the most likely candidate gene for QTL detected in regions homoeologous to the top of At V. Evidence for this hypothesis was based on comparative positional data for QTL and *Arabidopsis* candidates. In *B. nigra*, the likelihood peak of one QTL corresponded to the position of a *CO* homologue, and another QTL contained a *CO* homologue within the 90% confidence interval (Lagercrantz et al. 1996). In *B. oleracea*, the positions of *CO* homologues were inferred to be within the confidence intervals of three out of four QTL detected in three linkage groups (Bohuon et al. 1998). These studies did not include map position data for other candidate genes. In their analysis of *B. oleracea*, *B. rapa,* and *B. juncea* populations derived from annual parents, Axelsson et al. (2001) considered positional information for both *CO* and *FLC* and found that QTL likelihood peaks corresponded more closely to the position of *CO* homologues than *FLC* homologues for six of seven QTL. However, in many cases, positional data for candidate genes were inferred from other maps.

Although *CO* is a strong candidate for the flowering-time QTL detected in these studies, additional evidence would help support this hypothesis. First, segregation data for proximal candidate gene loci, such as *CO* and *FLC*, could be obtained for all QTL regions under study and the QTL effects re-estimated. This would provide better estimates of QTL effects with respect to the positions of candidate-gene loci. Second, given that the effects of *CO* are influenced by day length, if *CO* is allelic to a QTL, one would expect changes in day length to differentially modify the effects of early- and late-flowering QTL alleles. This could be tested by determining how plants segregating for the QTL are affected by long and short photoperiods.

Brassica napus genes homologous to *Arabidopsis CO* have been cloned and characterized for expression and function (Roberts et al. 1998). Four genes were isolated from an annual and a biennial oilseed rape cultivar corresponding to two alleles from two loci on linkage groups N10 and N19. All four genes were expressed in *B. napus*. One of the genes was transformed into a *co* mutant of *Arabidopsis* and it compensated the mutation causing even earlier flowering than in wild-type *Arabidopsis* under both long- and short-day conditions. It would be interesting to know if the annual and biennial alleles cloned from the two loci have different effects on flowering time when transformed into the *Arabidopsis co* mutant; however, results of such experiments have not been reported.

4.1.2 **FLC** *as a Candidate Gene*

Populations derived from annuals × biennials have been studied in *B. rapa* (Teutonico and Osborn 1995) and *B. napus* (Ferreira et al 1995; Butruille et al 1999b), and several QTL detected in those populations also were in regions having homoeology to the top of At V (Osborn et al. 1997). Based on results from these studies, *FLC* and not *CO* was favored as the most likely candidate gene for some QTL. First, it was known a priori that the QTL effects were due to natural allelic variation. *FLC* contributes to flowering time differences among ecotypes of *Arabidopsis*, and it seemed likely that genes that have been under natural selection also may have been targets of artificial selection. Second, like *FLC* but unlike *CO* in *Arabidopsis*, the effects of the late-flowering alleles were greatly diminished after vernalization treatment.

In a subsequent study, additional evidence was obtained that one of these QTL is homologous to *FLC*. This QTL, designated vernalization response flowering time from *B. rapa* 2 (VFR2), had been detected as the largest QTL effect in a segregating population of recombinant inbred (RI) lines, explaining 34% of the flowering time variation with an additive effect of 17 days (Osborn et al. 1997). The key to obtaining clear evidence that VFR2 corresponded to a specific candidate gene was the ability to have it segregate phenotypically as a single Mendelian factor. This was achieved by backcrossing the late-flowering allele at VFR2 from the biennial parent Per into the annual parent R500. After three generations of backcrossing, a BC_3S_1 population was evaluated for flowering time and leaf number in growth chamber and field experiments. In the growth chamber, one-quarter of the population had a distinctly earlier flowering time; and in the field, the population segregated into three distinct flowering time categories in a 1:2:1 ratio. The plants from these distinct flowering time categories were genotyped at RFLP loci near VFR2. DNA probes from *Arabidopsis*, including *CO* and *FLC*, were also used. The result from this analysis showed that VFR2 does not correspond to *CO* because these loci were 13 cM apart, but it cosegregated exactly (resolution of <0.24 cM) with an RFLP locus detected by *FLC* (Kole et al. 2001; Fig. 5.2).

The flowering time difference between homozygous VFR2 genotypes in this population was 95 days in the growth chamber and 43 days in the field, and the late-flowering effect of the Per allele was abolished by vernalization. Heterozygous genotypes were almost exactly intermediate to the homozygous genotypes, indicating complete additive gene action for the late-flowering allele. The delay in flowering was associated with an increase in the number of leaves produced before flowering (differences of 11 and 29 leaves between homozygous genotypes in the growth chamber and field, respectively) (Kole et al. 2001). *Arabidopsis* lines containing a functional *FRI* gene and segregating for early- and late-flowering *FLC* alleles have similar characteristics. Late-flowering alleles have additive gene action and confer vernalization-responsive phenotypes, and leaf number is highly correlated with flowering-time differences (Michaels and Amasino 2000).

In an *FRI*-containing *Arabidopsis*, natural early- and late-flowering alleles of *FLC* both produce RNA transcripts, although late-flowering alleles produce higher transcript levels, which may account for the phenotypic differences of the alleles (S. Michaels and R. Amasino, pers. comm.). In *Brassica*, the large effect of the Per allele at VFR2 on flowering time does not appear to be due to large differences in transcript levels. BC_3S_1 lines with different VFR2 alleles had similar levels of *FLC* RNA in gel-blot hybridizations (Kole et al. 2001). Thus, the flowering-time effects of VFR2 alleles may be due to differences in the accumulation or function of *FLC* protein rather than transcript levels. However, as discussed further below, the *Brassica* genome contains multiple copies of *FLC*, and expression of other *FLC* genes may have obscured potential differences in transcript levels between the early- and late-flowering VFR2 alleles. The biennial parent Per had higher levels of *FLC* transcripts than the BC_3S_1 line containing the Per allele, suggesting that other genes in Per promote expression of *FLC* and contribute to the biennial growth habit. As in *Arabidopsis*, vernalization reduced the levels of *FLC* transcripts.

4.2 Candidate Genes in Other Regions

Data from comparative mapping studies have provided evidence that some flowering time QTL in *Brassica* correspond to candidate genes other than those located at the top of At V. In a study of three segregating populations of *B. oleracea* derived from early- and late-flowering annual parents (Lan and Paterson 2000), several QTL were identified in genome regions that aligned to the top of At V, but QTL were also detected in regions that aligned elsewhere in the *Arabidopsis* genome, each of which contained candidate flowering-time genes. These QTL mapped to regions with homoeology to the top of *Arabidopsis* chromosome I, where *GI* and *FHA* are located; the bottom of chromosome I, where *EFS* is located; the top of chromosome III, where *HY2* and *VRN1* are located; and the bottom of chromosome V, where *EMF2* is located.

The QTL with the largest effect in a biennial × annual cross of *B. napus* (VFN1; 47% of the variation, additive effect of 22 days) was detected in a region of linkage group N2 that shows fractured collinearity with several regions of the *Arabidopsis* genome, but the QTL effect was most closely aligned with portions of At1 and At4. (Osborn et al. 1997). The aligned portion of At1 contains flowering time genes *FT* and *FLM*. *FT* belongs to a family of membrane-bound proteins, and it mediates signals for floral transition downstream of the photoperiod and autonomous pathways (Kardailsky et al. 1999; Kobayashi et al. 1999). *FLM* (syn. MAF1) is a MADS-box gene with homology to *FLC* (Ratcliffe et al. 2001; Scortecci et al. 2001). Late-flowering *FT* mutants and transgenics that over-express *FLM* do not respond to vernalization, unlike *B. napus* plants with the late-flowering VFN1 allele. Thus, neither of these candidate genes may correspond to VFN1, although the late-flowering allele of VFN1 may be unique form of one of these genes. The portion of At4 that aligned with the VFN1 region contains *FRI*, a gene that promotes expression of *FLC* (Johanson et al. 2000). *FRI* is a good candidate gene for VFN1 because allelic variation at *FRI* (along with *FLC*) is responsible for most of the flowering time differences among the earliest and latest flowering ecotypes of *Arabidopsis*, and late-flowering *FRI* lines are responsive to vernalization. This QTL region has a large effect on flowering time, and it should be possible to obtain Mendelian segregation after several generations of backcrossing to use for comparative fine-mapping, as described for *FLC* above.

5 Genetic Redundancy and Variation in Flowering Time

Cytological studies have long provided evidence that the genomes of diploid *Brassica* species (*B. rapa*, *B. oleracea,* and *B. nigra*) are extensively replicated (reviewed by Prakash and Hinata 1980). More recently, RFLP mapping studies also have provided evidence for replication within the diploid *Brassica* genomes (McGrath et al. 1990; Slocum et al 1990; Song et al. 1991; Chi et al. 1992; Hu et al. 1994; Kowalski et al. 1994; Ramsay et al. 1996; Lan et al. 2000). This genome replication is organized in blocks, consistent with a polyploid origin, and triplication of some of these blocks suggests that the diploid *Brassica* species may have evolved from an ancient hexaploid (Lagercrantz and Lydiate 1996; Lagercrantz 1998). Genome replication is compounded in *B. napus*, *B. juncea and B. carinata*, which are recent allopolyploids derived from hybridization of pairs of the diploid species. In these allopolyploid species, many genome regions are present in six copies. An example of *Brassica* genome replication is shown in Fig. 3 for three regions of *B. rapa* (R2, R3, and R10) and *B. oleracea* (O2, O3, and O9), and for the six homologous regions of *B. napus* (N2, N12, N3, N13, N10, and N19).

Polyploidy provides an attractive explanation for the occurrence of segregating QTL in replicated genome regions: these QTL may represent allelic variation at paralogous loci. In segregating populations of *B. oleracea*, Kennard et al. (1994) found that two regions representing a genome duplication each contained QTL for flowering, stem, and leaf traits; and Lan and Paterson (2000) reported that one-third of the QTL for curd traits in *B. oleracea* had possible paralogs. Some of the flowering time QTL detected by Lan and Peterson (2000) were in duplicated regions having homoeology to the top of

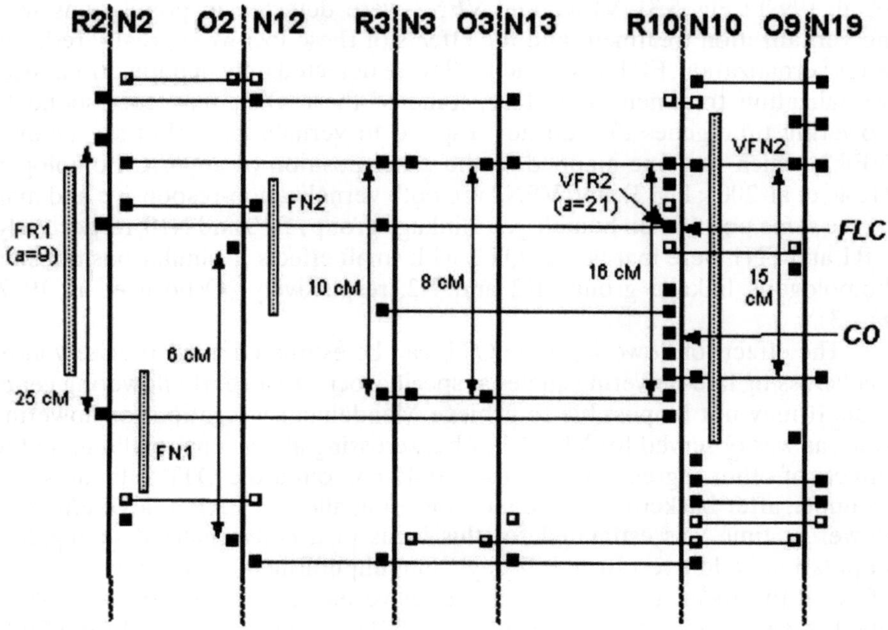

Fig. 3. Comparative map of *Brassica* linkage groups having homoeology to the top of chromosome V in *Arabidopsis thaliana*. Vertical lines represent homologous linkage groups in *B. rapa* (*R*) and *B. napus* (*N*), or *B. oleracea* (*O*) and *B. napus* (*N*) (Parkin et al. 1995). *Boxes* show the relative positions (but not relative distances) of RFLP loci on *R2*, *R3* and *R10* (*B. rapa* linkage groups from Kole et al. 1997 and Osborn et al. 1997, as converted to the Parkin et al. 1995 nomenclature); *O2*, *O3*, and *O9* (*B. oleracea* linkage groups from Camargo et al. 1997, as converted to the Parkin et al. 1995 nomenclature); and *N2*, *N12*, *N3*, *N13*, *N10*, and *N19* (*B. napus* linkage groups from Osborn et al. 1997). *Boxes connected with horizontal lines* are RFLP loci detected with the same probe. *Filled boxes* are loci known to have homologous sequences at the top of *A. thaliana* chromosome V based on comparative mapping (Osborn et al. 1997; Kole et al. 2001) or BLAST searches using DNA sequences of RFLP probes (our unpubl. data). *Open boxes* have not been tested for homology. *Shaded bars* indicate the positions of QTL for flowering time (Osborn et al. 1997; Butruille et al. 1999b; our unpubl. data). *VFR2* was mapped as a qualitative trait locus in a backcross population (Kole et al. 2001). Additive effects (*a*) in days are shown for alleles from *B. rapa* cv. Per after backcrossing to cv. R500. The map positions of homologues of the *A. thaliana* flowering time genes *CO* and *FLC* are shown, as are map distances of markers near these loci

At V, and Bohuon et al. (1998) also detected flowering time QTL in three regions with homoeology to top of At V. These loci may represent homologues of *CO*, *FLC*, or other flowering time genes in this region. Axelsson et al. (2001) hypothesized that all nine QTL they and Lagercrantz et al. (1996) detected in regions homoeologous to the top of At V in four *Brassica* species were the result of duplication of one *CO* ancestral gene.

Some of the QTL detected in populations derived from annuals × biennials of *B. rapa* and *B. napus* also may represent duplications of ancestral genes. VFR2 and FR1 in *B. rapa* and VFN2, FN1, and FN2 in *B. napus* all map to regions with homoeology to the top of At V (Osborn et al. 1997; Butruille et al. 1999b; Fig. 5.3). VFR2 and VFN2 were detected in populations with no vernalization treatment and the effects of these loci were greatly reduced after vernalization. FR1, FN1, and FN2 were detected only in populations after vernalization treatment; therefore, some of these QTL may correspond to flowering time genes that do not respond to vernalization (but see below). VFR2, which was fine-mapped to the exact position of an *FLC* homologue (Kole et al. 2001; Fig. 3), and VFN2 are both vernalization-responsive and map to the same position on homologous linkage groups R10 and N10, respectively. FR1 and FN1 were mapped as QTL with small effects to similar positions on homologous linkage groups R2 and N2, respectively (Osborn et al. 1997; Fig. 3).

The effects of flowering-time QTL can be estimated more precisely after backcrossing late-flowering alleles at specific loci into an early-flowering genotype. It may not be possible to achieve Mendelian segregation for flowering time, as was observed for VFR2, but backcrossing allows one to eliminate the effects of other segregating loci that could confound the QTL estimates. For example, after backcrossing the late-flowering allele at FR1, a large effect on flowering time was estimated for this locus in a nonvernalized segregating population (additive effect = 9 days; our unpublished data). Previously, an effect at this locus was detected only after vernalization (Osborn et al. 1997), and this had encouraged Axelsson et al. (2001) to propose that FR1 (and FN1) are *CO* homologues because these loci were apparently not responsive to vernalization. The detection of an effect in a nonvernalized population opens the possibility that FR1 may respond to vernalization. This needs to be tested, and the results could provide evidence favoring *FLC* as the more likely candidate.

FLC homologues have been mapped to R2, R3, and R10 in *B. rapa* (Fig. 3). The R10 position corresponds exactly to the position of VFR2, and the R2 map position of *FLC* corresponds generally to the position FR1 (Fig. 3). Thus, *FLC* loci may regulate flowering time in *Brassica* through the action of multiple gene copies. This model is consistent with that of Michaels and Amasino (2000), who proposed that *FLC* acts in a rheostat-like manner to control flowering time in *Arabidopsis*. *FLC* produces an inhibitor of flowering, and the delay in flowering is proportional to the copy number of late-flowering alleles. Naturally occurring *Arabidopsis* lines have a single *FLC* locus and the copy number late-flowering alleles is restricted to zero, one, or two, corresponding to homozygous for early-flowering alleles (producing ~12 leaves before flowering), heterozygous (producing ~46 leaves), or homozygous for late-flowering alleles (producing ~89 leaves) (Lee et al. 1994). Thus, late-flowering homozygotes behave as winter annuals, having much-delayed flowering but eventually flowering without vernalization. Transgenic lines containing additional functional copies of *FLC* do not flower without vernalization and behave as obligate biennials, like those found in the *Brassica* genus (Michaels and Amasino 1999; 2000).

Polyploidy has provided *Brassica* species with a potential mechanism to expand the rheostat-like effect of *FLC* on flowering time. In diploid species, such as *B. rapa* with three *FLC* loci on R2, R3, and R10, there are many more potential combinations of *FLC* alleles than in *Arabidopsis*. These combinations could contribute to the diversity in flowering time of these species, including the prevention of flowering through the fixation of multiple high-expressing alleles. Polyploid *Brassica* species have even more *FLC* loci. Tadege et al. (2001) cloned five copies from *B. napus*, and when transformed into *Arabidopsis* with a constitutive promoter, each gene significantly delayed flowering. Thus, multiple *FLC* loci produce functional gene products; however, it is not known if flowering-time differences in *B. napus* are due to allelic variation at multiple *FLC* loci.

Additional copies of *CO* also might expand genetic variation for flowering time, but perhaps in a more limited way. In *Arabidopsis*, *CO* is a promoter of flowering, and the difference in effects of functional and nonfunctional *CO* alleles is ~7 leaves (Putterill et al. 1995). Some mutant *co* lines expressing wild-type *CO* transgenes flower earlier under short days by reducing leaf number from 18 to 7 (Putterill et al. 1995). *CO* may be an important factor in flowering-time differences among annual forms of *Brassica* species, and it could explain some of the QTL effects observed in segregating populations derived from annuals × annuals.

6 Future Research Needs

Results from studies conducted thus far suggest that multiple gene copies contribute to the wide range of flowering time variation observed in *Brassica* species. However, additional data are needed to demonstrate exactly which loci are involved and to what extent they contribute to flowering time variation. Some further evidence could come from comparative fine-mapping experiments. If the allelic effects at a QTL are small, it may not be possible to map the loci as single Mendelian factors. However, loci with quantitative effects can still be further defined by selecting specific recombinants during backcrossing that isolate the candidate genes under study. It may even be possible to identify effects of candidate genes where QTL have not been mapped (e.g., *FLC* on linkage group R3 in Fig. 3) by backcrossing an allele at the candidate gene locus and analyzing its effects in isolation from other segregating loci. Gene expression analyses can provide further evidence that candidate genes affect flowering time if locus-specific probes or PCR primers can be designed to separately detect the transcripts of paralogous loci. Ultimately, gene transformation experiments may be needed to prove that specific candidate genes influence flowering-time variation in *Brassica*. For these experiments to contribute to our understanding of how replicated loci affect flowering variation, they must compare multiple alleles from each locus that is under the control of endogenous gene regulators.

A model including allelic variation at replicated loci may also apply to other components of morphological diversity in *Brassica* species. Variation for stem diameter, internode length, leaf shape and size, and other traits may be due, in part, to allelic variation at replicated loci for key genes regulating these developmental processes. These replicated loci have probably arisen by polyploid events; and thus, polyploidy may have been an important factor in the diversification of *Brassica* crops by providing the possibility of many more gene combinations for selective breeding. QTL mapping with candidate genes for these other components of morphology could be used to more broadly explore the contribution of replicated genes to phenotypic diversity in crop plants.

Acknowledgements. The authors thank Isobel Parkin and Derek Lydiate for providing unpublished DNA sequences of RFLP probes n102, o7 and o118, and Scott Michaels for reviewing the manuscript. Support for this work was provided by the USDA/NRI Competitive Grants Program, USDA/IFAFS Plant Genome Program, and by an NSF Bioinformatics Fellowship to L.L.

References

Araki T (2001) Transition from vegetative to reproductive phase. Curr Opin Plant Biol 4:63–68

Axelsson T, Shavorskaya O, Lagercrantz U (2001) Multiple flowering time QTLs within several *Brassica* species could be the result of duplicated copies of one ancestral gene. Genome 44: 856–864

Bohuon EJR, Ramsay LD, Craft JA, Arthur AE, Marshall DF, Lydiate DJ, Kearsey MJ (1998) The association of flowering time quantitative trait loci with duplicated regions and candidate loci in *Brassica oleracea*. Genetics 150:393–401

Butruille DV, Guries RP, Osborn TC (1999a) Increasing yield of spring oilseed rape hybrids (*Brassica napus* L.) through introgression of winter germplasm. Crop Sci 39:1491–1496

Butruille DV, Guries RP, Osborn TC (1999b) Linkage analysis of molecular markers and quantitative trait loci in populations of inbred backcross lines of *Brassica napus* L. Genetics 153: 949–964

Camargo LEA, Osborn TC (1996) Mapping loci controlling flowering time in *Brassica oleracea*. Theor Appl Genet 92:610–616

Camargo LEA,Savides L, Jung G, Nienhuis J, Osborn TC (1997) Location of theself-incompatibility locus in an RFLP and RAPD map of Brassicaoleracea. J Hered 88:57–60

Cavell AC, Lydiate DJ, Parkin IAP, Dean C, Trick M (1998) Collinearity between a 30-centimorgan segment of *Arabidopsis thaliana* chromosome 4 and duplicated regions within the *Brassica napus* genome. Genome 41:62–69

Chi Y-S, Hownecke ME, Sernyk JL (1992) A genetic linkage map of restriction fragment length polymorphism loci for *Brassica rapa* (syn. *campestris*). Genome 35:746–757

Detjen LR (1926) A preliminary report on cabbage breeding. Proc Am Soc Hortic Sci 23:325–332

Dickson MH (1968) Eight newly described genes in broccoli. Proc Am Soc Hortic Sci 93:356

Diers BW, Osborn TC (1994) Genetic diversity of oilseed *Brassica napus* germplasm based on restriction fragment length polymorphisms. Theor Appl Genet 88:662–668

Doebley J, Stec A, Gustus C (1995) *Teosinte branched 1* and the origin of maize: evidence for epistasis and the evolution of dominance. Genetics 141:333–346

Edwards MD, Helentjaris T, Wright S, Stuber CW (1992) Molecular-marker-facilitated investigations of quantitative trait loci in maize 4. Theor Appl Genet 83:765–774

Ferreira ME, Satagopan J, Yandell BS, Williams PH, Osborn TC (1995) Mapping loci controlling vernalization requirement and flowering time in *Brassica napus*. Theor Appl Genet 90:727–732

Hu J, Sadawski J, Osborn TC, Landry B, Quiros C (1998) Linkage group alignment from four independent *Brassica oleracea* RFLP maps. Genome 41:226–235

Johanson U, West J, Lister C, Michaels S, Amasino R, Dean C (2000) Molecular Analysis of *FRIGIDA* a major determinant of natural variation in *Arabidopsis* flowering time. Science 290:344–348

Kardailsky I, Shukla VK, Ahn JH, Dagenais N, Christensen SK, Nguyen JT, Chory J, Harrison MJ, Weigel D (1999) Activation tagging of the floral inducer *FT*. Science 286:1962–1965

Kennard WC, Slocum KM, Figdore SS, Osborn TC (1994) Genetic analysis of morphological variation in *Brassica oleracea* using molecular markers. Theor Appl Genet 87:721–732

Kobayashi Y, Kaya H, Goto K, Iwabuchi M, Araki T (1999) A pair of related genes with antagonistic roles in mediating flowering signals. Science 286:1960–1962

Koch MA, Haubold B, Mitchell-Olds T (2000) Comparative evolutionary analysis of chalcone synthase and alcohol dehydrogenase loci in *Arabidopsis, Arabis* and related genera (Brassicaceae). Mol Biol Evol 17:1483–1498

Kole C, Kole P,Vogelzang R, Osborn TC (1997) Genetic linkage map of a Brassicarapa recombinant inbred population. J Hered 88:553–557

Kole C, Quijada P, Michaels SD, Amasino RM, Osborn TC (2001) Evidence for homology of flowering-time genes *VFR2* from *Brassica rapa* and *FLC* from *Arabidopsis thaliana*. Theor Appl Genet 102:425–430

Kole C, Williams PH, Rimmer SR, Osborn TC (2002) Linkage mapping of genes controlling resistance to white rust (*Albugo candida*) in *Brassica rapa* (syn. *campestris*) and comparative mapping to *B. napus* and *Arabidopsis thaliana*. Genome 45:22–27

Koornneef M, Hanhart CJ, Van der Veen JH (1991) A genetic and physiological analysis of late flowering mutants in *Arabidopsis thaliana*. Mol Gen Genet 229:57–66

Koornneef M, Alonso-Blanco C, Peeters AJM, Soppe W (1998) Genetic control of flowering time in *Arabidopsis*. Annu Rev Plant Physiol Plant Mol Biol 49:345–370

Kowalski SP, Lan TH, Feldmann KA, Paterson AH (1994) Comparative mapping of *Arabidopsis thaliana* and *Brassica oleracea* chromosomes reveals islands of conserved organization. Genetics 138:499–510

Lagercrantz U (1998) Comparative mapping between *Arabidopsis thaliana* and *Brassica nigra* indicates that *Brassica* genomes have evolved through extensive genome replication accompanied by chromosome fusions and frequent rearrangements. Genetics 150:1217–1228

Lagercrantz U, Lydiate DJ (1996) Comparative genome mapping in *Brassica*. Genetics 144: 1903–1910

Lagercrantz U, Putterill J, Coupland G, Lydiate D (1996) Comparative mapping in *Arabidopsis* and *Brassica*, fine scale genome collinearity and congruence of genes controlling flowering time. Plant J 9:13–20

Lan TH, Paterson AH (2000) Comparative mapping of quantitative trait loci sculpting the curd of *Brassica oleracea*. Genetics 155:1927–1954

Lan TH, DelMonte TA, Reischmann KP, Hyman J, Kowalski SP, McFerson J, Kresovich S, Paterson AH (2000) An EST-enriched comparative map of *Brassica oleracea* and *Arabidopsis thaliana*. Genome Res 10:776–768

Lee I, Michaels SD, Masshardt AS, Amasino RM (1994) The late-flowering phenotype of *FRIGIDA* and *LUMINIDEPENDENS* is suppressed in the Landsberg *erecta* strain of *Arabidopsis*. Plant J 6:903–909

McGrath JM, Quiros CF, Harada JJ, Landry BS (1990) Identification of *Brassica oleracea* monosomic alien chromosome addition lines with molecular markers reveals extensive gene duplication. Mol Gen Genet 223:198–204

Michaels SD, Amasino RM (1999) *Flowering Locus C* encodes a novel MADS domain protein that acts as a repressor of flowering. Plant Cell 11:949–956

Michaels SD, Amasino RM (2000) Memories of winter: vernalization and the competence to flower. Plant Cell Environ 23:1145–1153

Osborn TC, Kole C, Parkin IAP, Sharpe AG, Kuiper M, Lydiate DJ, Trick M (1997) Comparison of flowering time genes in *Brassica rapa*, *B. napus* and *Arabidopsis thaliana*. Genetics 146: 1123–1129

Parkin IAP, Sharpe AG, Keith DJ, Lydiate DJ (1995) Identification of the A and C genomes of amphidiploid *Brassica napus* (oilseed rape). Genome 38:1122–1131

Parkin IAP, Lydiate DJ, Trick, M (2002) Assessing the level of collinearity between *Arabidopsis thaliana* and *Brassica napus* for *A. thaliana* chromosome 5. Genome 45:1–11

Prakash S, Hinata K (1980) Taxonomy, cytogenetics and origin of crop *Brassica*'s, a review. Opera Bot 55:1–57

Putterill J, Robson F, Lee K, Simon R, Coupland G (1995) The *CONSTANS* gene of *Arabidopsis* promotes flowering and encodes a protein showing similarities to zinc finger transcription factors. Cell 80:847–857

Ramsay LD, Jennings DE, Bohuon EJR, Arthur AE, Lydiate DJ (1996) The construction of substitution library of recombinant backcross lines in *Brassica oleracea* for the precision mapping of quantitative trait loci. Genome 39:558–567

Ratcliffe OJ, Nadzan GC, Reuber TL, Riechmann JL (2001) Regulation of flowering in *Arabidopsis* by an *FLC* homologue. Plant Physiol 126:122–132

Roberts LS, Robson F, Sharpe A, Lydiate D, Coupland G (1998) Conserved structure and function of the *Arabidopsis* flowering time gene *CONSTANS* in *Brassica napus*. Plant Mol Biol 37:763–772

Schmidt R, Acarkan A, Boivin K (2001) Comparative structural genomics in the Brassicaceae family. Plant Physiol Biochem 39:253–262

Scortecci KC, Michaels SD, Amasino RM (2001) Identification of a MADS-box gene *Flowering Locus M* that represses flowering. Plant J 26(2):229–236

Sheldon CC, Burn JE, Perez PP, Metzger J, Edwards JA, Peacock WJ, Dennis ES (1999) The *FLF* MADS box gene: a repressor of flowering in *Arabidopsis* regulated by vernalization and methylation. Plant Cell 11:445–458

Simpson GG, Gendall AR, Dean C (1999) When to switch to flowering. Annu Rev Cell Dev Biol 99:519–50

Slocum MK, Figdore SS, Kennard WC, Suzuki JY, Osborn TC (1990) Linkage arrangement of restriction fragment length polymorphism loci in *Brassica oleracea*. Theor Appl Genet 80: 57–64

Song KM, Suzuki JY, Slocum MK, Williams PH, Osborn TC (1991) A linkage map of *Brassica rapa* (syn. *campestris*) based on restriction fragment length polymorphism loci. Theor Appl Genet 82:296–304

Song K, Slocum MK, Osborn TC (1995) Molecular marker analysis of genes controlling morphological variation in *Brassica rapa* (syn. *campestris*). Theor Appl Genet 90:1–10

Tadege M, Sheldon CC, Helliwell CA, Stoutjesdijk P, Dennis ES, Peacock WJ (2001) Control of flowering time by FLC orthologues in *Brassica napus*. Plant J 28:545–553

Teutonico RA, Osborn TC (1995) Mapping loci controlling vernalization requirement in *Brassica rapa*. Theor Appl Genet 91:1279–1283

U N (1935) Genomic analysis in *Brassica* with special reference to the experimental formation of *B. napus* and peculiar mode of fertilization. Jpn J Bot 7:389–452

Williams PH, Hill C (1986) Rapid cycling populations of *Brassica*. Science 232:1285–1289

Yang YW, Lai KN, Tai PY, Li WH (1999) Rates of nucleotide substitution in angiosperm mitochondrial DNA sequences and dates of divergence between *Brassica* and other angiosperm lineages. J Mol Evol 48:597–604

6 Quantitative Trait Loci for Clubroot Resistance in *Brassica oleracea*

R.E. Voorrips, M.C. Jongerius, and H.J. Kanne

1 Introduction

Clubroot, caused by the fungus *Plasmodiophora brassicae* Wor., is one of the most damaging diseases of *Brassica oleracea* crops, including cauliflower, broccoli, Brussels sprouts, cabbage, and other cruciferous crops worldwide. The pathogen causes swelling of parts of the roots and sometimes of the stem base, giving rise to the characteristic "clubs" for which the disease is named (Fig. 1). These clubs inhibit nutrient and water transport, stunt the growth of the plant, and increase the susceptibility to wilting. Later on, the clubbed roots decay, weakening the support of the plant.

Clubroot is a widespread disease. Crête (1981) estimated that in northwestern Europe, Japan, North America, and Australia about 10% of a total of 660,000ha of *B. oleracea* crops was infested with clubroot. For all cruciferous crops, the infested area in those regions amounted to 196,000ha. The disease is not only widespread, but also very persistent in the soil. The spores can remain infectious for at least 15 years (Mattusch 1977). Therefore, crop rotation does not offer much promise, especially since most cruciferous crops and cruciferous weeds will maintain the disease presence. Cultural practices, especially the application of calcium and boron, and lime to decrease the pH of the soil, may reduce disease pressure but these measures are often not sufficient to keep the crop healthy. On the other hand, it is known that some soil types have a significant suppressive effect on the occurrence of clubroot (Murakami et al. 2000). Chemical control is limited, since the few effective treatments are either banned or too expensive.

The introduction of cultivars with resistance or tolerance to clubroot would therefore be desirable. However, only a very small number of resistant cultivars have been released. Breeding programs aimed at the introduction of resistance to clubroot have been few and the results often disappointing. This general lack of success is due to the relatively small number of known sources of resistance, the recessive and often apparently complex inheritance of resistance, and the genetic variability of the pathogen.

Plant Research International B.V., P.O. Box 16, 6700 AA Wageningen, The Netherlands

Biotechnology in Agriculture and Forestry, Vol. 52
Nagata/Tabata (Eds.) Brassicas and Legumes

Fig. 1. Roots of a cauliflower plant severely affected by clubroot

Tools now exist for analyzing genetic complexity and for introgressing the genes concerned that were not available only two decades ago. This chapter illustrates the advantages of using doubled haploids (DHs), molecular markers, and quantitative trait loci (QTL) mapping algorithms.

In *B. oleracea*, doubled haploids were first regenerated from haploid microspores through anther culture (Keller and Armstrong 1981). However, this procedure was laborious and the yield was low. The microspore culture technique, first developed for *B. napus* (Lichter 1982) and adapted for *B. oleracea* crops by Duijs et al (1992) was an improvement. Large numbers of microspores can be obtained by grinding complete flower buds, thus eliminating the isolation of individual anthers. Also, the average number of plants obtained per flower bud is larger with microspore culture. This made the generation of large numbers of doubled haploids a practical tool in both breeding and genetic studies. For genetic studies, the advantages of a doubled haploid population, compared with, e.g., an F2, are twofold: the individual DH genotypes can be maintained indefinitely through selfing, as they are fully homozygous; and because all heterozygous genotypes are missing the segregation patterns are less complicated. In the case of quantitative characters, there is an additional advantage: the absence of intermediate heterozygous genotypes makes it easier to discriminate between the different genotypic classes.

Molecular markers, as aids in breeding and genetic studies, have been used for a long time. A case in point concerns the resistance to *Meloidogyne incognita* in tomato, which was shown in 1974 to be linked to the isozyme *Aps-1*^{1} (Rick 1983). Nowadays, mostly genomic DNA markers are used to compile linkage maps. The first DNA markers to be used for genetic studies were restriction length fragment polymorphisms (RFLPs); subsequently, a multitude of different DNA markers techniques have been established. The essence of all molecular marker methods is that they yield information about the allelic composition at specific loci in different individual plants. The simultaneous segregation observed at many different loci in a population allows one to calculate recombination frequencies between loci. From these recombination frequencies, the most likely linkage map of these loci can be calculated.

The use of molecular markers is not restricted to the compilation of linkage maps. A very important application is the new possibility of locating chromosomal regions involved in the expression of quantitative traits, and the estimation of the effects of these regions. For these regions, the term quantitative trait loci (QTL) has been coined. Statistical methods have been developed to estimate the location, support intervals and effects of QTL for many different situations: different population types, dominant or codominant markers, dominant or additive action of QTL, mono-QTL or multiple-QTL models, etc.

For our study, we obtained a population of doubled haploid plants from the F1 of a cross between a clubroot-resistant cabbage and a susceptible broccoli parent. In this DH population, we scored two types of molecular markers: RFLPs and AFLPs (amplified fragment length polymorphisms). Although the principles of these two types of markers are very different, from a genetic viewpoint, they can be used in much the same way. One difference is that RFLPs can in general be scored codominantly, while AFLPs are scored dominantly, except when reliable quantitative scoring of band intensities on DNA gels is possible. For our study, the difference was not relevant since our DH population consisted solely of homozygotes. Finally, we developed a reliable quantitative scoring method for resistance with a distribution suitable for the application of a multiple QTL model. We then used this model to perform a QTL analysis of clubroot-resistance segregating in our DH population.

With the aid of these novel techniques, we were able to identify and characterize two previously unknown genes for clubroot resistance, as well as closely linked markers that can be used, directly or after conversion to more easily handled marker types, for marker-assisted selection (MAS).

2 Results

2.1 Resistance Tests

The severity of clubroot symptoms was measured by assigning symptom grades based on visual inspection, as well as by measurements of healthy and diseased root tissue, and calculating an R value (see Protocols, below). The mean symptom grade and the mean R value were calculated for all DH lines, parents and F1; the R values were corrected for test and block effects. A high proportion of the DH lines, as well as the F1 and the susceptible parent, scored the maximum disease symptom grade (Fig. 2, bottom), suggesting a recessive inheritance of resistance. In contrast, the distribution of the R values was much less skewed, and the F1 value was about halfway between the parental values (Fig. 2, top). Clearly, the grading scale is compressed at the susceptible side, and expanded at the resistant side compared with the R value scale.

2.2 Molecular Markers

Among the 107 DH lines of the mapping population, 92 polymorphic markers were scored: 66 RFLPs and 26 AFLPs.

RFLPs. Since almost all probe/enzyme combinations detected polymorphisms, no preliminary screening of parental blots was performed. Of 40 probes used, 37 detected a total of 66 polymorphisms using restriction endonuclease *Eco*R1. Three probes (1NH12, 2NA11, 3NG10) did not detect any RFLPs. Five RFLP groups without recombination were found (1NE1c/3ND12, 1NG9/2NA8a/3NB4b, 1NF2b/2NA1a, 1NF2a/1NH3a/2NF10, and 3NB4c/3NC3b). Since the *Eco*R1 restriction fragment sizes detected were unequal within all these groups, these markers detected different but closely linked loci.

AFLPs. With one set of selective primers 26 AFLPs were detected. Five AFLP pairs without recombination were found. In four pairs (A11/A12, A13/A14, A17/A18, and A22/A23), both amplified fragments were inherited from the same parent. Since the sizes of both fragments in these pairs were very similar, it is possible that they represent double bands as reported by Vos et al. (1995), and not different markers. In one pair (A04/A05), each of the two parents contributed one fragment.

2.3 Construction of the Linkage Map

Ninety-two markers were scored for 107 DH lines. We determined the grouping of the markers in linkage groups at a range of log of odds (LOD) scores and calculated preliminary maps for these linkage groups. If gaps larger than

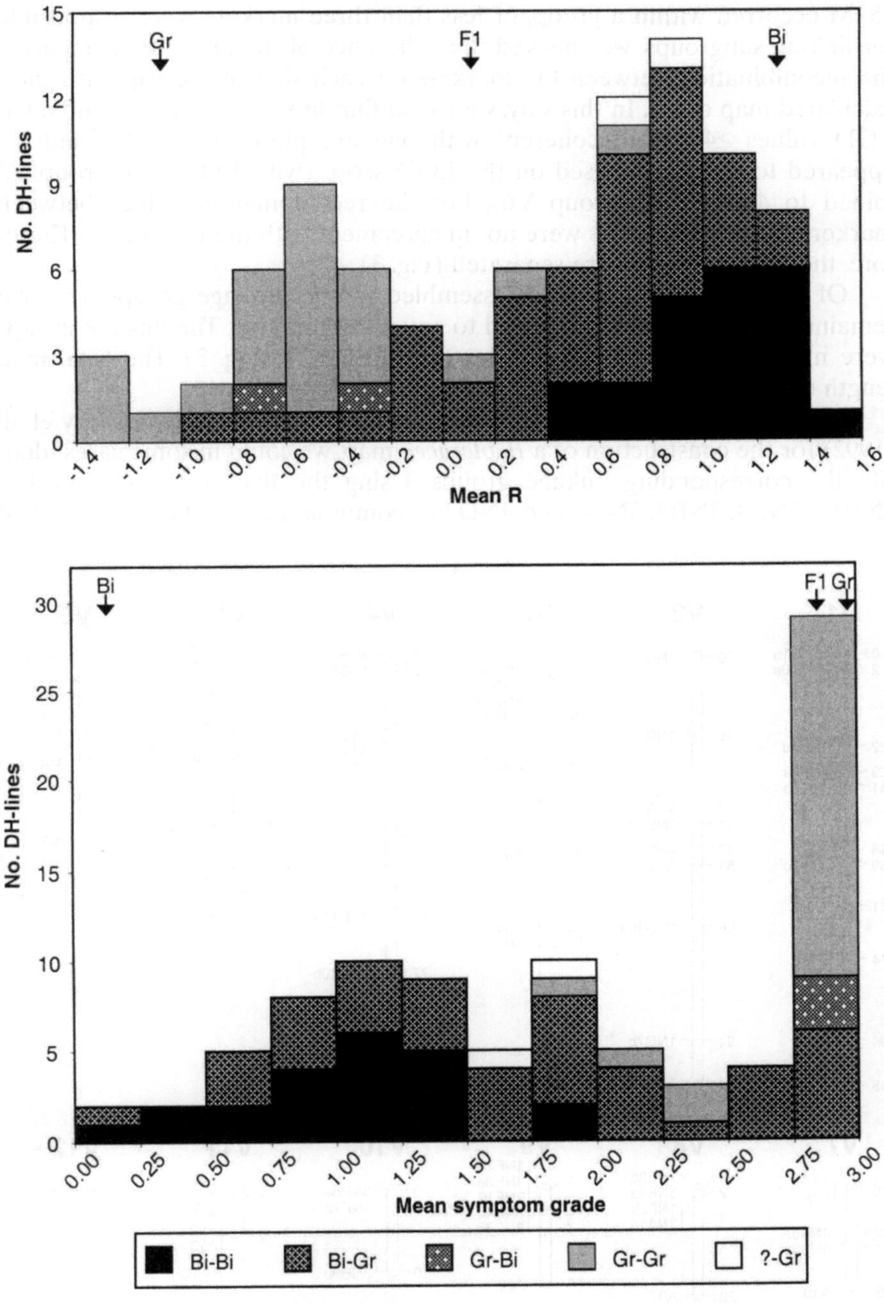

Fig. 2. Frequency distributions for the means of the measure of clubroot resistance R (*top*) and of the symptom grade (*bottom*) in a population of doubled haploid lines, in relation to the genotypic composition (Bi or Gr allele) for the two marker loci 4NE11a and 2NA8c most closely linked to clubroot resistance loci *pb-3* and *pb-4*, respectively

25 cM occurred within a group, or less than three markers were responsible for linking subgroups, we checked the coherence of the group by comparing the recombination between the markers on each side of the gap with their calculated map order. In this way, we found that linkage groups assembled at LOD values ≥4.0 were coherent with one exception: groups V3 and V6 appeared to be linked based on the LOD score (with 1NG1b on group V3 joined to 4NE11c on group V6), but the recombination values between markers in the two groups were not in agreement with the map order. Therefore, these two groups were separated (Fig. 3).

Of the 92 markers, 87 were assembled into 12 linkage groups. The five remaining markers were not linked to any other markers. The linkage groups were numbered in decreasing order of map length (Fig. 3). The total map length covered by the 12 linkage groups was 615 cM.

Since the RFLP probes used were a subset of those used by Landry et al. (1992) for the construction of a *B. oleracea* map, we could in some cases identify the corresponding linkage groups. Using the five single-copy probes 2NB10, 2NE4, 3NB3, 3NF4, and 4ND3 as common markers between the two

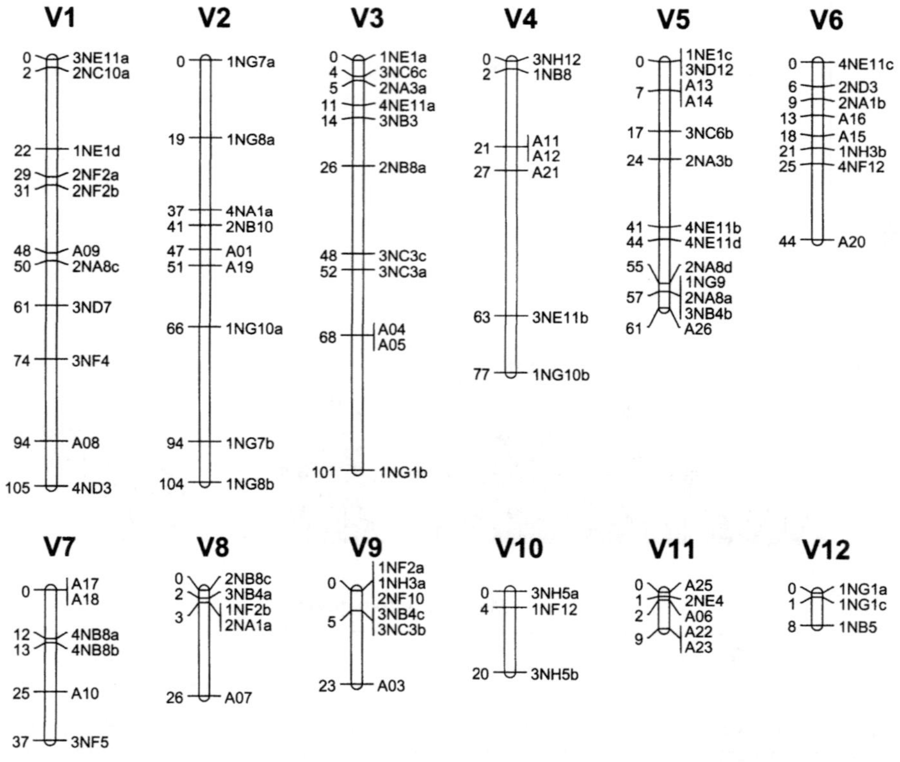

Fig. 3. Linkage map of *Brassica oleracea* based on 92 RFLP and AFLP marker loci segregating in a population of doubled haploid lines. Five markers (2NA8b, 2NB8b, 4NA1b, A02, and A24) were not linked to other loci

maps, our linkage groups V1, V2, and V3 could be identified with their groups 5, 4, and 1, respectively. From the relative map positions of multicopy probe RFLPs, our linkage groups V5 and V12 could further be tentatively associated with two regions of their linkage group 3.

We observed a high frequency of segregation distortion. The segregation of 59 markers (64%) deviated significantly ($P \leq 0.05$) from the 1:1 ratio expected in a doubled-haploid population. Almost all loci on linkage groups V3, V8, V9, and V12 showed an excess of Bi alleles, and all loci in group V11 an excess of Gr alleles. On all linkage groups except group V10, single loci or groups of loci showed segregation distortion. The most extreme segregation ratios were observed for markers 1NF2b and 2NA1a on linkage group V8 (93 Bi:9 Gr:5 missing) and unlinked marker 2NA8b (8 Bi:96 Gr: 3 missing).

2.4 Mapping QTLs for Clubroot Resistance

Means of the R values and the symptom grades were calculated for all lines. The number of tested plants per line varied considerably due to the often low fertility of the DH lines resulting in a very limited seed supply, as well as to a sometimes very slow or weak plant development. Both effects were presumably expressions of severe inbreeding depression, as might be expected from doubled haploids. Data from plants with a total root system weight <0.3 g were rejected, since such small plants generally lacked the resources to develop clubroot symptoms even if they were susceptible (Voorrips and Kanne 1997b). Symptom measurements were available for 91 DH lines. One very fertile DH line was used as a partially resistant standard genotype in the experiments; therefore, many data (179) were available for this line. For the other 90 DH lines, the numbers of evaluated plants ranged from 1 to 16, with an average of 7.4. The variance of 91 DH line means of R was 0.55. The error variance per plant, calculated from all DH lines, the original F_1, the two parents, and several other genetically uniform populations, was 0.19 (Voorrips and Kanne 1997b). The mean error variance of DH line means was calculated as 0.04, leaving an (additive) genetic variance of DH line means of 0.51. The narrow-sense heritability of the line means of R in the DH population was therefore $0.51/0.55 = 0.93$.

The mean R values of the DH lines were analyzed using the MQM method (Jansen 1994; Jansen and Stam 1994). In the first step, three markers were selected as cofactors: 4NE11a, 2NA8c, and 3NH5a on linkage groups V3, V1, and V10, respectively. From the total variance of DH line means of 0.55, a residual variance of 0.18 was not absorbed by the cofactors. In the second step of MQM analysis, the LOD scores of possible QTL throughout the genome were determined. Two QTL, designated *pb-3* and *pb-4* with LOD scores above the threshold of 2.7 were found near two markers used as cofactors: 4NE11a and 2NA8c (Fig. 4). Near the third cofactor (3NH5a) and over a large part of linkage group V6 between marker loci 2NA1b and A20, elevated LOD scores of about 2.0 were found, but these remained below the LOD threshold value.

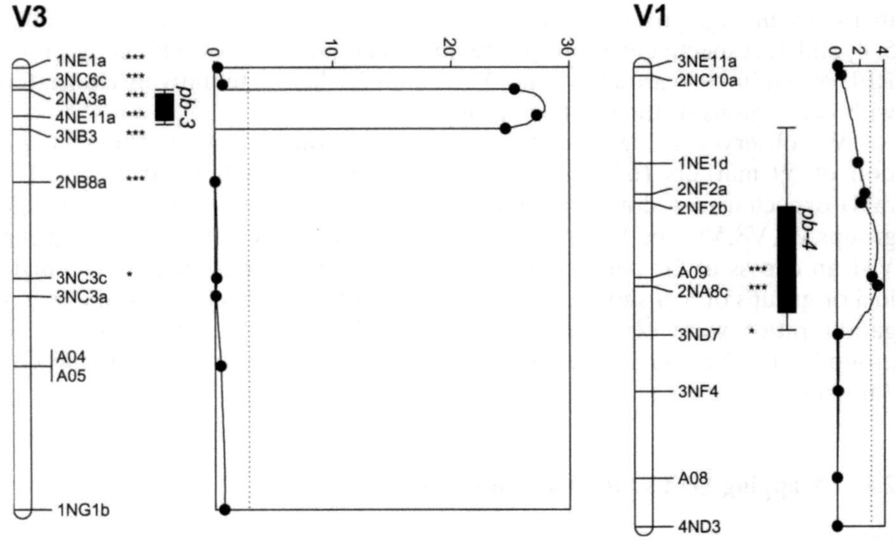

Fig. 4. QTLs for clubroot resistance on linkage groups *V3* (*pb-3*) and *V1* (*pb-4*). The results of MQM mapping of the mean measure of resistance R are shown as LOD graphs and as 1-LOD and 2-LOD support intervals. The *dotted line* indicates the threshold LOD value of 2.7. The significance of Kruskal-Wallis tests for mean symptom grades is indicated for each marker locus as * ($P < 0.01$), ** ($P < 0.001$) and *** ($P < 0.0001$)

The effects of substituting the homozygous resistant for the homozygous susceptible genotype were estimated as 1.17 and 0.41 R value units (50 and 18% of the difference between the parental line means), respectively, for *pb-3* and *pb-4*; these two QTL accounted for 54 and 6%, respectively, of the total genetic variance of the DH line means. As one of the four possible homozygous QTL genotypes (with the Gr allele of *pb-3* and the Bi allele of *pb-4*) was scarcely represented among the DH lines, a possible epistasis effect could not be estimated.

For *pb-3* and *pb-4*, and for the possible QTL near marker 3NH5 and on linkage group V6, the alleles conferring resistance were inherited from the resistant parent. The frequency distributions of mean R values of DH lines with the different genotypes at the peak marker loci near *pb-3* and *pb-4* (4NE11a and 2NA8c, respectively) are shown in Fig. 2 (top). Both 4NE11a and 2NA8c showed significantly distorted segregation ratios of 70:34:3 and 30:70 :7 (Bi:Gr:missing), respectively.

QTL analysis of mean symptom grades of the DH lines involved Kruskal-Wallis tests for the probability of finding a QTL linked with each marker locus. As expected, significant effects were found for markers closely linked to *pb-3* and *pb-4* (Fig. 4). The frequency distribution of mean visual symptom grades of DH lines with the different genotypes at loci 4NE11a and 2NA8c is shown in Fig. 2 (bottom). As in the analysis of the R values, notable but not quite significant effects were detected at some loci on linkage group V6. The possible

QTL near marker 3NH5a indicated by MQM mapping was not revealed by analysis of the symptom grades.

3 Discussion

Assigning symptom grades is the most commonly used method for measuring disease severity in clubroot research; it has the advantage of being fast and requiring relatively little labor. An alternative way to assess clubroot severity involves measuring the symptoms. The advantage of measurements is that they eliminate arbitrary distinctions between symptom grades and subjective scoring. Further, if a measure is found that shows a relatively uniform error variance over the whole range of values, then common statistical tools such as analysis of variance can be applied, that are not applicable to symptom grades.

The merit of using measurements (the R values, see Protocols, below) instead of symptom grades is clearly illustrated in Fig. 2: especially at the susceptible side of the scale, symptom grades allow much less precision than measurements, which results in a substantially skewed frequency distribution. Because the measurements allow for analysis of variance, correction for test and block effects is possible, reducing the error variance. Finally, MQM mapping is possible with the results of the measurements, while symptom grades only allow Kruskal-Wallis tests for QTL effects; this results in a much higher significance of QTL effects and a more precise mapping of QTL when measurements are available.

The genetic map presented here is based on RFLP as well as AFLP markers. For a DH line population as used in this study, the dominant marker type of AFLPs is as informative as the codominant type of most RFLPs, since heterozygosity does not occur. The RFLPs were detected with cDNA clones and thus show homology with expressed sequences, while the AFLPs presumably represent a random sample from the genome. There was no obvious difference in the distribution of both types of markers over the genome, although AFLPs and RFLPs never mapped at the same locus.

The length of the linkage map, 615 cM, was less than the length of other genomic maps published for *B. oleracea*. This is most likely due to the smaller number of markers (92) in comparison to other maps, *e.g.* 820 cM with 258 loci (Slocum et al. 1990), 747 cM with 108 loci (Kianian and Quiros 1992), 1,112 cM with 201 loci (Landry et al. 1992), 824 cM with 128 markers (Moriguchi et al. 1999), 893 cM in an integrated map with 547 markers (Sebastian et al. 2000).

Based on multicopy RFLP probes, our linkage groups V3 and V5 show homology over part of their length, presumably indicating an interchromosomal duplication. Recombination in the segment on linkage group V3 is apparently reduced in comparison with the homologous segment on group V5, as indicated by their different map lengths. Although these linkage groups are also homologous to linkage groups 1 and 3 of Landry et al. (1992), respectively, those authors found no evidence for this duplication due to an accidental lack of relevant polymorphisms in their mapping population (Fig. 5). Two other duplicated regions were found: 1NG7–1NG8 twice in linkage group V2, and 1NF2–3NB4 in groups V8 and V9. Also some closely linked duplications of the same probe were noted: 2NF2 in linkage group V1, 3NC3 in V3, 2NA8 in V5, 4NB8 in V7, 3NH5 in V10, and 1NG1 in V11.

The frequency of distorted segregation, affecting 65% of the loci, is much larger than that observed in intraspecific F_2 populations in *B. oleracea*, to wit, 5% (Slocum et al. 1990), 5–12% (Kianian and Quiros 1992), 12% (Landry et al. 1992), and 8% (Moriguchi et al. 1999) of the marker loci. In populations of DH lines of *B. napus*, 20–34% of the marker loci showed skewed segregation (Tanhuanpää et al. 1994; Cloutier et al. 1995; Ferreira et al. 1995). In contrast, Orton and Browers (1985) found significant deviations from the expected 1:1 ratios at all segregating isozyme loci in four DH populations of *B. oleracea* (respectively 3, 2, 1 and 1 loci). The most extreme segregation ratios mentioned

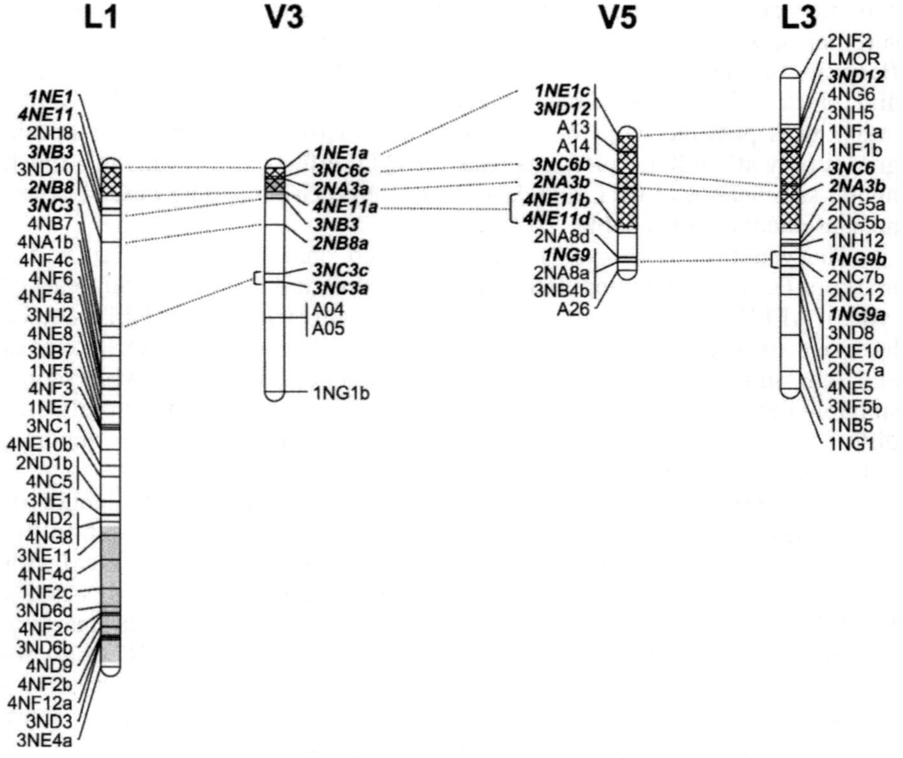

Fig. 5. Duplication between our linkage groups *V3* and *V5*, and the corresponding linkage groups *L1* and *L3* published by Landry et al. (1992) not showing this duplication. The homeologous regions of V3 and V5, and the corresponding regions on L1 and L3 are *cross-hatched*. The 2-LOD intervals of QTLs for clubroot resistance *pb-3* on V3 and *CR2b* on L1 are shown *in gray*

in Tanhuanpää et al. (1994), Ferreira et al. (1995) and Orton and Browers (1985) were 9:26, 17:81, and 24:76, respectively, while we found even more extreme ratios (9:93 and 96:8). Presumably, the increased frequency of distorted segregation ratios in *B. oleracea* DH populations reflects selection pressure due to the microspore culture and plant regeneration process, and possibly also an increased selection pressure against deleterious recessive alleles in a DH population compared to an F_2. In *B. oleracea*, the regeneration of plants from microspores generally occurs at a lower frequency than in *B. napus*; this may account for the higher frequency of segregation distortion found in DH populations in this species. The microspore regeneration capacity of the two accessions from which the parental lines were derived was similar: 10.9 and 7.4 embryos per flower bud for cv. Greenia and for the Bindsachsener accession (Duijs et al. 1992).

Voorrips and Kanne (1997a, b) studied segregation of clubroot resistance in populations descended from the same cross (Bi × Gr) that yielded the population of DH lines discussed here. Neither the analysis of symptom grades, nor that of R values determined for individual plants in F_2 and backcross populations showed evidence for simple inheritance of resistance. In retrospect, this is not surprising, since the QTL analyses presented here indicate that at least three QTL are involved in the resistance of DH line Bi. One QTL, *pb-3*, can be considered a major gene and is responsible for half of the total additive resistance effect in DH line Bi. However, apart from *pb-3* and *pb-4*, at least one further QT locus must be present to account for the unexplained 32% of the difference between the parental lines. Further, in the population of DH lines, both *pb-3* and *pb-4* showed significant segregation distortion. If this should also be the case in the F_2 or backcross populations, the genetic analyses carried out by Voorrips and Kanne (1997a,b) would be invalid. This illustrates two important advantages of QTL analysis: distorted segregation is easily observed and compensated for, and even a partial explanation of observed segregation of traits yields meaningful results.

Landry et al. (1992) were the first to report linkage of clubroot resistance genes with genetic markers. They studied an F_2 population from a cross between a cabbage breeding line resistant to race 2 (Williams 1966) and a susceptible rapid-cycling *B. oleracea* line. The parentage of the resistant line included *B. napus* cv. Wilhelmsburger, resistant to races 2, 3, 6, and 7 as well as cv. Badger Shipper, resistant to races 1, 3, and 6. The resistance tests were performed with a field isolate designated as race 2. Two QTL for clubroot resistance were mapped (*CR2a* and *CR2b*) in linkage groups 6 and 1, respectively. Intriguingly, *CR2a* was linked to marker 2NA8, while our gene *pb-4* was linked to marker 2NA8c detected with the same probe. However, *CR2a* probably originates from *B. napus* based on the resistance of the progenitors of the segregating population, while *pb-4* originates from *B. oleracea*. Although *CR2b* is located on the same linkage group as our major QTL *pb-3*, the 2-LOD support intervals do not overlap, thus indicating that these two QTLs are different (Fig. 5, gray segments on linkage groups L1 and V3). The magnitude of the gene effects could not be compared since a different measuring scale was used in their study. Landry et al. (1992) estimated that *CR2a* and *CR2b* together explained 61% of the variation for clubroot resistance in the F_2 population.

Figdore et al. (1993) studied the segregation of RFLP markers and clubroot resistance to race 7 (Williams 1966) in an F_2 from a cross between a susceptible cauliflower cv and broccoli line OSU CR-7, which was resistant to race 7. They found strong evidence for one resistance gene, and possibly spurious indications for two other resistance genes. Treating symptom grades as quantitative data, they estimated that this gene accounted for 12.6% of the variance in the F_2. Since different markers were used in their study, the map position of this resistance gene cannot be compared with the positions of *pb-3* and *pb-4*. Similarly, GrandClément and Thomas (1996) performed linkage studies of QTL for clubroot resistance with RAPD markers; their results also cannot be compared to ours because of a lack of common markers.

Moriguchi et al. (1999) mapped QTL for resistance based on field test results. Their field was infested with races 1 and 3 (Williams 1966). They found major QTL on linkage group LG3, which unfortunately cannot be identified with one of our linkage groups. It is certainly different from *pb-3* on our linkage group V3: this is the same as group 1 of Landry et al (1982), which corresponds to LG1 of Moriguchi et al. (1999; Hu et al 1998).

The nomenclature of clubroot resistance genes in *B. oleracea* is not standardized at this time. Chiang and Crête (1970) hypothesized the existence of two recessive genes for resistance to race 6 (Williams 1966), which they labeled *pb-1* and *pb-2*. However, in a later study (Chiang and Crête 1976), they found only evidence for one such gene. The work of Landry et al. (1992) was based on the same resistant parent used by those authors, but since they used another test method and another race of *P. brassicae,* it is not surprising that they discovered different (in their case dominant) genes for resistance. Other authors have also presented evidence for mono- and oligogenic inheritance of clubroot resistance in *B. oleracea*, but have not labeled those genes (reviewed in Voorrips 1995). The approach of Landry et al. (1992) to name the genes after the race to which they confer resistance is confusing, because there is not yet a universally recognized nomenclature for races, and because one gene is likely to confer resistance to multiple races. Therefore, we prefer to follow the nomenclature system used by Chiang and Crête (1970), who added sequential numbers to the prefix *pb-*. Since the R value of our F_1 was intermediate between the parents and the mean symptom grade of the F_1 close to the susceptible parent, we use the recessive symbol to indicate the resistant alleles.

4 Protocols

4.1 Plant Material

One of the accessions in which we found resistant plants was a line derived from an old cabbage landrace "Bindsachsener," originating in Germany.

From this accession we obtained doubled-haploid lines (DH lines) through microspore culture, using the method of Duijs et al. (1992). From this set of DH lines, one line "Bi" with a high level of resistance was selected. This resistant line Bi was crossed with susceptible DH line "Gr," obtained from broccoli cv. Greenia (Hammenhögs Frö AB, Hammenhög, Sweden). This susceptible line was selected for its high response in microspore culture. From three plants of the F_1 (Gr × Bi) a population of doubled haploids was obtained through microspore culture, again according to the procedure of Duijs et al. (1992). From about 2.2×10^7 microspores, isolated from 218 flowers, we obtained 390 embryos, 120 of which developed into flowering plants. Of these 120 flowering plants, 107 produced seeds upon selfing.

4.2 Pathogen

The field isolate of *P. brassicae* that was used in this study originated from a heavily infested field of the Experimental Station Brabant at Breda, The Netherlands, in 1986. Large clubs of an unknown, highly susceptible cauliflower cultivar were collected, washed, and stored at –20 °C. Spore suspensions were prepared by macerating frozen clubs in 4–5 volumes of demineralized water, using an electric blender and filtering the crude suspensions through four layers of cheesecloth. Prior to inoculation, the inoculum was diluted to 10^7 spores ml^{-1}. This isolate was characterized as ECD 16/3/30 (Buczacki et al 1975; Voorrips and Visser 1993).

4.3 Resistance Tests

Three resistance tests were carried out in a phytotron chamber, at 22 °C, with a photoperiod of 16 h at 110 $\mu E\ m^{-2} s^{-1}$ HPI-T illumination, according to Voorrips and Visser (1993). Disposable plastic sets of six $4.5 \times 4.5\,cm^2$ square, 110-ml pots were filled with potting compost (pH 6.0, sterilized by gamma irradiation) and placed on a tray in 1 cm water, refilled daily. One seed was sown per pot at a depth of 2.5 cm and covered with compost. Two ml of a suspension of 10^7 spores ml^{-1} was pipetted to each pot. After 6 weeks, the root systems of the plants were washed and evaluated.

For visual assessment of disease symptoms, we used a four-point scale (Voorrips and Kanne 1997a): grade 0: no symptoms; grade 1: small, separated globular or spindle-shaped clubs on tap root or side roots, not more than 2 mm in diameter; grade 2: intermediate symptoms; grade 3: clubs larger than 1 cm in diameter or affecting more than half of the root system (Fig. 6). Symptoms of grade 3 always affected the tap root and the proximal part of side roots, and sometimes extended into the hypocotyl as well.

In two of the three experiments, the fresh weights of the affected parts (the so-called clubs) and healthy parts of the root system of each plant were also measured. The compost was removed from the root system under running tap water. Care was taken to recover as much of the root system as possible,

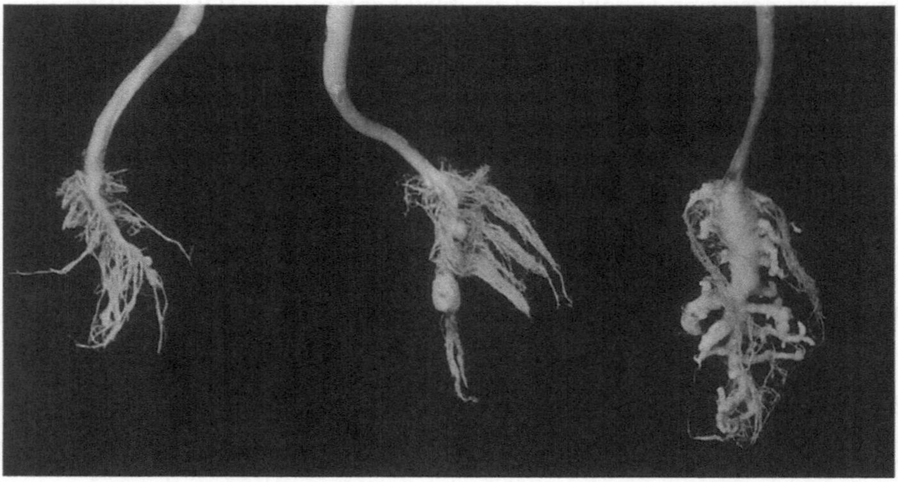

Fig. 6. Clubroot symptom grades 1 (small nodules), 2 (intermediate symptoms), and 3 (severe club formation) in young cabbage plants, 6 weeks after inoculation

but the finest roots were often lost, especially if decayed. The clubs were separated from the healthy roots; both were dried by squeezing between tissue paper and weighed separately on an electronic scale with 0.01 g precision. For statistical analysis, the function

$$R = \ln \frac{HRW + 0.5}{CW + 0.5}$$

was used as a measure of resistance, where HRW is healthy root weight and CW is club weight, both expressed in grams (g). Data of plants with a total root weight (HRW + CW) < 0.3 g were rejected. R was corrected for test and block effects before further analysis. This R value was shown to correlate well with the symptom grade, and to have a normally distributed error term with an approximately constant error standard deviation of 0.44 (Voorrips and Kanne 1997b)

4.4 RFLP markers

All RFLP procedures were as described by Van der Beek et al. (1992; 1994). Genomic DNA was digested with restriction endonuclease *Eco*R1. Probes sized 0.35–2.5 kB were obtained from a *B. napus* embryo cDNA library and cloned into the *Pst*1 site of vector pBR322 (Harada et al. 1988) and made available by B.S. Landry (Agriculture Canada Research Station, St-Jean-sur-Richelieu, Canada). Cloned inserts were PCR-amplified and radiolabeled (Bernatzky and Tanksley 1986).

RFLP markers were named after the probe (1NB8 to 4NF12) detecting them. If several polymorphic loci were detected with the same probe, the loci were designated by appending different lowercase letters to the probe code. This nomenclature is similar to that followed by Landry et al. (1992), but identically named markers in their work and ours do not necessarily refer to the same loci.

4.5 AFLP Markers

Genomic DNA was digested with restriction endonucleases *Eco*R1 and *Mse*1. After ligation of double-stranded adapters to the ends of the restriction fragments, preamplification was performed with primers specific for the *Eco*R1 and *Mse*1 adapters, including one selective nucleotide (underlined), followed by amplification with similar primers with three selective nucleotides (italics): 5′ GAC TGC GTA CCA ATT C*A*C *A* 3′ (E35), and 5′ GAT GAG TCC TGA GTA A*C*C *T* 3′ (M54). Depending on the amount of amplified DNA after preamplification, the reaction mix was diluted 1/20 or 1/40 in TE and 10 μl was used in the final amplification. The E35 primer was end-labeled with γ ^{33}P ATP (Feinberg and Vogelstein 1983). Amplification fragments were separated on 5% denaturing polyacrylamide gels. Autoradiography was carried out with Kodak XOMAT AR X-ray film. Segregating AFLPs were labeled A01 to A26 in order of decreasing fragment size.

4.6 Linkage Analysis

We used the JoinMap 2.0 package (Stam 1993; Stam and Van Ooijen 1995) to assign markers to linkage groups and to calculate the probable order and distances of markers within each linkage group. Map distances were based on Kosambi's (1944) mapping function. The basis for assigning markers to linkage groups was an LOD score ≥4.0 with one or more other markers in the group. Charts of linkage groups were prepared using the MapChart 2.0 program (Voorrips 2002).

4.7 Quantitative Trait Loci (QTL) Analysis

Since the error terms of the R value were approximately normally distributed (Voorrips and Kanne 1997b), the mean R values of the DH lines could be analyzed with the Multiple QTL Model (MQM) method (Jansen 1994; Jansen and Stam 1994). In short, markers near putative QTL are used as cofactors in a model used to estimate QTL effects in other parts of the genome. Because the cofactors absorb a large part of the effect of the QTL in their vicinity, the power of QTL detection in the remaining parts of the genome increases considerably.

The MQM method is a two-step method, where the first step consists of the selection of markers to be used as cofactors, and the second consists of the estimation of the effects of putative QTL throughout the genome after correction for the QTL effects at the selected markers. In the first step, cofactors were selected from 38 markers distributed over the linkage map at approximately 25 cM intervals, the criterion being a significance of $P \leq 0.02$ for the marker effect on R. In the second step, an LOD threshold of 2.7 was used to obtain a confidence level of 0.95 throughout the genome. This threshold was determined based on 1,000 simulations with the actual molecular marker data, according to the method of Jansen (1994). The MQM analysis was carried out with a Genstat5 program developed by R.C. Jansen (Plant Research International, Wageningen, The Netherlands).

The error terms of the mean symptom grades of the DH lines were not normally distributed and the mean symptom grades were therefore not suitable for MQM analysis. A nonparametric Kruskal-Wallis test for significance of the effect on the quantitative trait was performed at each marker locus. In order to obtain a confidence level of 0.95 throughout genome, the tests per marker locus were considered significant if $P \leq 0.001$. The Kruskal-Wallis analysis was carried out using the MapQTL 3.0 package (Van Ooijen and Maliepaard 1995). Since the segregating population consisted of DH lines, no dominance effects could be assessed using either analysis method.

Acknowledgements. Dr. R.C. Jansen (Plant Research International) wrote the Genstat5 program used for the MQM analysis. The RFLP probes were kindly supplied by Dr. B.S. Landry (Agriculture Canada).

References

Bernatzky R, Tanksley SD (1986) Toward a saturated linkage map in tomato based on isozymes and random cDNA sequences. Genetics 112:887–898

Buczacki ST, Toxopeus H, Mattusch P, Johnston TD, Dixon GR, Hobolth LA (1975) Study of physiologic specialization in *Plasmodiophora brassicae*: proposals for attempted rationalization through an international approach. Trans Br Mycol Soc 65:295–303

Chiang MS, Crête R (1970) Inheritance of clubroot resistance in cabbage (*Brassica oleracea* L. var. *capitata* L.). Can J Genet Cytol 12:253–256

Chiang MS, Crête R (1976) Diallele analysis of the inheritance of resistance to race 6 of *Plasmodiophora brassicae* in cabbage. Can J Plant Sci 56:865–868

Cloutier S, Cappadocia M, Landry BS (1995) Study of microspore-culture responsiveness in oilseed rape (*Brassica napus* L.) by comparative mapping of a F_2 population and two microspore-derived populations. Theor Appl Genet 91:841–847

Crête R (1981) Worldwide importance of clubroot. Clubroot Newsl 11:6–7

Duijs JG, Voorrips RE, Visser DL, Custers JBM (1992) Microspore culture is successful in most crop types of *Brassica oleracea* L. Euphytica 60:45–55

Feinberg AP, Vogelstein B (1983) A technique for radiolabeling DNA restriction endonuclease fragments to high specific activity. Anal Biochem 132:6–13

Ferreira ME, Williams PH, Osborn TC (1995) RFLP mapping of *Brassica napus* using doubled haploid lines. Theor Appl Genet 89:615–621

Figdore SS, Ferreira ME, Slocum MK, Williams PH (1993) Association of RFLP markers with trait loci affecting clubroot resistance and morphological characters in *Brassica oleracea* L. Euphytica 69:33–44

GrandClément C, Thomas G (1996) Detection and analysis of QTLs based on RAPD markers for polygenic resistance to *Plasmodiophora brassicae* Woron in *Brassica oleracea* L. Theor Appl Genet 93:86–90

Harada JJ, Baden CS, Comai L (1988) Spatially regulated genes expressed during seed germination and postgerminative development are activated during embryogeny. Mol Gen Genet 212: 466–473

Hu J, Sadowski J, Osborn TC, Landry BS, Quiros CF (1998) Linkage group alignment from four independent *Brassica oleracea* RFLP maps. Genome 41:226–235

Jansen RC (1994) Controlling the type I and type II errors in mapping quantitative trait loci. Genetics 138:871–881

Jansen RC, Stam P (1994) High resolution of quantitative traits into multiple loci via interval mapping. Genetics 136:1447–1455

Keller WA, Armstrong KC (1981) Production of anther-derived dihaploid plants in autotetraploid marrowstem kale (*Brassica oleracea* var. *acephala*). Can J Genet Cytol 23:259–265

Kennard WC, Slocum MK, Figdore SS, Osborn TC (1994) Genetic analysis of morphological variation in *Brassica oleracea* using molecular markers. Theor Appl Genet 87:721–732

Kianian SF, Quiros CF (1992) Generation of a *Brassica oleracea* composite RFLP map – linkage arrangements among various populations and evolutionary implications. Theor Appl Genet 84:544–554

Kosambi DD (1944) The estimation of map distances from recombination values. Ann Eugenics 12:172–175

Landry BS, Hubert N, Crête R, Chiang M, Lincoln SE, Etoh T (1992) A genetic map for *Brassica oleracea* based on RFLP markers detected with expressed DNA sequences and mapping of resistance genes to race 2 of *Plasmodiophora brassicae* (Woronin). Genome 35:409–420

Lichter R (1982) Induction of haploid plants from isolated pollen of *Brassica napus*. Z Pflanzenphysiol 105:427–434

Mattusch P (1977) Epidemiology of clubroot of crucifers caused by Plasmodiophora brassicae. In: Buczacki ST, Williams PH (eds) Woronin +100 International Conference on Clubroot. University of Wisconsin, Madison, WI, pp 24–28a

Moriguchi K, Kimizuka-Takagi C, Ishii K, Nomura K (1999) A genetic map based on RAPD, RFLP, isozyme, morphological markers and QTL analysis for clubroot resistance in *Brassica oleracea*. Theor Appl Genet 93:86–90

Murakami H, Tsushima S, Shishido Y (2000) Soil suppressiveness to clubroot disease of Chinese cabbage caused by *Plasmodiophora brassicae*. Soil Biol Chem 32:11–12

Orton TJ, Browers MA (1985) Segregation of genetic markers among plants regenerated from cultured anthers of broccoli (*Brassica oleracea* var. *italica*). Theor Appl Genet 69:637–643

Rick CM (1983) Tomato (*Lycopersicon*). In: Tanksley SD, Orton TJ (eds) Isozymes in plant genetics breeding, part B. Elsevier, Amsterdam, pp 147–165

Sebastian RL, Howell EC, King GJ, Marshall DF, Kearsey MJ (2000) An integrated AFLP and RFLP *Brassica oleracea* linkage map from two morphologically distinct doubled-haploid mapping populations. Theor Appl Genet 100:75–81

Slocum MK, Figdore SS, Kennard WC, Suzuki JY, Osborn TC (1990) Linkage arrangement of restriction fragment length polymorphism loci in *Brassica oleracea*. Theor Appl Genet 80:57–64

Stam P (1993) Construction of integrated genetic linkage maps by means of a new computer package: JoinMap. Plant J 3:739–744

Stam P, Van Ooijen JW (1995) JoinMap(TM) version 2.0: software for the calculation of genetic linkage maps. CPRO-DLO, Wageningen, Netherlands

Tanhuanpää PK, Vilkki JP, Vilkki HJ (1994) Segregation and linkage analysis of DNA markers in microspore derived and F_2 populations of oilseed rape (*Brassica napus* L). Euphytica 74:59–65

Van der Beek JG, Verkerk R, Zabel P, Lindhout P (1992) Mapping strategy for resistance genes in tomato based on RFLPs between cultivars: *Cf9* (resistance to *Cladosporium fulvum*) on chromosome 1. Theor Appl Genet 84:106–112

Van der Beek JG, Pet G, Lindhout P (1994) Resistance to powdery mildew (*Oidium lycopersicum*) in *Lycopersicon hirsutum* is controlled by an incompletely dominant gene *Ol-1* on chromosome 6. Theor Appl Genet 89:467–473

Van Ooijen JW (1992) Accuracy of mapping quantitative trait loci in autogamous species. Theor Appl Genet 84:803–811

Van Ooijen JW, Maliepaard C (1995) MapQTL version 3.0: software for the calculation of QTL positions on genetic maps. CPRO-DLO, Wageningen, Netherlands

Voorrips RE (1995) *Plasmodiophora brassicae*: aspects of pathogenesis and resistance in *Brassica oleracea*. Euphytica 83:139–146

Voorrips RE (2002) MapChart: software for the graphical presentation of linkage maps and QTLs. J Hered 93:77–78

Voorrips RE, Kanne HJ (1997a) Genetic analysis of resistance to *clubroot (Plasmodiophora brassicae*) in *Brassica oleracea*. 1. Analysis of symptom grades. Euphytica 93:31–39

Voorrips RE, Kanne HJ (1997b) Genetic analysis of resistance to clubroot (*Plasmodiophora brassicae*) in *Brassica oleracea*. 2. Quantitative analysis of root symptom measurements. Euphytica 93:41–48

Voorrips RE, Visser DL (1993) Examination of resistance to clubroot in accessions of *Brassica oleracea* using a glasshouse seedling test. Neth J Plant Pathol 99:269–276

Vos P, Hogers R, Bleeker M, Reijans M, Van de Lee T, Hornes M, Frijters A, Pot J, Peleman J, Kuiper M, Zabeau M (1995) AFLP: a new technique for DNA fingerprinting. Nucleic Acids Res 23:4407–4414

Williams PH (1966) A system for the determination of races of *Plasmodiophora brassicae* that infect cabbage and rutabaga. Phytopathology 56:624–626

7 Structural Differences of *S* Locus Between *Brassica oleracea* and *Brassica rapa*

T. NISHIO and K. SATO

1 Introduction

Self-incompatibility as well as male sterility is successfully used for the production of F_1 hybrid seeds in *Brassica* vegetables. Since the first commercial F_1 hybrid cultivars in cabbage and Chinese cabbage were released by a Japanese seed company using a self-incompatibility system in 1950, many F_1 hybrids of various *Brassica* vegetables have been developed by the same breeding methodology. Although the use of self-incompatibility has a problem of a contamination of selfed seeds in the F_1 seeds due to instability of self-incompatibility, many seed companies still adopt self-incompatibility because of their higher yield of the F_1 hybrid seeds than the use of the male sterility.

Self-incompatibility in *Brassica* is controlled by multiple alleles in a single locus, *S*, and the recognition specificity of the alleles in *S* locus is expressed sporophytically. That is, the recognition specificity of pollen is determined by a genotype of a plant bearing the pollen, not by an allele in a pollen grain itself (Bateman 1955). Fifty and 30 alleles have been identified in the *S* locus of *Brassica oleracea* and *Brassica rapa*, respectively (Ockendon 1974, 2000; Nou et al. 1993). Dominant-recessive relationships between the alleles in pollen do not necessarily correspond to those in stigma, and the relationships are non-linear (Thompson and Taylor 1966; Hatakeyama et al. 1998).

Allele-specific stigma proteins encoded by *S* locus have been identified first in *Brassica* by an immunological technique (Nasrallah and Wallace 1967) and by isoelectric focusing (Nishio and Hinata 1977). The *S*-specific protein in *Brassica* is a secreted glycoprotein on the stigma surface, and named *S*-locus glycoprotein (SLG). The finding of SLG has accelerated intensive studies on *S*-specific stigma proteins in various plant species, and has led to the isolation of cDNA clones of *S*-locus genes in Brassicaceae, Solanaceae, Papaveraceae, and Rosaceae (Nasrallah et al. 1985; Anderson et al. 1986; Stein et al. 1991; Foote et al. 1994; Sassa et al. 1996). However, *S*-specific proteins in pollen had not been identified for a long time after the finding of SLG. Recently, the gene for the *S*-specific pollen protein that is the recognition molecule of self-incompatibility in pollen has been elucidated in *Brassica* (Schopfer et al. 1999; Suzuki et al. 1999; Takayama et al. 2000).

Graduate School of Agricultural Science, Tohoku University, 1-1 Tsutsumidori-Amamiyamachi, Aoba-ku, Sendai 981-8555, Japan

Biotechnology in Agriculture and Forestry, Vol. 52
Nagata/Tabata (Eds.) Brassicas and Legumes

Brassica is a good material for the molecular biological study of self-incompatibility because: first, one plant bears large number of flowers of relatively large size; second, the age of a flower can be easily determined by the position of the flower in an inflorescence; third, inbred lines can be obtained by bud pollination or anther culture; and, finally, each generation lasts no more than 1 year. *Brassica* is closely related to self-compatible *Arabidopsis thaliana*, in which the genome sequence has been determined (The Arabidopsis Genome Initiative 2000). The finding by molecular biology of self-incompatibility can clearly contribute to F_1 hybrid breeding.

2 *S*-Locus Genes in *Brassica*

The *SLG* gene was the first to be identified as an *S*-locus gene (Nasrallah et al. 1988). Subsequently, another gene *SRK* (*S*-locus receptor kinase), tightly linked to *SLG,* was isolated as a homologue of *SLG* (Stein et al. 1991). *SRK* encodes a receptor protein kinase composed of an outer-membrane receptor domain similar to SLG (S-domain), the transmembrane domain, and the cytoplasmic serine-threonine protein kinase domain. *SLG* and *SRK* are transmitted to progeny as one set together with the gene for the *S*-specific pollen protein; therefore, a set of the alleles of *SLG, SRK*, and the *S*-specific pollen protein is referred to as *S* haplotype.

S haplotypes are classified into two groups, class I and class II, based on their *SLG* and *SRK* sequences. The deduced amino acid sequences of SLG and SRK have more than 72% identity between different *S* haplotypes within the classes, and less than 70% between the two classes (Chen and Nasrallah 1990; Nishio and Kusaba 2000). Class II *S* haplotypes are generally recessive to the class I *S* haplotypes in pollen.

Mutation in *SRK* resulted in a self-compatibility phenotype (Nasrallah et al. 1994), while there are self-incompatible *S* homozygotes lacking the *SLG* gene (Okazaki et al. 1999). The *SLG* genes in some *S* haplotypes have a frameshift or a nonsense mutation in the coding region, and cannot synthesize a full-length protein (T. Suzuki et al. 2000). These results suggest that SLG is not necessary for self-incompatibility. By introducing the *SLG* and *SRK* genes into a plant having a different *S* haplotype, Takasaki et al. (2000) demonstrated that SRK is the determinant of self-recognition specificity in stigma and SLG enhances self-incompatibility response.

The determinant of self-recognition specificity in pollen has long been searched for, but most of the trials have been unsuccessful. Doughty et al. (1998) identified cysteine-rich protein in pollen-coat protein extracted by hexane and found that the application of the pollen-coat protein to the stigma modified recognition specificity of pollen. In a ca. 80-kb fragment harboring *SLG* and *SRK* from *Brassica rapa* S^9 homozygote, a gene encoding a cysteine-rich protein, *SP11*, was found by Suzuki et al. (1999) among 11 newly identified genes expressed in pollen. In a plant transformation experiment

Schopfer et al. (1999) demonstrated an *S*-locus gene encoding a cysteine-rich protein to be the determinant of *S*-haplotype specificity, naming it *SCR* (*S*-locus cysteine-rich protein). *SP11* and *SCR* were soon found by Takayama et al. (2000) to be the same gene (*SP11* is used hereafter).

SLG and the receptor domain of *SRK* in the same *S* haplotype have highly similar sequences (Nishio and Kusaba 2000). A part of the sequence of *SLG* is the same as that of *SRK* in some *S* haplotypes, suggesting that gene conversion occurred in the past. Distributions of *S* haplotypes in phylogenetic trees were found to be similar between the tree of *SP11* and that of the hypervariable region of *SRK* (Sato et al. 2002).

3 Genomic Organization of the *S* Locus

SP11, SRK, and *SLG* are closely linked to each other in the *S* locus. In many *S* haplotypes so far analyzed, *SP11* is located between *SRK* and *SLG*, while *SRK* is between *SP11* and *SLG* in S^{910} (Fig. 1; Suzuki et al. 1999; Brugiere et al. 2000; Takayama et al. 2000, Kimura et al. 2002). Distances between these three genes are largely different among *S* haplotypes. Directions of transcrip-

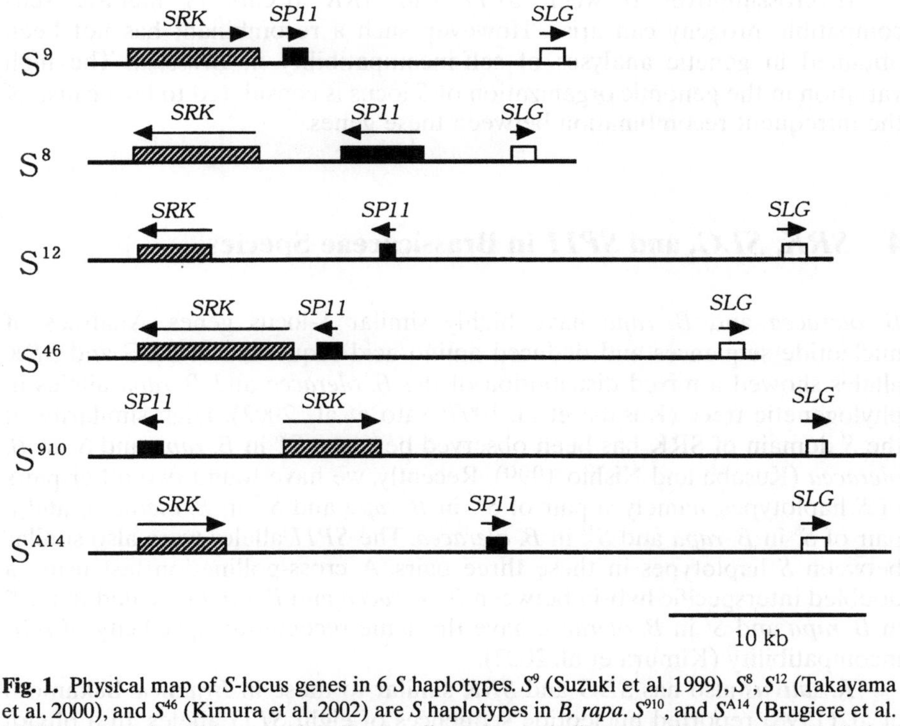

Fig. 1. Physical map of *S*-locus genes in 6 *S* haplotypes. S^9 (Suzuki et al. 1999), S^8, S^{12} (Takayama et al. 2000), and S^{46} (Kimura et al. 2002) are *S* haplotypes in *B. rapa*. S^{910}, and S^{A14} (Brugiere et al. 2000) are those in self-incompatible *B. napus* lines. *Arrows* indicate the direction of transcription

tion of these genes are also different among *S* haplotypes. These differences suggest that insertion, deletion, and inversion of DNA fragments occurred in the evolution of *S* haplotypes.

Several genes closely linked to the *S*-locus genes have been reported. Some genes are located near *SRK* and *SLG*, and had been expected to be the determinant of self-recognition specificity in pollen. The *S* locus anther gene (*SLA*) is located near the *SLG* gene, ca. 1 kb downstream of *SLG*, in the S^2 haplotype in *B. oleracea*. This gene is expressed in developing microspores and anthers. *SLA* has two open reading frames that predict proteins of 10 and 7.5 kDa (Boyes and Nasrallah 1995). This gene cannot be detected in other *S* haplotypes or another line of the S^2 haplotype (Kusaba et al. 2000). *SLL1* and *SLL2* (*S*-locus-linked genes 1 and 2) are between *SRK-910* and *SLG-910* (Yu et al. 1996). *SLL1* is expressed specifically in anthers, while *SLL2* is in anthers and stigmas. In *SLL1* cDNA, there are two small open reading frames that can be translated to 2- and 3-kDa proteins. The *SLL1* gene is not present in the *S* locus of the S^9 haplotype in *B. rapa*. In the S^9 of *B. rapa*, *SAE1* (*S*-locus anther-expressed gene) is located near *SLG*, ca. 3 kb downstream. Other *S* haplotypes do not have this gene except for S^3 in *B. oleracea* (Watanabe et al. 1999). These studies indicate that different *S* haplotypes have different genes in the *S* locus. Although the function of these genes has not yet been elucidated, it can be inferred that *S* heterozygotes possessing various dominant genes may have an advantage over *S* homozygotes.

If crossing-over between *SP11* and *SRK* occurs in meiosis, self-compatible progeny can arise. However, such a recombinant has not been obtained in genetic analyses of self-incompatibility in *Brassica*. The high variation in the genomic organization of *S* locus is considered to be a cause of the infrequent recombination between these genes.

4 *SRK*, *SLG*, and *SP11* in Brassicaceae Species

B. oleracea and *B. rapa* have highly similar *S*-locus genes. Analyses of nucleotide sequences and deduced amino acid sequences of *SLG* and *SRK* alleles showed a mixed distribution of the *B. oleracea* and *B. rapa* alleles in phylogenetic trees (Kusaba et al. 1997; Sato et al., 2002). High similarity of the *S* domain of SRK has been observed between S^{46} in *B. rapa* and S^7 in *B. oleracea* (Kusaba and Nishio 1999). Recently, we have found two other pairs of *S* haplotypes, namely, a pair of S^{47} in *B. rapa* and S^{12} in *B. oleracea,* and a pair of S^8 in *B. rapa* and S^{32} in *B. oleracea*. The *SP11* alleles were also similar between *S* haplotypes in these three pairs. A cross-pollination test using a doubled interspecific hybrid between *B. oleracea* and *B. rapa* revealed that S^{46} in *B. rapa* and S^7 in *B. oleracea* have the same recognition specificity of self-incompatibility (Kimura et al. 2002).

R. sativus also has *SLG* and *SRK* similar to those in *Brassica*. Sakamoto et al. (1998) reported nucleotide sequences of eight *SLG* alleles. In a phylogenetic tree, *R. sativus SLG* alleles did not form an independent cluster, and were intermingled with *Brassica SLG*, suggesting that diversification of *S* haplotypes predates the divergence of genus *Brassica* and *Raphanus*. The *SRK* and *SP11* alleles in *Raphanus sativus* are also similar to those in *Brassica* (Okamoto unpubl. result).

Artificially synthesized allopolyploids of AC genome show self-incompatibility, while an AC genome species, i.e., *Brassica napus*, is self-compatible. The self-compatibility of *B. napus* cv. Wester is considered to be due to a frameshift mutation in *SRK* (Goring et al. 1993). However, *B. napus* may have various *S* haplotypes, because lines in *B. rapa* have been used as breeding materials of oil seed rape in *B. napus*. Investigation of *S* haplotypes in various cultivars in *B. napus* is necessary to understand the genetic basis of the self-compatibility in *B. napus*.

Recently, *S*-locus genes in self-incompatible *Arabidopsis lyrata* have been isolated (Kusaba et al. 2001). Self-recognition molecules in stigma and pollen of *A. lyrata* are similar to SRK and SP11 in *Brassica*, respectively. There is no *SLG* in the *A. lyrata S* locus. In self-compatible *Arabidopsis thaliana*, *SRK* and *SP11* orthologs have been found on chromosome IV. This *SRK* ortholog encodes a truncated open reading frame (ORF) due to the use of a splice site different from that of *Brassica* and *A. lyrata SRK* genes. Three *SP11* orthologs identified downstream from the *A. thaliana SRK* ortholog do not encode full-length protein, either. These findings suggest that *A. thaliana* has lost the self-incompatibility of the ancestral species by the inactivation of these *S*-locus genes.

5 Difference of *S* Locus Between *B. oleracea* and *B. rapa*

Using pulsed-field gel electrophoresis, G. Suzuki et al. (2000) analyzed large DNA fragments harboring *S*-locus genes of seven *S* haplotypes in *B. oleracea* and six *S* haplotypes in *B. rapa*. *Bss*HII fragments and *Mlu*I fragments in *B. oleracea S* haplotypes detected by an *SLG* probe, which is hybridized with both *SLG* and the *S* domain of *SRK*, were larger than those in *B. rapa S* haplotypes. Another *S* locus probe, *SP4*, also detected larger DNA fragments in *B. oleracea* than in *B. rapa*. The fragment sizes were more variable in *B. oleracea* than in *B. rapa*. From these results, it is inferred that *S* locus region in *B. oleracea* is larger and more variable than that in *B. rapa*.

The materials used in the study of G. Suzuki et al. (2000) were *S* haplotypes selected arbitrarily. Since there is a large variation of the *S*-locus structure among different *S* haplotypes in a species (Fig. 1), it is difficult to distinguish the size difference between the two species from that among *S* haplotypes in a species. Therefore, Kimura et al. (2002) compared the pairs of *S* haplotypes between the two species, i.e., *B. oleracea* S^{7} and *B. rapa* S^{46}, *B. oleracea* S^{12} and *B. rapa* S^{47}, and *B. oleracea* S^{32} and *B. rapa* S^{8}, using the same method. The lengths of the *Bss*HII fragments detected by the *SLG* probe and

SP11 probes were longer in *B. oleracea* than in *B. rapa*. Two *Mlu*I fragments were detected in *B. oleracea* S^7, while a single *Mlu*I fragment was in the other five *S* haplotypes. The lengths of the *Mlu*I fragments (in S^7, the sum of the two fragments) were also longer in *B. oleracea* than those in *B. rapa*.

One possible explanation of the presence of *S* haplotypes having the same recognition specificity in *B. oleracea* and *B. rapa* is a recent introgression of *S* haplotypes from one species to the other. However, the electrophoretic analysis of the *S* haplotype pairs in *B. oleracea* and *B. rapa* indicated large structural differences between these pairs. Nucleotide sequences of the neighboring regions of *SRK* and *SP11* were also highly different between the *S* haplotype pair despite of the high similarity of *SRK*, *SLG*, and *SP11* sequences (Kimura et al. 2002). These results suggest that the similarity of the nucleotide sequences of the *S*-locus genes and the commonality of the function among the *S* haplotype pairs is not due to the recent introgression of *S* haplotypes, but due to trans-specific evolution, where each allele in an ancestral species is inherited by two daughter species (Fig. 2).

Although the neighboring regions of *SRK* and *SP11* in the *S* locus have been largely changed after speciation of *B. oleracea* and *B. rapa*, the sequence important for the recognition specificity in *SRK* and *SP11* has been maintained. This fact suggests that their recognition specificity cannot be changed easily. If the recognition specificity of either *SRK* or *SP11* is changed, a self-compatible *S* haplotype arises. In *B. oleracea* and *B. rapa*, self-compatible wild plants are rare, suggesting a disadvantage of self-compatibility in these species. It is considered that they have maintained self-incompatibility without changing recognition specificity of *S* haplotypes after speciation.

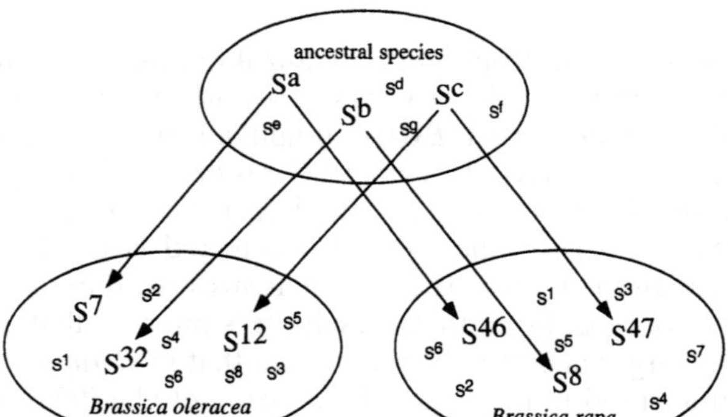

Fig. 2. Trans-specific evolution of *S* haplotypes in *Brassica*. It is considered that the S^a haplotype in an ancestral species has been inherited by *B. oleracea* as S^7 and by *B. rapa* as S^{46}; and S^b and S^c haplotypes by *B. oleracea* as S^{32} and S^{12}, respectively, and by *B. rapa* as S^8 and S^{47}, respectively. *B. oleracea* S^7 and *B. rapa* S^{46} have the same recognition specificity in pollen and stigma but the genomic organization of *S* locus is highly different between these *S* haplotypes, suggesting that these *S* haplotypes were split at the time of speciation and still have the same recognition specificity as that of the ancestral *S* haplotype

6 Conclusion and Prospects

In the study of self-incompatibility, one interesting question is how a new *S* haplotype has been generated. The number of *S* haplotypes in a population tends to decrease by genetic drift. In order to maintain the large number of *S* haplotypes in a species, a high mutation rate at the *S* locus has been postulated (Mayo 1966). However, generation of a new *S* haplotype is not a simple mutation event. In order to generate a new *S* haplotype, both *SRK* and *SP11* have to be changed simultaneously. Matton et al. (1999) produced an *S*-RNase gene having dual specificity in petunia with site-directed mutagenesis; they presented a dual-specificity model as a mechanism for generating a new *S* haplotype. In *Brassica*, allele-specific binding between the SRK protein and the SP11 protein has been demonstrated (Kachroo et al. 2001; Takayama et al. 2001). Nucleotide sequences of a large number of the *SRK* and *SP11* alleles were determined in *B. oleracea* and *B. rapa* (Sato et al. 2002). *Brassica* self-incompatibility would be the best system for the study of the generation mechanism of a new *S* haplotype.

Molecular genetic study of *S* haplotypes in *Brassica* can contribute to the development of the breeding methodology for *Brassica* vegetables. Because of the difficulty of *S* haplotype identification, breeders of *Brassica* vegetables have performed F_1 hybrid breeding without information of *S* haplotypes in their breeding lines. By identifying *S* haplotypes, information on the characteristics of each *S* haplotype, e.g., rank in the dominance relationships and stability of self-incompatibility, can be put to account. Identification of *S* haplotypes in breeding lines became much easier by analyzing DNA polymorphism of *S*-locus genes than by a traditional cross-pollination test. In the DNA polymorphism analysis of the *S*-locus genes, conserved sequences in *SLG* or *SRK* are used (Brace et al. 1994; Nishio et al. 1996, 1997). Since nucleotide sequences of most of *SLG, SRK*, and *SP11* alleles in *B. oleracea* and *B. rapa* have been determined, more reliable and simple techniques for *S* haplotype identification will be developed.

References

Anderson NA, Cornish EC, Mau S-L, Williams EG, Hoggart R, Atkinson A, Bonig I, Grego B, Simpson R, Roche PJ, Haley JD, Penschow JD, Niall HD, Tregear GW, Coghlan JP, Crawford RJ, Clarke AE (1986) Cloning of cDNA for a stylar glycoprotein associated with expression of self-incompatibility in *Nicotiana alata*. Nature 321:38–44

Bateman AJ (1955) Self-incompatibility systems in angiosperms (III). Cruciferae. Heredity 9:52–68

Boyes DC, Nasrallah JB (1995) An anther-specific gene encoded by an *S* locus haplotype of *Brassica* produces complementary and differentially regulated transcripts. Plant Cell 7:1283–1294

Brace J, King GJ, Ockendon DJ (1994) A molecular approach to the identification of *S*-alleles in *Brassica oleracea*. Sex Plant Reprod 7:203–208

Brugiere N, Cui Y, Rothstein SJ (2000) Molecular mechanisms of self-recognition in *Brassica* self-incompatibility. Trends Plant Sci 5:432–438

Chen C-H, Nasrallah JB (1990) A new class of *S* sequences defied by a pollen recessive self-incompatibility allele of *Brassica oleracea*. Mol Gen Genet 222:241–248

Doughty J, Dixon S, Hiscock SJ, Willis AC, Parkin IAP, Dickinson HG (1998) PCP-A1, a cysteine-rich *Brassica* pollen coat protein that binds the *S* locus glycoprotein, is the product of gametophytic gene expression. Plant Cell 10:1333–1347

Foote HCC, Ride JP, Franklin-Tong VE, Walker EA, Lawrence MJ, Franklin CH (1994) Cloning and expression of a distinctive class of self-incompatibility (*S*) gene from *Papaver rhoeas* L. Proc Natl Acad Sci USA 91:2265–2269

Goring DR, Glavin TL, Schafer U, Rothstein SJ (1993) An S receptor kinase gene in self-compatible *Brassica napus* has a 1-bp deletion. Plant Cell 5:531–539

Hatakeyama K, Watanabe M, Takasaki T, Ojima K, Hinata K (1998) Dominance relationships between *S*-alleles in self-incompatible *Brassica campestris* L. Heredity 79:241–247

Kachroo A, Schopfer CR, Nasrallah ME, Nasrallah JB (2001) Allele-specific receptor-ligand interactions in *Brassica* self-incompatibility. Science 293:24–26

Kimura R, Sato K, Fujimoto R, Nishio T (2002) Recognition specificity of self-incompatibility maintained after the divergence of *Brassica oleracea* and *Brassica rapa*. Plant J 29:215–223

Kusaba M, Nishio T (1999) Comparative analysis of *S* haplotypes with very similar *SLG* alleles in *Brassica rapa* and *Brassica oleracea*. Plant J 17:83–91

Kusaba M, Nishio T, Satta Y, Hinata K, Ockendon D (1997) Striking sequence similarity in inter- and intra-specific comparisons of class I *SLG* alleles from *Brassica oleracea* and *Brassica campestris*: implications for the evolution and recognition mechanism. Proc Natl Acad Sci USA 94:7673–7678

Kusaba M, Matsushita M, Okazaki K, Satta Y, Nishio T (2000) Sequence and structural diversity of the *S* locus genes from different lines with the same self-recognition specificities in *Brassica oleracea*. Genetics 154:413–420

Kusaba M, Dwyer K, Hendershot J, Vrebalov J, Nasrallah JB, Nasrallah ME (2001) Self-incompatibility in the genus *Arabidopsis*: characterization of the *S* locus in the outcrossing *A. lyrata* and its autogamous relative *A. thaliana*. Plant Cell 13:627–643

Matton DP, Luu DT, Xike Q, Laublin G, O'Brien M, Maes O, Morse D, Cappadocia M (1999) Production of an *S*-RNase with dual specificity suggests a novel hypothesis for the generation of new *S* alleles. Plant Cell 11:2087–2097

Mayo O (1966) On the problem of self-incompatibility. Biometrics 22:111–120

Nasrallah JB, Kao T-H, Goldberg ML, Nasrallah ME (1985) A cDNA clone encoding an *S*-locus-specific glycoprotein from *Brassica oleracea*. Nature 318:263–267

Nasrallah JB, Yu SM, Nasrallah ME (1988) Self-incompatibility genes of *Brassica*: expression, isolation, and structure. Proc Natl Acad Sci USA 85:5551–5555.

Nasrallah JB, Rundle SJ, Nasrallah ME (1994) Genetic evidence for the requirement of the *Brassica S*-locus receptor kinase gene in the self-incompatibility response. Plant J 5:373–384

Nasrallah ME, Wallace DH (1967) Immunogenetics of self-incompatibility in *Brassica oleracea*. Heredity 22:519–527

Nishio T, Hinata K (1977) Analysis of S-specific proteins in stigma of *Brassica oleracea* L. by isoelectric focusing. Heredity 38:391–396

Nishio T, Kusaba M (2000) Sequence diversity of *SLG* and *SRK* in *Brassica oleracea* L. Ann Bot 85 (Suppl A):141–146

Nishio T, Kusaba M, Watanabe M, Hinata K (1996) Registration of *S* alleles in *Brassica campestris* L. by the restriction fragment sizes of *SLG*s. Theor Appl Genet 92:388–394

Nishio T, Kusaba M, Sakamoto K, Ockendon DJ (1997) Polymorphism of the kinase domain of the S-locus receptor kinase gene (*SRK*) in *Brassica oleracea* L. Theor Appl Genet 95, 335–342

Nou IS, Watanabe M, Isogai A, Hinata K (1993) Comparison of *S*-alleles and *S*-glycoproteins between two wild populations of *Brassica campestris* in Turkey and Japan. Sex Plant Reprod 6:79–86

Ockendon DJ (1974) Distribution of self-incompatibility alleles and breeding structure of open-pollinated cultivars of Brussel sprouts. Heredity 33:159–171

Ockendon DJ (2000) The *S*-allele collection of *Brassica oleracea*. Acta Hortic 539:25–30

Okazaki K, Kusaba M, Ockendon DJ, Nishio T (1999) Characterization of *S* tester lines in *Brassica oleracea*: polymorphism of restriction fragment length of *SLG* homologues and isoelectric points of *S*-locus glycoproteins. Theor Appl Genet 98:1329–1334

Sakamoto K, Kusaba M, Nishio T (1998) Polymorphism of the *S*-locus glycoprotein gene (*SLG*) and the *S*-locus related gene (*SLR1*) in *Raphanus sativus* L. and self-incompatible ornamental plants in the Brassicaceae. Mol Gen Genet 258:397–403

Sassa H, Nishio T, Kowyama Y, Hirano H, Koba T, Ikehashi H (1996) Self-incompatibility (S) alleles of the Rosaceae encode members of a distinct class of the T2/S ribonuclease superfamily. Mol Gen Genet 250:547–557

Sato K, Nishio T, Kimura R, Kusaba M, Suzuki T, Hatakeyama K, Ockendon DJ, Satta Y (2002) Co-evolution of the S-locus genes, *SRK*, *SLG*, and *SP11/SCR* in *Brassica oleracea* and *Brassica rapa*. Genetics 162:931–940

Schopfer CR, Nasrallah ME, Nasrallah JB (1999) The male determinant of self-incompatibility in *Brassica*. Science 286:1697–1700

Stein JC, Howlett B, Boyes DC, Nasrallah ME, Nasrallah JB (1991) Molecular cloning of a putative receptor protein kinase gene encoded at the self-incompatibility locus of *Brassica oleracea*. Proc Natl Acad Sci USA 88:8816–8820

Suzuki G, Kai N, Hirose T, Fukui K, Nishio T, Takayama S, Isogai A, Watanabe M, Hinata K (1999) Genomic organization of the *S*-locus: identification and characterization of genes in *SLG/SRK* region of S^9 haplotype of *Brassica campestris* (syn. *rapa*). Genetics 153:391–400

Suzuki G, Watanabe M, Nishio T (2000) Physical distances between *S*-locus genes in various *S* haplotypes of *Brassica rapa* and *B. oleracea*. Theor Appl Genet 101:80–85

Suzuki T, Kusaba M, Matsushita M, Okazaki K, Nishio T (2000) Characterization of *Brassica S*-haplotypes lacking *S*-locus glycoprotein. FEBS Lett 482:102–108

Takasaki T, Hatakeyama K, Suzuki G, Watanabe M, Isogai A, Hinata K (2000) The *S* receptor kinase determines self-incompatibility in *Brassica* stigma. Nature 403:913–916

Takayama S, Shiba H, Iwano M, Shimosato H, Che F-S, Kai N, Watanabe M, Suzuki G, Hinata K, Isogai A (2000) The pollen determinant of self-incompatibility in *Brassica campestris*. Proc Natl Acad Sci USA 97:1920–1925

Takayama S, Shimosato H, Shiba H, Funato M, Che F-S, Watanabe M, Iwano M, Isogai A (2001) Direct ligand-receptor complex interaction controls *Brassica* self-incompatibility. Nature 413:534–538

Tantikanjana T, Nasrallah ME, Stein JC, Chen C-H, Nasrallah JB (1993) An alternative transcript of the *S* locus glycoprotein gene in a class II pollen-recessive self-incompatibility haplotype of *Brassica oleracea* encodes a membrane-anchored protein. Plant Cell 5:657–666

The Arabidopsis Genome Initiative (2000) Analysis of the genome sequence of the flowering plant *Arabidopsis thaliana*. Nature 408:796–815

Thompson KF, Taylor JP (1966) Non-linear dominance relationships between *S*-alleles. Heredity 21:345–362

Watanabe M, Suzuki G, Toriyama K, Takayama S, Isogai A, Hinata K (1999) Two anther-expressed genes downstream of SLG^9: identification of a novel *S*-linked gene specifically expressed in anthers at the uninucleate stage of *Brassica campestris* (syn. *rapa*) L. Sex Plant Reprod 12: 127–134

Yu K, Schafer U, Glavin TL, Goring DR, Rothstein SJ (1996) Molecular characterization of the *S* locus in two self-incompatible *Brassica napus* lines. Plant Cell 8:2369–2380

8 *Brassica* and Legume Chromosomes

K. FUKUI

1 Historical Perspectives

The numbers of the chromosomes of *Brassica* and legumes were identified at an early stage of chromosome research. The earliest report on the chromosome numbers of *Brassica* species was back in 1916 when Takamine precisely determined the chromosome number of *B. rapa* (*B. campestris*, A genome) as 20. Trouard-Riolle also determined the chromosome number of *Raphanus sativus* as 18 in 1916. The chromosome numbers of *B. oleracea* and *B. nigra*, which contain the C and B genomes, respectively, were also determined as 18 and 16, respectively, by Karpechenko in 1924 (Gomez-Campo and Hinata 1980). By 1980, chromosome numbers of more than 240 taxa of the tribe Brassicaceae had been determined and listed (Gomez-Campo and Hinata 1980). Chromosome research into the legumes has a similar history to that of *Brassica*. The chromosome number of *Glycine max*, with typical small type (S-type; Fukui 1996; Fukui et al. 2000) chromosomes, was reported by Karpechenko in 1925 and by Veatch in 1934 (Singh et al. 2001). The chromosome number of *Lotus japonicus* was reported as $2n = 24$ in 1930 (Kawakami 1930). The fact, however, that all the chromosomes are similar in both size and morphology at mitotic metaphase, made it difficult to identify individual chromosomes. Although Yanagisawa et al. (1991) found that there are marker chromosomes commonly observed in the species of the genus *Glycine*, it is far from complete identification of all the chromosomes. The situation of chromosome identification for the species in the genus *Brassica* was more or less similar to the case of S-type legume chromosomes represented by *Glycine max* until the 1980s. All the chromosomes are evenly condensed at mitotic metaphase without conspicuous morphological landmarks. Thus it was a rather common opinion of cytological researchers that *Brassica* and S-type legume chromosomes were not suitable for cytological studies.

Hu (1964) and Kurata and Omura (1978) identified 7 or 8 different chromosomes among the 12 rice chromosomes which are the typical S-type chromosomes, based on the appearance of uneven condensation of the rice chromosomes at the prometaphase stage. Although the appearance of the uneven condensation of the chromosomes during mitotic prometaphase was

Department of Biotechnology, Graduate School of Engineering, Osaka University. Suite 565-0871, Osaka, Japan

Biotechnology in Agriculture and Forestry, Vol. 52
Nagata/Tabata (Eds.) Brassicas and Legumes

known empirically, many researchers thought that the pattern was too unstable to be useful in morphological landmarking for identification purposes. The pattern progressively changes during the progress of the cell cycle, but the dynamic changes in the patterns and the pattern at each substage are characteristic. The acquisition of numerical data from uneven condensation along the prometaphase chromosome, which is rather difficult by visual inspection, is attained by the introduction of image-analysis methods (Fukui 1986; Fukui and Iijima 1991; Kato and Fukui 1998). The density profile at the mid-rib of each chromatid is measured and the gray values of each pixel comprising the density profile are satisfactorily obtained. Now the condensation pattern is recognized as an effective morphological marker, and moreover, quantitative chromosome maps have been developed based on the density profile for several *Brassica* and legume plants (Fukui et al. 1998; Kamisugi et al. 1998; Ito et al. 2000a,b). Various recent cytological methods are also effective in characterizing and identifying *Brassica* and legume chromosomes. Fuchs et al. (1998b) revealed the structural differentiation among large *Vicia faba* chromosomes using, among other methods, Giemsa staining, fluorescent dye staining, and restriction enzyme treatment. The data obtained provided the fundamental information on the structure of *Vicia faba* chromosomes. The fluorescence in situ hybridization (FISH) method has also played a major role in the assignment of linkage groups to pea chromosomes (Fuchs et al. 1998a).

In this chapter, since the two are often used in combination to identify and characterize *Brassica* and legume chromosomes, the method based on the condensation pattern is described, as are those molecular cytological methods represented by FISH.

2 Identification and Characterization of Plant Chromosomes Based on Condensation Pattern

2.1 Two Types of Plant Chromosomes

Plant chromosomes are basically classified into large type (L-type) chromosomes and small type (S-type) chromosomes (Fukui 1996). Prominent differences between L- and S-type chromosomes lie not only in their sizes, but also in the appearance of the condensation pattern of their chromosomes at the prometaphase stage as shown in Fig. 1. Fukui and Mukai (1988) confirmed the reproducible appearance of an uneven condensation pattern at prometaphase of S-type chromosomes in *Atriplex rosea*, the pattern of which has come to be referred to simply as the "condensation pattern". Subsequently, Kamisugi et al. (1993) revealed that the numerical data from the patterns could be used for automatic identification of rice chromosomes. Thus the condensation pattern has been established as a useful morphological landmark for the characterization and identification of plant chromosomes in addition to the two

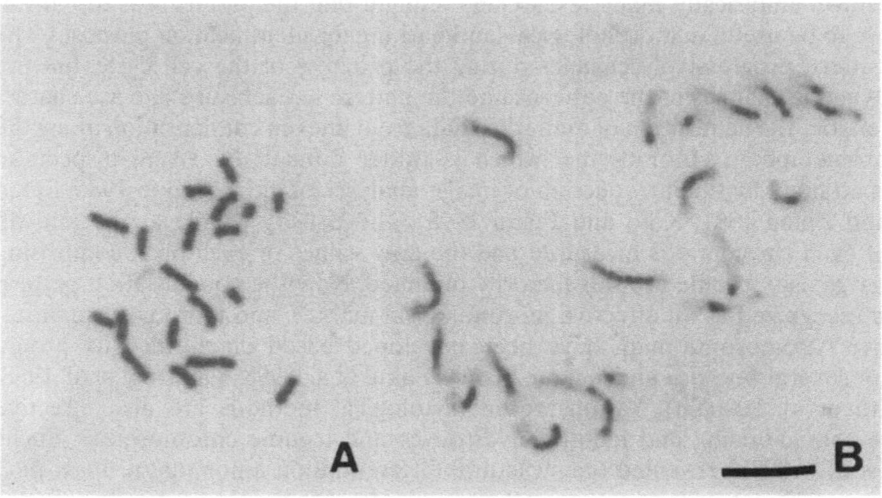

Fig. 1. Haploid *Brassica napus* chromosomes ($n = 19$) at **A** metaphase and **B** prometaphase; Giemsa stained. The *scale bar* indicates 5 μm. (Kamisugi et al. 1998)

conventional parameters of length and arm ratio. Condensation pattern is especially valuable in characterization and identification of S-type chromosomes where the difference in morphology is almost negligible at mitotic metaphase. *Brassica* and many legume chromosomes share typical S-type chromosome characteristics, being 1 to 3 μm in length at the metaphase stage and having similar morphology. These similarities explain why more than 70 years were required to identify the chromosomes of *Brassica* species.

Although the concept of condensation pattern was first developed in *Atriplex* chromosomes, *Brassica* chromosomes have also been characterized and identified based on uneven condensation at prometaphase (e.g., Cheng et al. 1995a; Olin-Fatih 1996). The unevenness is not caused by differential affinity of the staining dye to different chromosomal regions, as is the case with G-band manifestation after Giemsa staining, but is proved to be caused by the differential condensation of the chromosomes at the prometaphase stage (Fukui and Iijima 1991). In contrast, L-type plant chromosomes have been identified by several banding methods such as the C-banding method (Endo 1986) or simply by their morphological features (Fukui and Kamisugi 1995) without a prominent condensation pattern at prometaphase.

2.2 Image Analysis Methods for Condensation Pattern

Imaging methods are indispensable for the analysis of density which is difficult to quantify by visual inspection alone. Image technology now supports researchers' decisions by reproducibly and reliably digitizing the color, shade, volume, and even shape of complex biological objects. The first comprehensive image analysis system for plant chromosomes was developed in 1985 and named the Chromosome Image Analyzing System, CHIAS (Fukui 1986). Now the third generation CHIAS (CHIAS 3 is released via CHAIS 2; Fukui and Nakayama 1996) and all the necessary software and its manual can be obtained via the Internet (Kato et al. 1997; Kato and Fukui 1998; Ha et al. 1999). After identification of each individual chromosome by its condensation pattern, it is image analyzed using CHIAS 3. The numerical data of its density profile obtained from the condensation pattern becomes the image parameter CP. The image analyses proceed basically following three steps. First, the original image is image-analyzed to obtain source images for data acquisition by correction of light distortion, elimination of noise, normalization of gray values, and removal of other artifactual elements. Second, the density profile (CP) is obtained from the condensation pattern in each chromatid of individual chromosomes identified, and the gray values of the profile are recorded and averaged to give the standard CP (stCP). Third, the stCP is delimited by one or two gray values to discriminate the condensed and dispersed regions – or the heavily condensed, condensed, and dispersed regions (Kamisugi et al. 1993). Based on the stCPs for individual chromosomes, divided into dispersed, condensed, and heavily condensed regions by gray-value thresholds, quantitative chromosome maps for several *Brassica* and legume plants, such as *B. napus* (Kamisugi et al. 1998), *B. rapa, B. nigra, B. oleracea* (Fukui et al. 1998), *A. thaliana* (Ito et al. 2000a), *G. max* (Yanagisawa et al. 1991), and *L. japonicus* (Ito et al. 2000b) have been developed successfully.

3 Case Histories of Chromosome Identification and Development of Quantitative Chromosome Maps: *Brassica* and *Arabidopsis*

3.1 *Brassica* Chromosomes

Figure 2 depicts the chromosomes of three *Brassica* species with basic genomes of A (*B. rapa*), B (*B. nigra*), and C (*B. oleracea*), respectively (Fukui et al. 1998). Prominent condensation is visualized at both the proximal regions of all the chromosomes. Some chromosomes also have faint, unstable, and small condensations (FUSCs) at the end of the chromosomes. Based on the overall morphology, all the chromosomes within a complement are classified into three major categories as: type I (median and submedian), type II (subterminal and terminal) and type III (satellite chromosome), by visual inspection first. Then each chromosome within the type is identified by considering its condensation pattern, FUSCs, and the other morphological characteristics as described in the case of rice chromosomes (Iijima et al. 1991). Detailed inspection of the condensation pattern enables identification of individual chromosomes of even similar length and arm ratio.

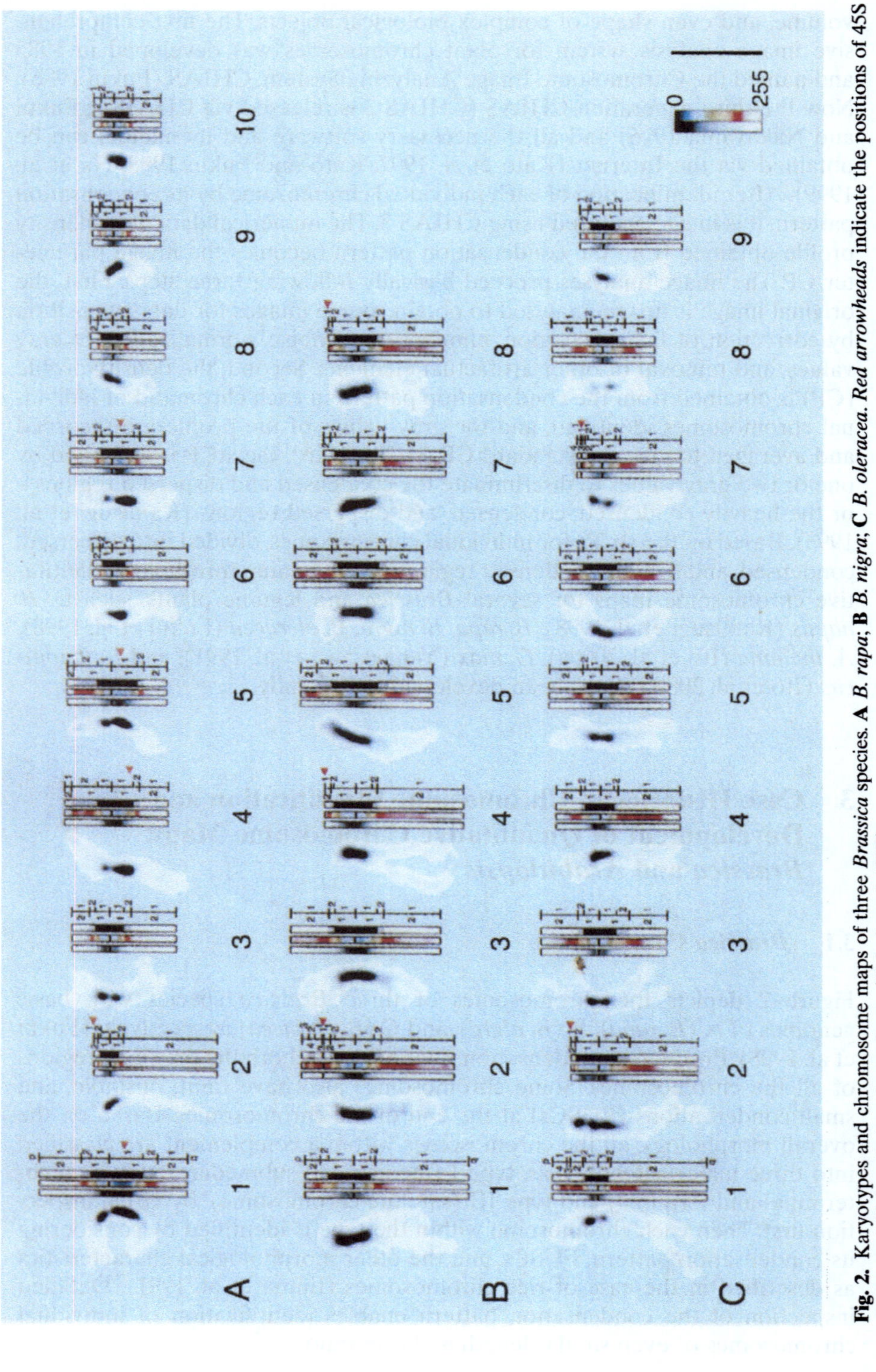

Fig. 2. Karyotypes and chromosome maps of three *Brassica* species. **A** *B. rapa*; **B** *B. nigra*; **C** *B. oleracea*. *Red arrowheads* indicate the positions of 45S rDNA loci. The *scale bar* shows 5 μm. (Fukui et al. 1998)

Although the overall chromosomal characteristics are similar within the three species demonstrating prominent condensed regions at the centromeres, species-specific characteristics are also revealed. These characteristics enable identification of the genomic origin of individual chromosomes even in the tetraploid species. The most prominent species-specific characteristic that appears consistently is the difference in the relative size of the condensed regions at the centromeric regions. In the chromosomes of *B. rapa*, the larger heterochromatic regions usually occur at the proximal region of the long arms as shown in Fig. 2. In the case of *B. oleracea*, the opposite tendency, i.e., the largest heterochromatic regions occur at the proximal region of the short arms, is observed in some chromosomes. The chromosomes of *B. nigra* have almost even-sized heterochromatic blocks on both arms.

Figure 3 shows the karyotype and quantitative chromosome maps of *B. napus* ($n = 19$). The species-specific condensation at the proximal regions is consistently observed among the chromosomes of *B. napus*, which together with the parameters of length and arm ratio, enable the identification and classification of the entire *B. napus* chromosome complement into ten *rapa*-type chromosomes (A genome origin, black capitals) and nine *oleracea* types (C genome origin, red capitals). In Fig. 3, three images of each chromosome are presented. The left-hand image is a karyogram which demonstrates the uneven chromatin condensation within chromosomes. In the center is a chromosome map representing the condensation pattern after image analysis of the CP. The density distribution is presented in pseudocolor for greater visual contrast. Each color corresponds to a certain gray value as indicated on the gray scale/color table. The chromosome map on the right hand side is constructed by applying threshold values (168 and 219) to the standard CP or the averaged CP. The borders between heavily condensed and dispersed regions observed in the chromosome spreads correspond to the borders between gray and white regions in the chromosome map. The 5S and 45S rDNA loci are also mapped onto the chromosome map.

3.2 *Arabidopsis* Chromosomes

Ito et al. (2000a) stained *A. thaliana* chromosomes with the DNA-binding fluorochrome DAPI (diaminidophenylindole) instead of using ordinary dyes. It is not possible to identify each chromosome at metaphase because of the smallness of the chromosomes with similar morphology, as is often the case in S-type chromosomes. DAPI staining, however, enables the detailed comparison of the condensation pattern of *A. thaliana* chromosomes at prometaphase. Figure 4 depicts *A. thaliana* chromosomes at metaphase and prometaphase in addition to the results of multicolor FISH using 5S and 45S rDNA as probes. Based on the unique condensation pattern characteristic to each homologous chromosome pair, all the chromosomes are identifiable and the chromosome number is designated in accordance with the numbers of the linkage map (Koornneef et al. 1983; Murata et al. 1997). The uneven condensation pattern

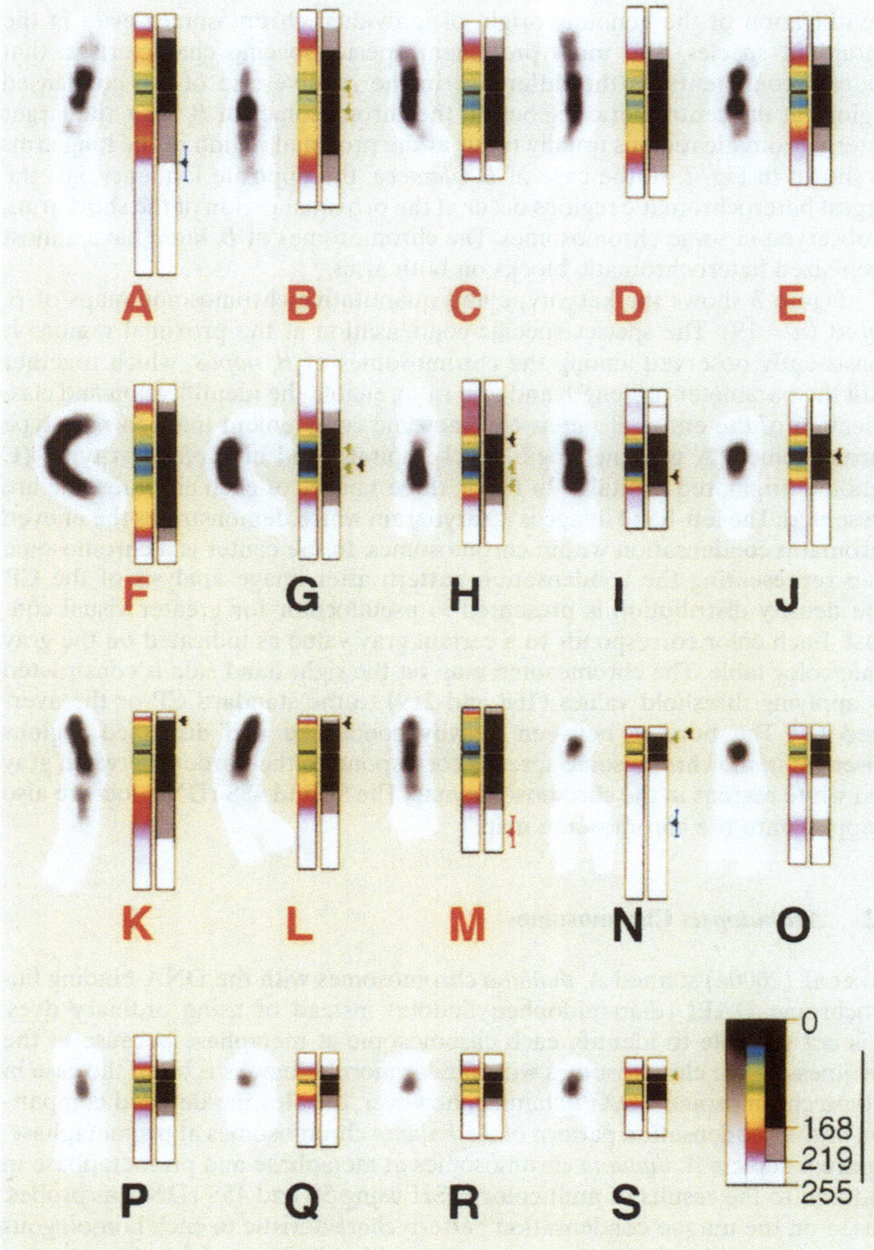

Fig. 3. Karyotypes and chromosome maps of *Brassica napus* ($n = 19$). *Solid* and *open circles* show 5S and 45S rDNA loci, respectively. *Solid* and *open triangles* indicate *SLR1* and *SLG* genes with the standard deviation. The *scale bar* shows 5 μm. (Fukui et al. 1998)

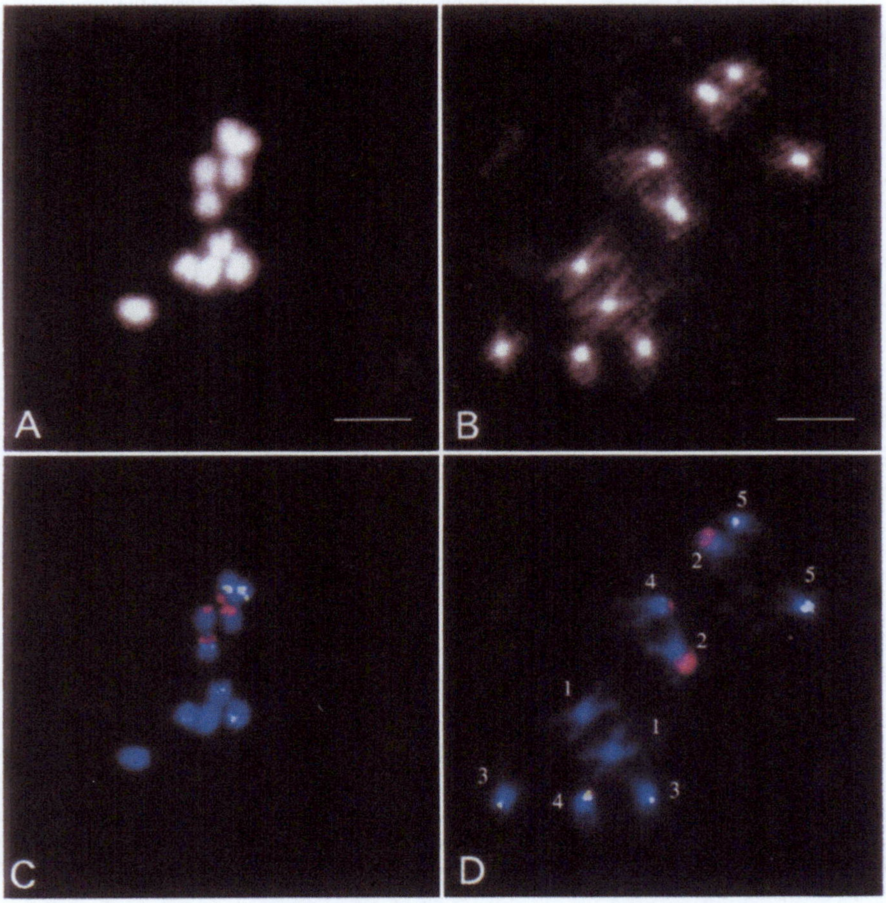

Fig. 4. *Arabidopsis thaliana* chromosomes at metaphase (**A**, **C**), prometaphase (**B**, **D**), with 45S (*red*) and 5S ribosomal DNA (*green*) signals after multicolor FISH (**C**, **D**). The *scale bar* shows 5 μm. (Ito et al. 2000a)

of each chromosome is quantified using CHIAS 3, and CP is digitally obtained for each chromatid. Sixteen CPs from eight chromosomes from four chromosomal spreads are averaged and stCP is obtained for each of five *Arabidopsis* chromosomes. Then a single gray value was determined in order to discriminate the condensed and dispersed regions. It is not possible to determine the second threshold as in the case of *G. max* chromosomes, mainly because of smallness of the chromosomes.

Figure 5 depicts quantitative chromosome maps developed for individual chromosomes. They satisfactorily represent morphological characteristics of each chromosome at the prometaphase stage shown in Fig. 4. Figure 5 also depicts the CPs and density profiles for FISH signals. The vertical and horizontal axes represent the number of pixels along the chromosomes

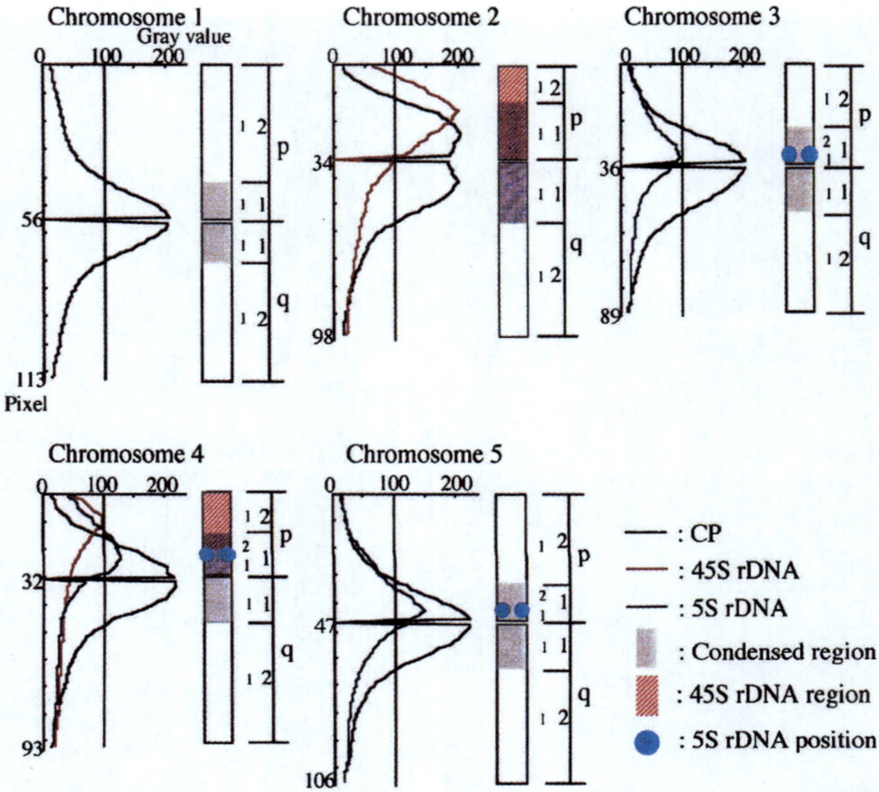

Fig. 5. CPs and density profiles for FISH signals, and quantitative chromosome maps of *Arabidopsis thaliana* with the addresses and locations of 45S and 5S rDNA loci. *Vertical* and *horizontal axes* represent the number of pixels along the chromosomes and the gray value, respectively. (Ito et al. 2000a)

and the gray value, respectively. The proximal condensations determined by the threshold (gray value = 100) are shown by gray regions in the quantitative chromosome maps. The 45S and 5S rDNA loci are designated by the red regions and solid blue circles on the chromosome map, respectively. The 5S rDNA loci map to *p*1.1 of chromosome 3, 4, and 5. Short and long arms are indicated by *p* and *q*, respectively.

Table 1 shows the morphological data and the positions of rDNA loci, which are averages based on eight prometaphase chromosomes (Ito et al. 2000a). The total length of prometaphase chromosomes is 16.5 μm, which is about 1.8 times the size of metaphase chromosomes. The characters of each chromosome and locations of the rDNA loci are summarized as follows: chromosome 1 is metacentric and the longest chromosome. It shows condensation only in the proximal regions of the centromere. Chromosome 2 is submetacentric with a 45S rDNA locus in the telomeric region of the short arm. It has a longer condensed region than chromosome 4, which is a key feature dis-

Table 1. Numerical data of somatic chromosomes of *A. thaliana*. (Ito et al. 2000a)

Chromosome No.	Short arm		Long arm		Arm ratio	Total length			45S rDNA Region (μm)	5S rDNA Region (μm)
	Length (μm)	SD	Length (μm)	SD		Length (μm)	SD	Relative length (%)		
1	1.83	0.18	1.90	0.29	1.04	3.73	0.42	22.68		
2	1.10	0.11	2.15	0.28	1.96	3.24	0.36	19.69	0–1.10	
3	1.16	0.13	1.76	0.36	1.15	2.92	0.46	17.74		1.00
4	1.02	0.09	1.96	0.32	1.99	3.51	0.61	18.57	0–0.65	0.76
5	1.54	0.30	1.96	0.32	1.27	3.51	0.61	21.32		1.36

criminating chromosome 2 from chromosome 4. Chromosome 3 is metacentric, is the shortest chromosome, and has a 5S rDNA locus at the proximal region on the short arm. Chromosome 5 is metacentric with a 5S rDNA locus close to the centromere on the short arm.

4 Case Histories of Chromosome Identification and Development of Quantitative Chromosome Maps: *Glycine* and *Lotus*

4.1 *Glycine* Chromosomes

It has been difficult to identify all the chromosomes of legume S-type chromosomes represented by *G. max* in spite of many long trials. Figure 6 shows

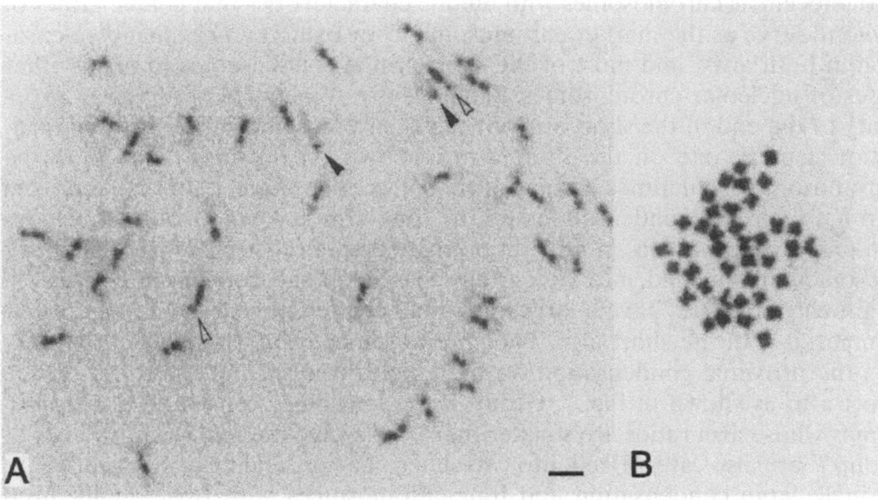

Fig. 6. *Glycine max* chromosomes at metaphase (*A*) and prometaphase (*B*). *Solid* and *open arrowheads* indicate marker chromosomes and chromosomes with the distinct characters. The *scale bar* shows 5 μm. (Yanagisawa et al. 1991)

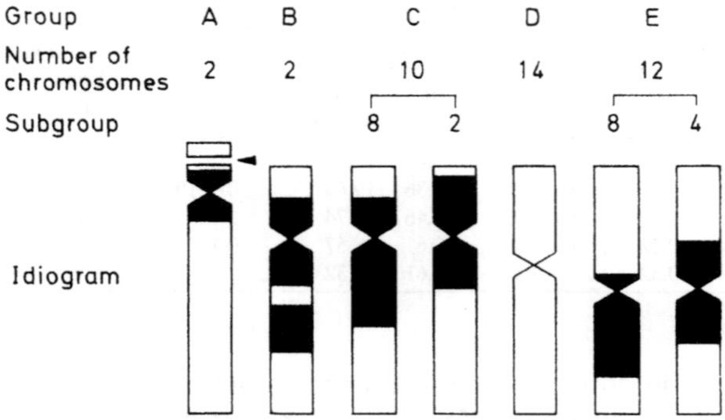

Fig. 7. Five groups of the *Glycine max* chromosomes ($2n = 40$) based on their condensation patterns. *Solid arrowhead* indicates the position of nucleolar organizing regions. Condensation pattern is not represented in group D because a single pattern is not defined among the chromosomes within group D. (Yanagisawa et al. 1991)

G. max chromosomes at prometaphase and metaphase stages. Typical condensation patterns are demonstrated on the prometaphase chromosomes indicating the possibility of identification of the chromosomes. Among 40 chromosomes, 2 show an extremely high arm ratio of 5.7, although the arm ratios of the rest of the chromosomes range from 1.3 to 2.5. A pair of chromosomes exists in all the nine species of the genus *Glycine* and they are referred to as the marker chromosomes of the genus *Glycine*. The marker chromosomes have been proven to be satellite chromosomes or nucleolar chromosomes by the N-banding method (Yanagisawa et al. 1991).

Application of the identification method based on condensation pattern, however, enables categorization of *G. max* chromosomes into five groups (Yanagisawa et al. 1991). Figure 7 depicts the five groups based on the condensation patterns of 40 *G. max* chromosomes. Group A chromosomes include two telocentric chromosomes with unique characteristics that enable either of them to serve as the marker chromosome. They exhibit a proximal condensation in both arms, and most of the short arm is condensed as in many other cases of nucleolar chromosomes. Satellites are sometimes observed in proximity to the end of the short arm. Group B chromosomes have three condensation centers, one on the short arm and two on the long arm. Thus, the chromosomes sometimes display trimodal condensation patterns when the gap between two condensations on the long arm is clear. Group C includes ten chromosomes with an arm ratio larger than 1.6. These submedian chromosomes are divided into two groups based on the condensation pattern. Eight chromosomes have a larger proximal condensation on the long arm as compared with the short arm. Two chromosomes exhibit a reversed pattern, i.e., the proximal condensation on the long arm is smaller than that on the short arm as shown in Fig. 7. Group E chromosomes consist of 12 chromosomes whose arm ratios are smaller than 1.2. Twelve median chromosomes of group E are also categorized into two subgroups, i.e., eight chromosomes with a single large condensation and four chromosomes with two equally sized proximal condensations. Group D chromosomes include the remaining 14 chromosomes whose arm ratios are neither large nor small enough to classify into group C or E. The condensation pattern of group D chromosomes is not

Table 2. Average arm ratio and total length of *G. max* chromosomes within each group. (Yanagisawa et al. 1991)

Group	Arm ratio		Total length		No. of Chromosomes
	Mean	SD	Mean	SD	
A	7.22	2.13	2.29	0.47	2
B	2.23	0.29	2.01	0.17	2
C	2.37	0.38	1.95	0.36	10
D	1.56	0.20	2.00	0.39	14
E	1.17	0.08	1.89	0.33	12

depicted because they show several patterns. Table 2 shows the average values and the standard deviation of the arm ratio and the total length of the chromosomes for each group based on 200 *G. max* chromosomes from five prometaphase chromosomal spreads. The largest values in arm ratio and total length are obtained in group A chromosomes due to their rapid contraction as the cell cycle progresses.

4.2 *Lotus* Chromosomes

Legumes are important for their ability to fix nitrogen by symbiosis with rhizobia. *L. japonicus* is now used as one of the model legumes for the study of nitrogen fixation (Handberg and Stougaard 1992) because of its small genome size, short life cycle, diploidy with 12 chromosomes (Kawakami 1930), self-fertility, and availability of transformation. Cheng and Grant (1973) reported the karyotype of *L. japonicus* as five pairs of submetacentric and one pair of metacentric chromosomes. Jiang and Gresshoff (1997) identified molecular polymorphism in genomic DNA. *Lotus japonicus* has the typical S-type chromosomes, small and evenly condensed at metaphase, as shown in Fig. 8. Because it is difficult to visually identify all the *Lotus* chromosomes at metaphase, prometaphase chromosomes demonstrating prominent condensation patterns were used for identification. In addition to proximal condensation, small condensations were observed at terminal regions in one accession, namely, Miyakojima. Six pairs of prometaphase chromosomes in *L. japonicus* were thus successfully identified by condensation patterns. Two accessions of Miyakojima and Gifu were used for chromosome identification and characterization. Three Miyakojima and seven Gifu chromosomal spreads were statistically confirmed to be at the same mitotic stage. Then their CPs were averaged for the same chromosomes, and quantitative chromosome maps were developed by setting two thresholds as shown in Fig. 9. The gray value that delimited heavily condensed and condensed regions was 83, and that between condensed and dispersed regions was 148 in the case of Miyakojima. The corresponding gray values of Gifu were 58 and 131. Faint, unstable, and small condensations (FUSCs) were observed, mainly at the terminal regions, in Miyakojima only. To differentiate FUSCs from the ordinary condensed

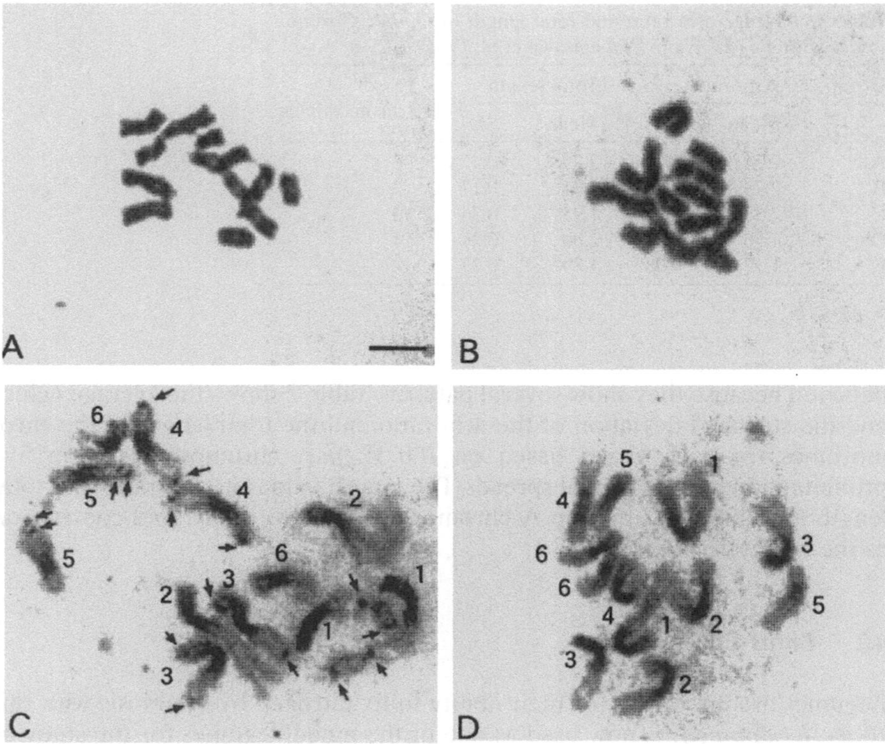

Fig. 8. Giemsa stained *Lotus japonicus* chromosomes at metaphase (**A** Miyakojima; **B** Gifu) and prometaphase (**C** Miyakojima; **D** Gifu). Accession Miyakojima shows prominent FUSCs at dispersed regions as indicated by *arrows*. *Numbers 1* to *6* indicate the chromosome number according to length order. The *scale bar* shows 5 μm. (Ito et al. 2000b)

regions which always appeared in the proximal regions, small solid circles are placed on the chromosome map of Miyakojima. The letters A to F are used to indicate the corresponding chromosomes of Miyakojima and Gifu, based on their condensation patterns and rDNA loci. As a result, the order of chromosome length in Gifu is chromosome B, C, A, D, E, and F, indicating a large difference in size of chromosome A between Miyakojima and Gifu, probably due to chromosome aberration.

Figure 9 depicts the quantitative chromosome map and the characteristics of the chromosomes. Chromosome A: Miyakojima chromosome 1 is the

Fig. 9. Chromosome maps of *L. japonicus*, accession Miyakojima (*A*) and Gifu (*B*). *CEN*, *p*, and *q* show centromeric position, short and long arms, respectively. *Numbers 1* to *6* are the chromosome numbers based on their lengths, and *A to F* indicate corresponding chromosomes between the two accessions. *Gray bars* represent the condensed chromosomal regions, while *solid bars* represent the heavily condensed regions. The *open bars* correspond to dispersed regions of the chromosomes. The positions of 45S, 5S rDNA loci and FUSCs are shown by *double*, *single open*, and *solid circles* on the maps, respectively. (Ito et al. 2000b)

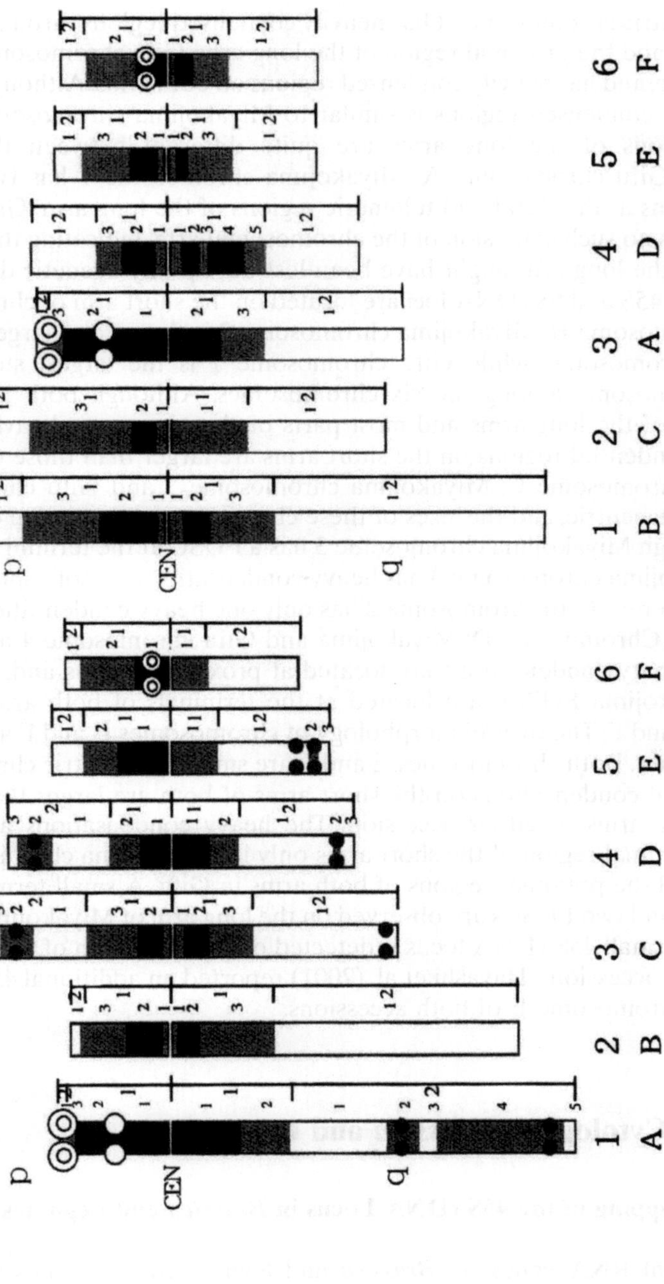

p
CEN
q
1 2 3 4 5 6
A B C D E F
Miyakojima
p
CEN
q
1 2 3 4 5 6
B C A D E F
Gifu
: Heavily condensed region
: Condensed region
: Dispersed region
: 45S rDNA locus
: 5S rDNA locus
: FUSCs

longest subtelocentric chromosome. It has heavily condensed regions throughout the short arm and the proximal region of the long arm. Gifu chromosome 3 is submetacentric, and has heavily condensed regions on both arms. Although the pattern of the condensed regions is similar to Miyakojima chromosome 1, the distal regions of the long arms are quite different between the Miyakojima and Gifu chromosome A. Miyakojima chromosome 1 has two more condensations at the distal and telomeric regions of the long arm. Gifu chromosome 3 has no such extension of the chromosomal arm, indicating that the distal part of the long arm might have been lost during phylogenetic differentiation. Both 45S and 5S rDNA loci are located on the short arm of chromosome A. Chromosome B: Miyakojima chromosome 2 is the second largest subtelocentric chromosome, while Gifu chromosome 1 is the largest submetacentric chromosome among the six chromosomes. Although both the proximal regions of the long arms and most parts of short arms are heavily condensed, the condensed regions on the short arms are larger than those on the long arms. Chromosome C: Miyakojima chromosome 3 and Gifu chromosome 2 are metacentric, and the sizes of these chromosomes are similar to each other, although Miyakojima chromosome 3 has a FUSC at the termini of both arms. Miyakojima chromosome 3 has heavy condensations on both short and long arms, whereas Gifu chromosome 2 has only one heavy condensation on the short arm. Chromosome D: Miyakojima and Gifu chromosome 4 are metacentric. Ordinary condensations are located at proximal regions and, in the case of Miyakojima FUSCs, are located at the terminals of both arms. Chromosomes E and F: The overall morphology of chromosomes E and F are similar to each other. Both chromosomes E and F are small metacentric chromosomes. Proximal condensations on the short arms of both are larger than those on the long arms in either accession. The heavy condensations are located at the proximal region of the short arms only in Miyakojima chromosomes 5 and 6 and the proximal regions of both arms in Gifu. A small terminal condensation and two FUSCs are observed on the long arm of Miyakojima chromosome 5. A small 45S rDNA locus is detected on the short arm of chromosome F in both accessions. Hayashi et al. (2001) reported an additional 45S rDNA locus on chromosome E of both accessions.

5 Molecular Cytology of *Brassica* and Legumes

5.1 Physical Mapping of the 45S rDNA Locus in *Brassica* and Legumes

Mapping ribosomal RNA genes on *Brassica* and legume chromosomes by FISH have been reported, among others, for *Lens* and *Cicer* (Abbo et al. 1994), *Phaseolus* and *Vigna* (Zheng et al. 1994), *Glycine* (Singh et al. 2001), *Arabidopsis* (Maluszynska and Heslop-Harrison 1993a), *Brassica* (Maluszynska and Heslop-Harrison 1993b; Cheng et al. 1995b; Snowdon et al. 1997; Fukui et al. 1998). FISH experiments reveal variability in the number of the 45S rDNA loci among three *Brassica* species with each of four genomes of A, B, C and AC. Five rDNA loci are visualized in *B. rapa*, three in *B. nigra*, two in *B. alboglabra* and *B. oleracea*, and seven in *B napus* (Cheng et al. 1995b; Fukui et al. 1998; Kamisugi et al. 1998). A different number of 45S rDNA loci has been reported in *B. oleracea* and *B. napus* (Maluszynska and Heslop-Harrison 1993b; Snowdon et al. 1997). The 45S rDNA loci are classified into two groups based upon where they map, regardless of their fluorescence intensity. One group is located at the end of short arm and the other at proximal regions, which latter are observed only in *B. rapa* chromosomes. The terminal positions are usual for the 45S rDNA locus and a chromosome with a terminal 45S rDNA locus is referred to as a satellite chromosome, because the tip of the chromosomal end is sometimes observed separately when the 45S rDNA is activated and extended.

Two pairs of intense and three pairs of weak fluorescence signals are clearly observed in *B. rapa*. Among five loci, four are located at the proximal regions and one at the end of a short arm. Both the 45S and 5S rDNA loci at the proximal regions are the main causes of the unbalanced heavy condensation at the proximal condensations between the short and long arms. One pair of large and two pairs of small signals detected in *B. nigra* are mapped at the end of the short arms of the chromosomes. Two pairs of medium-sized signals are observed at the end of the short arms of *B. oleracea* chromosomes. The mechanism is not known to explain the diverged locations and number of 45S rDNA loci among closely related species. Based on collinearity of gene locations among closely related species, 45S rDNA loci should be at similar places on corresponding chromosomes. One possible explanation for this peculiar positioning of 45S rDNA loci detected among the species in the genus *Brassica* is that by unknown mechanisms they move around the genome like transposons (Schubert 1984; Dubcovsky and Dvorak 1995; Shishido et al. 2000).

B. napus was reported to have either six or seven 45S rDNA loci (Maluszynska and Heslop-Harrison 1993b; Snowdon et al. 1997; Kamisugi et al. 1998). Ten pairs of chromosomes originated from the ancestor of *B. rapa* and nine pairs of those derived from *B. oleracea* in *B. napus* are identified by their characteristic condensation patterns as already described (Kamisugi et al. 1998). Some *rapa*-type chromosomes have larger proximal condensations at one arm than the other. FISH using 45S and 5S rDNA as probes reveal that eight of ten metacentric or submetacentric chromosomes showing unbalanced condensation between two arms harbor 5S and/or 45S rDNA loci in the heavily condensed regions. As a result, five and two 45S rDNA loci are located on the *rapa*-type and *oleracea*-type chromosomes, respectively. Their positions are conserved among *B. rapa, B. oleracea,* and *B. napus*. Thus it is clear that the difference in the number of 45S rDNA loci between the *rapa* and *oleracea* types was established before the origin of *B. napus* through natural hybridization between the ancestral *B. rapa* and *B. oleracea* genomes (Kamisugi et al. 1998).

Skorupska et al. (1989) and Griffor et al. (1991) detected a 45S rDNA locus in interphase cells and on metaphase chromosomes of *G. max*. Singh et al. (2001) further explored the distribution of 45S rDNA loci using 23 accessions from 16 species of the genus *Glycine*. As a result, one 45S rDNA locus was detected in 17 accessions of 14 diploid species including *G. max* and *G. soja*. Two 45S rDNA loci were detected in two diploid species of *G. curvata* and *G. cyrtoloba*. Two tetraploid species of *G. tabacina* and *G. tomentella* had two 45S rDNA loci, although they had one locus in different diploid accessions. These results support the tetraploid hypothesis of the 40-chromosome *Glycine* species. Cytogenetic and molecular evidence further supports the allotetraploid origin of the 40-chromosome *Glycine* (e.g., Shoemaker et al. 1996; Xu et al. 2000).

5.2 Mapping Genes on *Brassica* and Legume Chromosomes by Advanced FISH Methods

Based on the technical development of FISH, localizing single copy genes on plant chromosomes is now practicable. The ability to map single copy genes facilitates a wide range of genetic studies. Fuchs and Schubert (1995) detected the seed protein genes of legumin B4 at the proximal region of a long arm of chromosome 3 and short arm of chromosome 2. Legumin A gene is also located at the interstitial region of chromosome 5. Figure 10 depicts the mapping of the single copy *S*-locus glycoprotein (*SLG*) and *S*-locus-related (*SLR1*) genes involved in the *Brassica* self-incompatibility system by FISH together with 5S and 45S rDNA loci (Kamisugi et al. 1998). The *SLR1* probe is detected on the long arm of an *oleracea*-type (chromosome A, Fig. 3) and a *rapa*-type chromosome (chromosome N, Fig. 3). However, the *SLG* probe produced a clear doublet of fluorescent signals on the long arm of only one *oleracea*-type chromosome (chromosome M, Fig. 3). It is observed that the fluorescent signals from the *SLG* are more frequent after FISH than those from *SLR1*. This would indicate physical proximity of *SLG* and *SRK* genes which have high nucleotide sequence similarity over a common duplicated domain. These factors thus combine to confer an effective copy number of two on any *SLG* probe. The observation that the *SLG* and *SLR1* genes are located on different chromosomes visually confirmed the nonlinkage of these loci previously shown by RFLP analysis (Lalonde et al. 1989).

The spatial distribution of *SLG* and *SRK* genes on a DNA molecule is further visualized by the advanced FISH method using extended DNA fibers (EDF-FISH). The EDF-FISH method in plants was first developed by Fransz et al. (1996) using *Arabidopsis* and tomato DNA. The spatial relationship of some cosmid clones of *Arabidopsis* was confirmed visually by EDF-FISH and the distance between two unit sequences of 25S and 18S ribosomal RNA genes was also determined by this method. EDF-FISH enables much higher spatial resolution than chromosomal FISH because it employs DNA fibers extended on glass slides as the targets (Fransz et al. 1996; Zhong et al. 1998; Ohmido et al. 2001). Suzuki et al. (1999) visually mapped *SLG* and *SRK* genes cloned in the 76-kb fragment derived from *B. rapa* genomic DNA using a modified EDF-FISH as shown in Fig. 11. As a result, the hypothesis that explains the more

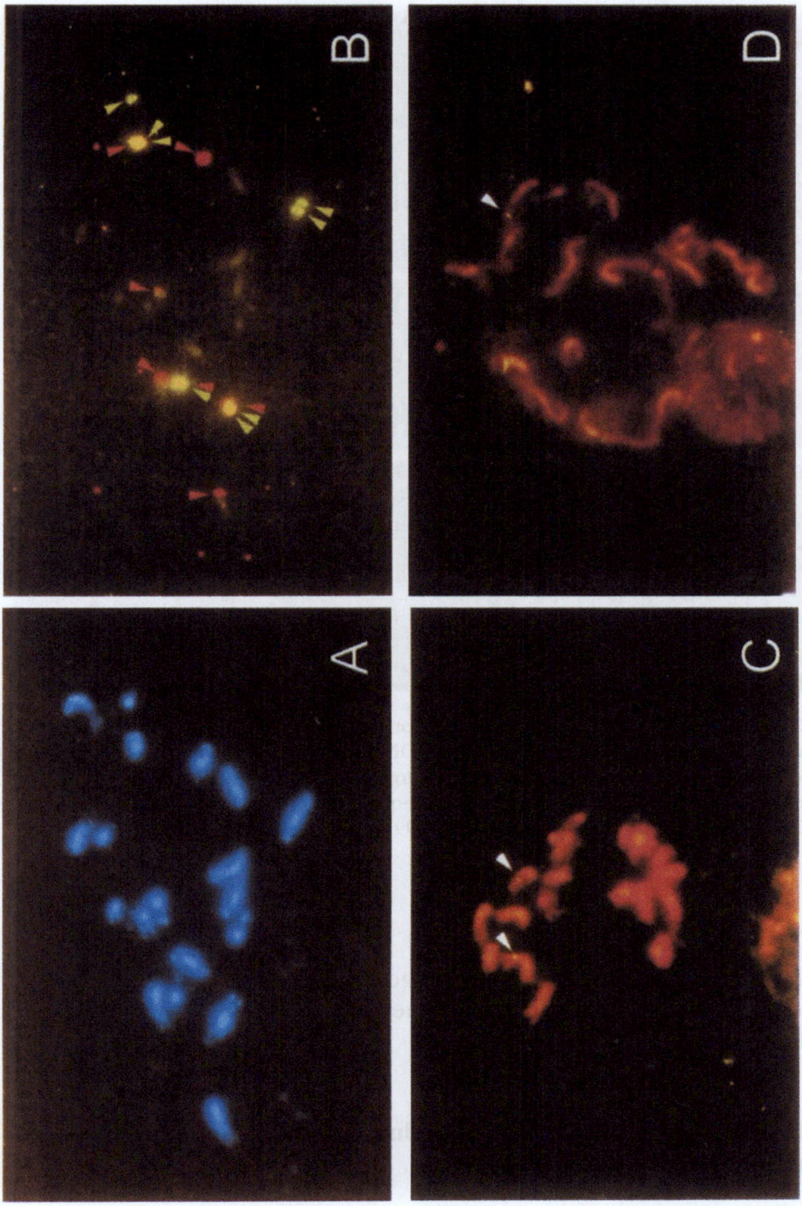

Fig. 10. Detection of 45S rDNA and single copy genes on *Brassica napus* chromosomes by FISH. **A** Haploid chromosomes stained with DAPI. **B** Multicolor FISH detecting 45S rDNA (red signals, *red arrowheads*), 5S rDNA (green signals, *green arrowheads*). Detection of **C** *SLR1* and **D** *SLG* genes as indicated by *arrowheads*. (Kamisugi et al. 1998)

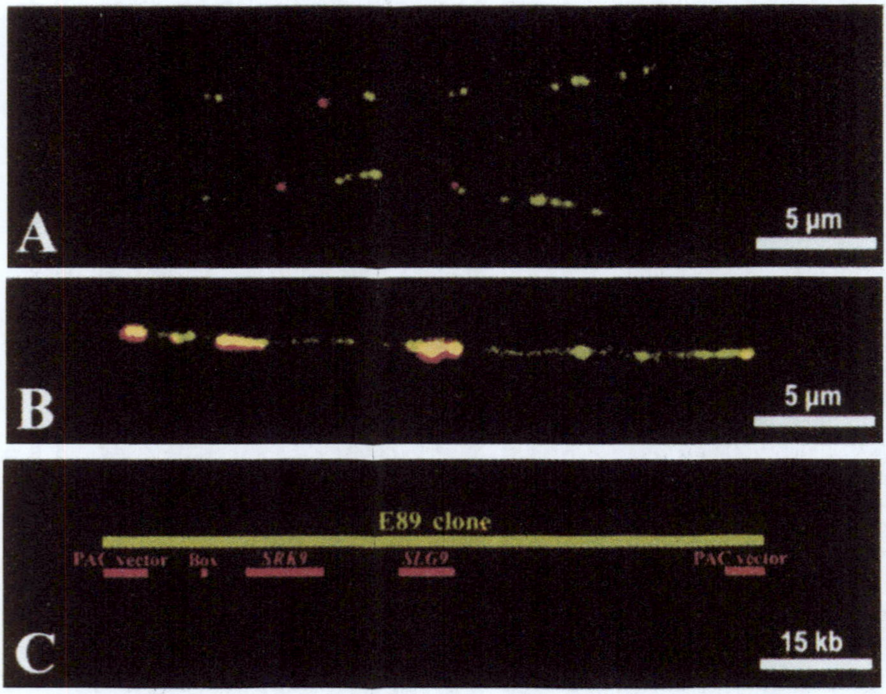

Fig. 11. EDF-FISH detecting *SLG* and *SRL* genes on a cloned genomic DNA fragment of *Brassica napus*. **A** Two green signal tracks of extended DNA fibers with *SLG* and *SRK* signals (*red*). **B** *Red fluorescent signals* at both ends indicate vector DNA sequences. Other major *red signals* indicate the positions of *SLG/SRK* promoter-like sequences, *SRK* and *SLG* sequences from *left* to *right*. The *scale bar* shows 5 μm. (Suzuki et al. 1999)

frequent appearance of *SLG* than *SLR* genes is visually proven by the two fluorescent signals of *SLG* and *SRK* genes located closely on the extended DNA fiber.

5.3 Identification of the Genomic Origin of *Brassica* Chromosomes by Genomic in Situ Hybridization

Figure 12 depicts the biased spreading of *B. rapa*- and *B. oleracea*-type chromosomes within a two dimensional chromosomal spread revealed by genomic in situ hybridization (GISH, Schwarzacher et al. 1989). GISH uses total genomic DNA as the probe for the FISH experiment and enables painting whole chromosomes if the chromosomes belong to the same genome. Leitch et al. (1991) reported the spatial differentiation of the different genomes within a single nucleus after distant hybridization between barley and wild rye. They also pointed out that the expression of certain phenotypes was affected

Fig. 12. Two-dimensional spreading of *oleracea*- and *rapa*-type chromosomes of *Brassica napus*. **A** Chromosome spread of *Brassica napus* stained with DAPI. **B** Chromosome spread of *Brassica napus* with signals from genomic DNA extracted from *Brassica rapa*

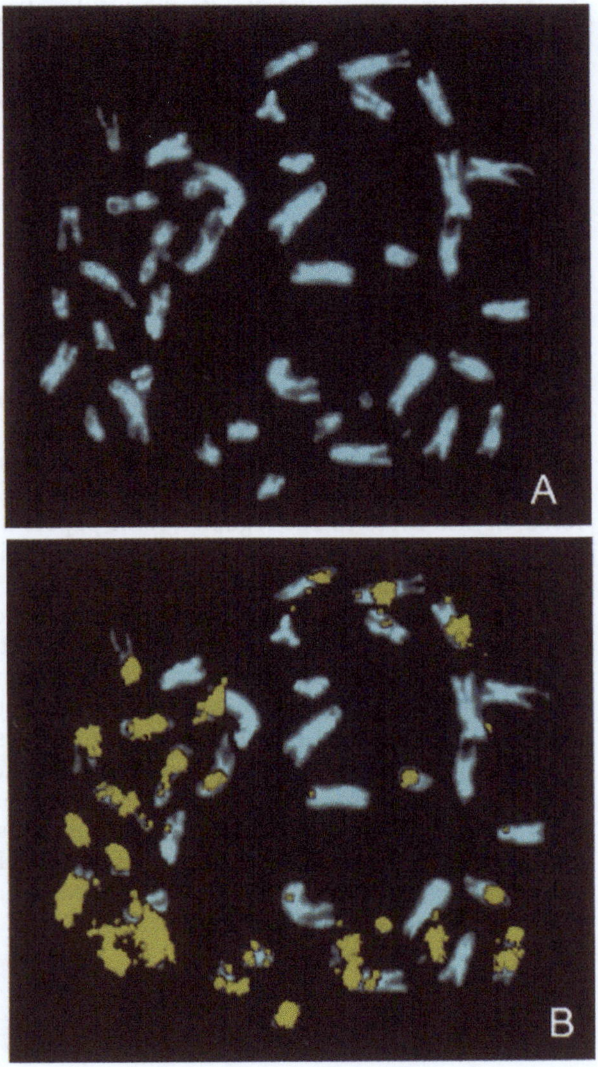

by the positioning of the genome that contained the genes responsible for the phenotype. In *Brassica* species, distant hybridizations between species belonging to different genera and/or species have been made often by embryo-rescue techniques. Somatic hybridization has also been tried and several somatic hybrids have been produced. Production of artificial *B. napus* by somatic hybridization between the protoplasts obtained from *B. rapa* and *B. oleracea* is a typical example, although *B. napus* can be produced by sexual hybridization as well. Production of various types of *B. rapa* is known empirically, and the cause of the phenomenon has been explained in various ways. The evidence for different spatial distribution of the chromosomes originating from two different genomes of *B. napus* is shown in Fig. 12. *Rapa*-type chromosomes are distributed peripherally in the spread with signals, and the three dimensional compartment may be one of the causes of variability among the plants obtained by somatic hybridization.

6 Future Perspectives

Chromosome studies of *Brassica* and legumes have rapidly been developing as reviewed above. The recent papers on the draft maps of whole genomes of two rice subspecies, *O. sativa* ssp. *indica* and ssp. *japonica,* have served as evidence of the opening of a new era in post-genome research in plants, in which proteins and chromatin/chromosome structures play a major role in understanding and thus regulating genetic and biological functions. In chromosome research into *Brassica* and legumes, some papers have already been published in this field. Li et al. (2001) reviewed the relationship between chromatin structure and phaseolin gene regulation. Fuchs et al. (1998) revealed that distribution of phosphorylated serine 10 at the N-terminal of histone H3 corresponded to the positions of the heterochromatic regions with high-copy sequence pVf7 in *Vicia faba*. Chua et al. (2001) reported that histone acetylation altered accessibility of micrococcal endonuclease to the promoter/enhancer regions of pea plastocyanin gene. Post-translational modifications, especially of histone proteins, are now intensively studied in relation to the higher regulation of gene expression, chromosome condensation, sister-chromatid cohesion (Strahl and Allis 2000). It is anticipated that chromosome research targeting chromosomal proteins will become ever more important. Balanced understanding of chromosomes in both the DNA and the protein aspects promises to yield a total view of chromosome structure and function.

7 Protocols

7.1 Preparations of *Brassica* and Legume Chromosomes

For chromosome sample preparation, an enzymatic maceration and air-drying (EMA) method is effective. Root tips are excised from 10–20 mm roots of seeds. They are immediately fixed with a fixative (ethanol:acetic acid 3:1) and stored in a freezer at –20°C before use. For longer storage, the fixative may be replaced with 70% ethanol to avoid degradation of DNA. The fixed roots are washed with distilled water for 5–10 min and dipped in an enzymatic mixture. The enzymatic mixture consists of 4% Cellulase Onozuka RS, 1% Pectolyase Y-23 and is adjusted to pH 4.2. Based on the condition of the root tips, 20–40 min of maceration is applied. Duration and concentration should be adjusted according to the condition of the materials. The root tips are then carefully picked up by a Pasteur pipette and carefully released into distilled water. After thorough washing, the root tips are picked up again and placed on a glass slide. The meristem regions only are excised and tapped by fine forceps until they become invisible cell clusters; fresh fixative is added (methanol:acetic acid 3:1). The air-dried samples are Giemsa stained or used for FISH experiments. The success of the EMA method is dependent upon the duration of digestion and concentration of the enzymatic mixture in order to remove as much cytoplasmic debris as possible (Fukui 1996; Kamisugi et al. 1998).

7.2 Preparation of Extended DNA Fibers (EDFs)

One or two grams of young leaves taken from plants are frozen in liquid nitrogen and ground to fine powder with a mortar and pestle. After gently adding 20 ml ice-cold nuclear isolation buffer [NIB; 10 mM Tris-HCl, pH 9.5, 10 mM EDTA, 100 mM KCl, 0.5 M sucrose, 4.0 mM spermidine, 1.0 mM spermine, and 0.1% (v/v) 2-mercapto ethanol], the homogenate is consecutively filtered through 165-, 50-, 20-, and 10-μm nylon mesh filters. Nuclei are then centrifuged at 2000 *g* for 3–5 min and resuspended in 200 μl NIB. A 3-μl solution of nuclear suspension is put on a microscopic slide and left to dry for 10 min. Twenty microliters of lysis buffer (0.5% w/v SDS, 5 mM EDTA, and 100 mM Tris-HCl, pH 7.0) is put on the nuclei to disrupt the nuclear matrixes. After a 5-min incubation, slides are tilted to produce extended fibers along the meniscus of the droplet and are left to dry. The quality of EDFs is checked using a fluorescence microscope by staining with an aqueous solution of 1 μg/ml YOYO-1 iodide (Molecular Probes; Fransz et al. 1996; Ohmido et al. 2001).

7.3 Characterization of Condensed and Dispersed Chromosomal Regions

Condensation patterns are clearly observed after Giemsa staining, although without such stain they could be observed by phase contrast microscopy. Because it is not possible to stop the cell cycle at a suitable stage in prometaphase, selection of a suitable stage of the chromosome spreads is important. It is known empirically that most chromosome spreads are at prometaphase in the S-type chromosomes, indicating that this stage occupies the longest period during the M-stage of mitosis. The statistical selection even among chromosomal spreads showing prometaphase characteristics is essential in order to develop the chromosome maps and consistent descriptions of chromosome characteristics, to map genes by FISH, and to accomplish whatever else may be required. Prometaphase is estimated by the total length of the chromosomes within the spread. The total length of the chromosomes within the spread is measured and the data is statistically analyzed. If the total length of the chromosomes is either significantly longer or shorter (at the 1% level by *t*-test) than the average of the total length obtained from chromosomal spreads used, the spread is either at the advanced stage close to metaphase or at the earlier stage close to prophase. These chromosomal spreads must be omitted from the further analysis.

Determination of the gray level threshold for discrimination of heavily condensed and condensed regions, and condensed and dispersed regions is based on the following two criteria. The threshold that divides condensed and dispersed regions should directly reflect the border between condensed and dispersed regions visually observed; and one value is easily determined for all the chromosomes of the chromosome spread. On the other hand, the threshold that divides condensed and heavily condensed regions is difficult to set because there is no very clear border in gray values between heavily condensed and condensed regions. The density threshold that divides condensed and heavily condensed regions could rather be set up to highlight the characteristics of chromosomal condensation giving the heavily condensed regions within and chromosome map. For example, rice chromosome 11 has three heavily condensed regions, one on the short arm and two on the long arm, and the characteristic is best represented by setting the second threshold between condensed and heavily condensed regions. The first threshold that divides the condensed and dispersed regions cannot represent this pattern of the three condensation states, which is unique to this particular chromosome.

7.4 Image Analysis of the Condensation Patterns for Construction of Quantitative Chromosome Maps

For quantification of the condensation pattern, the chromosome spreads suitable for image analysis are selected by statistical tests for the total chromosome length. Five to ten chromosomal spreads are appropriate for obtaining standard CP (stCP). The chromosome images selected are digitally captured via a CCD camera. The images stored can be analyzed by CHIAS 3 (Kato et al. 1997; Kato and Fukui 1998). Imaging procedures basically consist of the following six steps: (1) elimination of optical distortion in the captured image, (2) enhancement of image contrast, (3) discrimination of the chromosomal regions from the background field, (4) determination of homologous chromosomes and extraction of chromosomal characteristics based on the pseudocolor representation of condensation patterns, (5) interactive drawing of the mid-rib line on each of the chromatids after enlargement of the chromosome image and normalization of the density distribution, (6)and CP data acquisition. Detailed explanation and a manual for the image analysis can be obtained both in written form (Kato et al. 1997) and via the Internet (http://www.bio.eng.osaka-u.ac.jp/cell/).

The method for development of quantitative chromosome maps based on the CP is also included in a program within CHIAS 3. As in the case of CP analysis, the program can be downloaded free on the Internet ,as can the written manual. The crucial point for the development of the quantitative map lies in the appropriate determination of the second threshold to discriminate the heavily condensed and condensed regions. The second threshold should highlight the chromosome characteristics at the prometaphase stage. Thus a single gray value that is applicable to all the chromosomes is preferable for the second threshold, as in rice chromosomes (Fukui and Iijima 1991). However, it is sometimes not possible to set up a single gray value for the second threshold. In this case, multiple gray values can be employed for the second gray values as in the chromosomes of wild sugarcane (Ha et al. 1998).

7.5 Fluorescence in Situ Hybridization

7.5.1 Probe Preparation

Various sizes of DNAs in the range 1–400 kb can be used as probes in FISH experiments such as genomic DNAs in yeast artificial chromosomes, bacterial artificial chromosomes, and cosmid and plasmid clones (Ohmido and Fukui 1997; Ohmido et al. 1998). Short DNA sequences of a few kb such as molecular markers and genes cloned in ordinary cloning vectors or amplified by PCR methods can be used in FISH as well. The cloned DNA sequences are labeled by nick translation using biotin-16-dUTP. The probe DNAs are also amplified and labeled simultaneously by PCR, using the RV and M4 primer set (Roche). In both cases, the length of the probe is critical. It should be 200–400 bp, adjusted by the concentration of DNA polymerase and DNase, in the case of nick translation, and by primers in the case of PCR. Longer probes result in high nonspecific background noise after FISH.

7.5.2 Hybridization

The hybridization mixture contained 5 ng/μl biotin-labeled probe, 10% dextran sulfate (*Mr* 500,000), 50% formamide, 2 × SSC, and 500 ng/μl sheared salmon sperm DNA. In the case of the long genomic probes with long sequences, a 50-fold excess of unlabeled total genomic DNA plus a 100-fold excess of salmon sperm DNA is added as blocking DNAs. Genomic DNA is sheared to an average size of ca. 200–400 bp by autoclaving at 105 °C for 5 min or by digestion with DNase. In the case of hybridization mixtures with small probe DNAs like molecular markers or cDNA clones, addition of genomic DNA and salmon sperm DNA can be avoided. After hybridization overnight, the chromosome spreads are washed at 42 °C three times in 50% formamide/2×SSC, and three times in 0.1 × SSC at 60 °C. To detect the probes on chromosomes, two layers of Fluorescein Avidin DCS (Vector Lab, USA) and one layer of biotinylated anti-avidin D (Vector Lab) are applied alternately in a blocking solution (1% BSA) at 37 °C for 30–60 min. After each immunological detection step, the slides are washed three times with 2 × SSC at room temperature. Chromosomes are counterstained with propidium iodide (PI). Green fluorescent signals from fluorescein, and the red fluorescence of chromosomes stained with PI, are detected separately using different filter sets (B10 and G15, Zeiss). The two images are captured independently using a cooled CCD camera (PXL 1400, Roper) and the fluorescence images are merged into single images by imaging software (IP-Lab, Adobe Photoshop) (Ohmido et al. 1998).

7.5.3 Image Analysis

The FISH signals detected on the chromosomes are analyzed by CHIAS3, FISH version, to determine accurate localization of the clones. An adequate number of chromosomes at the same stage (mainly prometaphase for *Brassica* and legume chromosomes) with FISH signals are digitally captured. Then they are normalized to enhance the contrast, and the chromosome images are enlarged. The distribution of both the intensities of the fluorescent signal and chromosome are measured along the mid-rib line of the chromatids on which the FISH signals are localized. The length of the density profile lines are measured and averaged for both chromosomes and FISH signals. Note that the lengths of the profile lines are not equal to the number of pixels of the lines. Then both profiles are superimposed to form a single image in order to localize the precise position of the FISH signal on the chromosome (Ohmido et al. 1998).

Acknowledgements. The author thanks Nobuko Ohmido (Kobe University, Kobe, Japan) and Glyn Jenkins (University of Wales, Aberystwyth, UK) for their critical reading of the manuscript and valuable comments.

References

Abbo S, Miller TE, Reader SM, Dunford RR, King IP (1994) Detection of ribosomal DNA sites in lentil and chickpea by fluorescent in situ hybridization. Genome 37:713–716

Cheng BF, Heneen WK, Chen BY (1995a) Mitotic karyotypes of *Brassica campestris* and *Brassica alboglabra* and identification of the *B. alboglabra* chromosome in an addition line. Genome 38:313–319

Cheng BF, Heneen WK, Pedersen C (1995b) Ribosomal RNA gene loci and their nucleolar activity in *Brassica alboglabra* Bailey. Hereditas 123:169–173

Cheng RI, Grant WF (1973) Species relationships in the *Lotus corniculatus* group as determined by karyotype and cytophotometric analysis. Can J Genet Cytol 15:101–105

Chua YL, Brown APC, Gray JC (2001) Targeted histone acetylation and altered nuclease accessibility over short regions of the pea plastocyanin gene. Plant Cell 13:599–612

Dubcovsky J, Dvorak J (1995) Ribosomal RNA multigene loci: nomads of the Triticeae genomes. Genetics 140:1367–1377

Endo TR (1986) complete identification of wheat chromosomes by means of a C-banding technique. Jpn J Genet 61:89–96

Fransz RF, Alonso-Blanco C, Liharska TB, Peeters AJM, Zabel P et al. (1996) High-resolution physical mapping in *Arabidopsis thaliana* and tomato by fluorescence in situ hybridization to extended DNA fibers. Plant J 9:421–430

Fuchs J, Schubert I (1995) Localization of seed protein genes on metaphase chromosomes of *Vicia faba* via fluorescence in situ hybridization. Chromosome Res 3:94–100
Fuchs J, Kuhne M, Schubert I (1998a) Assignment of linkage groups to pea chromosomes after karyotyping and gene mapping by fluorescent in situ hybridization. Chromosoma 107:272–276
Fuchs J, Strehl S, Brandes A, Schweizer D, Schubert I (1998b) Molecular-cytogenetic characterization of the *Vicia faba* genome–heterochromatin differentiation, replication patterns and sequence localization. Chromosome Res 6:219–230
Fukui K (1986) Standardization of karyotyping plant chromosome by a newly developed chromosome image analyzing system (CHIAS). Theor Appl Genet 72:27–32
Fukui K (1996) Plant chromosomes at mitosis. In: Fukui K, Nakayama S (eds) Plant chromosomes: laboratory methods. CRC Press, Boca Raton, pp1–17
Fukui K, Iijima K (1991) Somatic chromosome map of rice by imaging methods. Theor Appl Genet 81:589–596
Fukui K, Kamisugi Y (1995) Mapping of C-banded *Crepis* chromosomes by imaging methods. Chromosome Res 3:79–86
Fukui K, Mukai Y (1988) Condensation pattern as a new image parameter for the identification of small chromosomes in plants. Jpn J Genet 63:59–366
Fukui K, Nakayama S (1996) Analysis of chromosome information. In: Fukui K, Nakayama S (eds) Plant chromosomes: laboratory methods. CRC Press, Boca Raton, pp 241–255
Fukui K, Nakayama S, Ohmido N, Yoshiaki H, Yamabe M (1998) Quantitative karyotyping of three diploid *Brassica* species by imaging methods and localization of 45S rDNA loci on the identified chromosomes. Theor Appl Genet 96:325–330
Fukui K, Ohmido N, Wako T (2000) Smallness: gain and loss in plant chromosome research. In: Olmo E, Redi A (eds) Chromosome today, vol 13. Birkhˇuser Verlag, Basel, Switzerland, pp 287–301
Griffor MC, Vodkin LO, Singh RJ, Hymowitz T (1991) Fluorescent in situ hybridization to soybean metaphase chromosomes. Plant Mol Biol 17:101–109
Gomez-Campo C, Hinata K (1980) A check list of chromosome numbers in the tribe Brassicaceae. In: Tsunoda S, Hinata K, Gomez-Campo C (eds) *Brassica* crops and wild allies. Japan Scientific Societies Press, Tokyo, pp 51–63
Ha S, Moore PH, Heinz D, Kato S, Ohmido N, Fukui K (1999) Quantitative chromosome map of the polyploid *Saccharum spontaneum* by multicolor fluorescence in situ hybridization and imaging methods. Plant Mol Biol 39:1165–1173
Handberg K, Stougaard J (1992) *Lotus japonicus*, an autogamous, diploid legume species for classical and molecular genetics. Plant J 2:487–496
Hayashi M, Miyahara M, Sato S, Kato T, Yoshikawa M, Taketa M, Hayashi M, Pedrosa A, Onda R, Imaizumi-Anraku H, Bachmair A, Sandal N, Stougaard J, Murooka Y, Tabata S, Kawasaki S, Kawaguchi M, Harada K (2001) Construction of a genetic linkage map of the model legume *Lotus japonicus* using an intraspecific F_2 population. DNA Res 8:301–310
Hu CH (1964) Further studies on the chromosome morphology of *Oryza sativa* L. In: Rice genetics and cytogenetics. Elsevier, Amsterdam, pp 51–61
Iijima K, Kakeda K, Fukui K (1991) Identification and characterization of somatic rice chromosomes by imaging methods. Theor Appl Genet 81:597–605
Ito M, Ohmido N, Akiyama Y, Fukui K (2000a) Quantitative chromosome map of *Arabidopsis thaliana* L. by imaging methods. Cytologia 65:325–331
Ito M, Miyamoto J, Mori Y, Fujimoto S, Uchiumi T, Abe M, Suzuki A, Tabata S, Fukui K (2000b) Genome and chromosome dimensions of *Lotus japonicus*. J Plant Res 113:435–442
Jiang Q, Gresshoff PM (1997) Classical and molecular genetics of the model legume *Lotus japonicus*. Mol Plant Microbe Interact 10:59–68
Kamisugi Y, Furuya N, Iijima K, Fukui K (1993) Computer-aided automatic identification of rice chromosomes by image parameters. Chromosome Res 1:189–196
Kamisugi Y, Nakayama S, O'Neill CN, Mathias RJ, Trick M, Fukui K (1998) Visualization of the *Brassica* self-incompatibility *S*-locus on identified oilseed rape chromosomes. Plant Mol Biol 38:1043–1052

Kato S, Fukui K (1998) Condensation pattern (CP) analysis of plant chromosomes using a newly developed chromosome image analyzing system, CHIAS III. Chromosome Res 6:473–479

Kato S, Hirose T, Akiyama Y, O'Neill CM, Fukui K (1997) Manual on the chromosome image analyzing system III, CHIAS III. Res Rep Dev Agric Hokuriku Area 36:1–76

Kawakami J (1930) Chromosome numbers in Leguminosae. Bot Mag (Tokyo) 44:319–328

Koornneef M, wan Eden J, Hanhart CJ, Stam CJ, Branksma FJ, Feenstra WJ (1983) Linkage map of *Arabidopsis thaliana*. J Hered 74:265–272

Kurata N, Omura T (1978) Karyotype analysis in rice I. A new method for identifying all chromosome pairs. Jpn J Genet 53:251–255

Lalonde BA, Nasrallah ME, Dwyer KG, Chen CH, Barlow B, Nasrallah JB (1989) A highly conserved *Brassica* gene with homology to the *S*-locus specific glycoprotein structural gene. Plant Cell 1:249–258

Leitch AR, Schwarzacher T, Mosgoller W, Bennett MD, Heslop-Harrison JS (1991) Parental genomes are separated throughout the cell cycle in a plant hybrid. Chromosoma 101:206–213

Li G, Chandrasekharan MB, Wolffe AP, Hall TC (2001) Chromatin structure and phaseolin gene regulation. Plant Mol. Biol 46:121–129

Malusynska J, Heslop-Harrison JS (1993a) Molecular cytogenetics of the genus *Arabidopsis*: in situ localization of rDNA sites, chromosome numbers and diversity in centromeric heterochromatin. Ann Bot 71:479–484

Malusynska J, Heslop-Harrison JS (1993b) Physical mapping of rDNA in *Brassica* species. Genome 36:774–781

Murata M, Heslop-Harrison JS, Motoyoshi F (1997) Physical mapping of the 5S ribosomal RNA genes in *Arabidopsis thaliana* by multi-color fluorescence in situ hybridization with cosmid clones. Plant J 12:31–37

Ohmido N, Fukui K (1997) Visual verification of close disposition between a rice A genome-specific DNA sequence (*TrsA*) and the telomere sequence. Plant Mol Biol 35:963–968

Ohmido N, Akiyama Y, Fukui K (1998) Systematic mapping of unique nucleotide sequences on identified rice chromosomes. Plant Mol Biol 38:1081–1087

Ohmido N, Kijima K, Ashikawa I, de Jong JH, Fukui K (2001) Visualization of the terminal structure of rice chromosomes using multicolor FISH on extended DNA fibers. Plant Mol Biol 47:413–421

Olin-Fatih M (1996) The morphology, cytology, and C-banded karyotypes of *Brassica campestris, B. oleracea* and *B. napus* plants regenerated from protoplasts. Theor Appl Genet 93:414–420

Schubert I (1984) Mobile nucleolus organizing regions (NORs) in *Allium* (Liliaceae s. lat.) – inferences from the specificity of silver staining. Plant Syst Evol 144:291–305

Schwarzacher T, Leitch AR, Bennett MD (1989) In situ localization of parental genomes in a wide hybrid. Ann Bot 64:315–324

Shishido R, Sano Y, Fukui K (2000) Ribosomal DNAs: an exception to the conservation of gene order in rice genomes. Mol Gen Genet 263:586–591

Shoemaker RC, Polzin K, Labate J, Specht J, Brummer EC, Olson T, Young N, Concibido V, Wilcox J, Tamulonis JP, Kochert G, Boerma HR (1996) Genomic duplication in soybean (*Glycine* subgenus *soja*). Genetics 144:329–338

Singh RJ, Kim HH, Hymowitz T (2001) Distribution of rDNA loci in the genus *Glycine* Wild. Theor Appl Genet 103:212–218

Skorupska H, Albertsen MC, Longholz KD, Palmer RG (1989) Detection of ribosomal RNA genes in soybean, *Glycine max* (L.) Merr., by in situ hybridization. Genome 32:1091–1095

Snowdon RJ, Koehler W, Koehler A (1997) A chromosomal localization and characterization of rDNA loci in the *Brassica* A and C genomes. Genome 40:582–587

Strahl BD, Allis CD (2000) The language of covalent histone modifications. Nature 403:41–45

Suzuki G, Kai N, Hirose T, Fukui K, Nishio T, Takayama S, Isogai A, Watanabe M, Hinata K (1999) Genomic organization of the *S* locus: Identification and characterization of genes in *SLG/SRK* region of S^9 haplotype of *Brassica campestris* (syn. *rapa*). Genetics 153:391–400

Xu SJ, Singh RJ, Hymowitz T (2000) Monosomics in soybean: origin, identification, cytology, and breeding behavior. Crop Sci 40:985–989

Yanagisawa T, Tano S, Fukui K, Harada K (1991) Marker chromosomes commonly observed in the genus *Glycine*. Theor Appl Genet 81:606–612

Zheng J, Irifune K, Hirai K, Nakata M, Tanaka R, Morikawa H (1994) In situ hybridization to metaphase chromosomes in six species of *Phaseolus* and *Vigna* using ribosomal DNA as the probe. J Plant Res 107:365–369
Zhong X, Fransz PF, Wennekes-van Eden J, Ramanna MS, van Kammen A, Zabel P (1998) FISH studies reveal the molecular and chromosomal organization of individual telomere domains in tomato. Plant J 13:507–517

9 Sequence Analysis of the *Lotus japonicus* Genome

S. SATO, E. ASAMIZU, and S. TABATA

1 Introduction

One of the most epoch-making accomplishments in plant genetics in the 20th century was the completion of genome sequencing of *Arabidopsis thaliana* (The Arabidopsis-Genome Initiative 2000). As a consequence, an enormous amount of information on gene structures and their functions have been and are still being accumulated in this organism. Nevertheless, other plant species have their own characteristics and advantages for the study of individual biological phenomena. Further, comparison of knowledge from *A. thaliana* and that from other plant species is a promising approach for obtaining universal knowledge on the genetic systems in all plants.

Legumes comprise 18,000 diverse species with a variety of characteristics, and many of them have long been utilized in agriculture in broad regions of the world. A few species among them have been chosen as "model legumes" for genetic and physiological studies on legume-specific phenomena. *Lotus japonicus* is a typical model legume with characteristics of short life cycle (3 months on average), self-fertility, diploidy ($n = 6$), and small genome size (472.1 Mb) (Handberg and Stougaard 1992; Ito et al. 2000; Kawaguchi 2000). A variety of mutants in various steps of symbiosis and nitrogen fixation have been isolated (Imaizumi-Anraku et al. 1997; Schauser et al. 1998; Szczyglowski et al. 1998), and genes expressed during these steps have been characterized utilizing the established transformation system (Martirani et al. 1999; Schauser et al. 1999; Webb et al. 2000). Characteristics of *L. japonicus*. are also referred in Chapter 10 by Kawaguchi and Nishimura.

To understand the genetic system of legume species, we initiated a large-scale genome analysis of *L. japonicus*. The project comprises three approaches: EST collection, genome sequencing, and linkage mapping. In this report, the current status of the project will be summarized.

Kazusa DNA Research Institute, 2-6-7 Kazusa-kamatari, Kisarazu, Chiba 292, Japan

Biotechnology in Agriculture and Forestry, Vol. 52
Nagata/Tabata (Eds.) Brassicas and Legumes

2 EST Analysis

Expressed sequence tag (EST) analysis is performed to make a catalogue of expressed portions of the genome by developing anonymous partial cDNA sequences. In an EST analysis, synthesized cDNA is directionally cloned into a vector and a single-pass sequence is read either from the 5′- or 3′-end of the cDNA. A 300–500 bp-long sequence thus obtained is sufficient to identify a gene by similarity search of the public databases. The EST information is useful in understanding the expression levels or expression properties of certain genes. Since the efficacy of EST analysis was obvious, rapidly generating transcribed sequence data, this method has been applied in many other organisms. As of January 2002, the EST database in GenBank (http://www.ncbi.nlm.nih.gov/dbEST/) contains 10,067,985 entries from more than 380 different organisms, and the number is continuously growing. This includes a large number of EST sequences from several model plant species such as *A. thaliana* (Höfte et al. 1993; Newman et al. 1994; Cooke et al. 1996; Asamizu et al. 2000b), *Medicago truncatula* (Covitz et al. 1998; Gyorgyey et al 2000) and *Oryza sativa* (Yamamoto and Sasaki 1997). It is notable that accumulation of EST sequences is predominant in important crop plants such as *Glycine max*, *Lycopersicon esculentum,* and *Zea mays*.

More than 31,000 EST sequences of *L. japonicus* have been deposited in the public EST database as of January 2002. EST sequences numbering 5,495 from mature and immature nodules have been deposited by M.K. Udvardi's group at the Max Planck Institute for Molecular Plant Physiology, Golm, Germany. L. Poedenphant's group at the University of Aarhus in Denmark deposited 2,399 ESTs from whole roots and nodule primordia.

Two projects in Japan have also made a significant contribution. By taking advantage of the relatively large floral organ of *L. japonicus*, dissection of floral reproductive organs of this plant with gene expression has been carried out. In order to analyze expression patterns of genes related to the differentiation and function of reproductive organs, ESTs from flower buds were collected at Iwate University. A total of 842 5′-end sequences were generated (Endo et al. 2000). Large-scale EST sequencing is in progress at Kazusa DNA Research Institute. As the first part of the project, young plants (2 weeks old) of *L. japonicus* accession Miyakojima MG-20 were used as material and two types of cDNA libraries, normalized and size-selected libraries, were constructed (Asamizu et al. 2000a). Normalization was performed as described by Bonaldo et al. (1996). Briefly, a single-stranded cDNA library was prepared by the combined action of gene II endonuclease of phage F1 and exonuclease III. Self-hybridization was performed with an excess amount of PCR-amplified inserts, and nonhybridized, single-stranded circles were purified by hydroxyapatite chromatography. Single-stranded circles were converted to double strands, transformed into a host *Escherichia coli*, and served as a template for the sequence generation. To overcome the elimination of longer cDNA species during the normalization step, synthesized cDNA was separated on agarose gel, and fragments longer than 3 kb were recovered from the gel and cloned into a vector. This library was called the size-selected library, and subjected to sequencing without normalization. A total of 22,983 5′-end EST sequences were generated, and these sequences were clustered into 7,137 nonredundant groups by assembly (Asamizu et al. 2000a).

Nodule development, secondary metabolism, and seed development are examples of important biological processes of leguminous plants from both a biological and an agricultural point of view. The accumulation of EST sequences derived from pods, roots, nodules, flower buds, and flowers is underway as the next step of this project. The sequence information and search results of clones generated in these two projects are available at the website http://www.kazusa.or.jp/en/plant/lotus/EST/. The EST sequence information and the corresponding cDNA clones will no doubt serve as a valuable source for creating a cDNA array in the future genome-wide gene expression studies.

3 Genome Sequencing

3.1 Genomic Libraries

To obtain a source of genome sequencing clones, we have constructed a genomic library of *L. japonicus* accession Miyakojima MG-20 in a transformation-competent artificial chromosome (TAC) vector. The TAC vector accepts and maintains large genomic DNA fragments stably in both *E. coli* and *Agrobacterium tumefaciens*, and contains the *cis* sequence required for *Agrobacterium*-mediated gene transfer into a plant genome (Liu et al. 1999). Therefore, the TAC clone is a useful resource for further functional analysis. For the construction of a genomic library, two restriction enzymes, *Hin*d III and *Mbo* I, were used for partial digestion of genomic DNA to avoid the cloning bias. The average insert sizes were 87, 96, 105, and 106 kb for four independent preparations. In totality, these TAC libraries are equivalent to a 7.7-haploid genome. The TAC libraries thus generated were arrayed in 93 384-well-microtiter plates, and 48 DNA pools, each containing 384 clones, were subjected to PCR screening (Sato et al. 2001).

Because of their importance for physical mapping and positional cloning, large insert genomic libraries of *L. japonicus* have also been constructed by several other groups. Gresshoff's group in Australia constructed a bacterial artificial chromosome (BAC) library of *L. japonicus* accession Gifu B-129 with an average insert size of 94 kb, covering six-sevenfold genome equivalents (Men et al. 2000). This library was constructed using vector V41, which allows direct *Agrobacterium*-mediated plant transformation. Sandal et al. in Denmark produced a BAC library of *L. japonicus* accession Gifu B-129, consisting of 32,000 clones with an average insert size of 70 kb (Sandal et al. 2000). DNA pools for PCR screening of this library have been made available. Kawasaki's group in Japan also constructed a large insert BAC library of

L. japonicus accession Gifu B-129 with the average insert size of about 140 kb, covering 8 genome equivalents (Kawasaki and Murakami 2000).

3.2 Clone Selection and Sequencing

TAC clones for sequencing were isolated by screening the three-dimensional DNA pools of *L. japonicus* TAC genomic libraries by PCR with oligonucleotide primers designed on the basis of nucleotide sequences of the public ESTs and cDNAs from *L. japonicus*.

The nucleotide sequence of each TAC insert, which is approximately 100 kb in length, was determined according to the bridging shotgun method (Sato et al. 1997). In brief, the TAC DNAs were subjected to sonication followed by size-fractionation by agarose gel electrophoresis. Fractions of approximately 1.0 and 2.5 kb were cloned into M13mp18 and named libraries of element clones and bridge clones, respectively. Purified single-stranded DNA or PCR-amplified inserts were subjected to sequence analysis according to the standard procedure. The single-pass sequences of one-strand of element clones and those of both ends of bridge clones, a total of which correspond to an approximately sixfold equivalent of an insert, were assembled using Phred-Phrap programs (P. Green, University of Washington, Seattle, USA). After extension of the termini of each contig by the primer extension method followed by re-connection, most of the TAC inserts were reconstructed into one to three contigs with more than 95% coverage of both strands. A lower threshold of acceptability for generation of consensus sequences was set at Phred score 20 for each base.

3.3 Gene Assignment

Assignment of the protein-coding regions and gene modeling were performed by combining similarity search and computer prediction. Similarity search against the nonredundant protein sequence database nr (compiled by NCBI) was carried out using the BLASTX program (Altschul et al. 1990). In parallel, positions of potential protein-encoding regions were deduced by combining the predictions with computer programs Grail (Uberbacher and Mural 1991), GENSCAN (Burge and Karlin 1997), and NetGene2 (Hebsgaard et al. 1996). The transcribed regions of the genome were assigned by comparison of the nucleotide sequences with *L. japonicus* ESTs using the BLASTN program (Altschul et al. 1990). All the results obtained were compiled with the aid of our new web-based tool, KAPSEL (Kazusa Annotation Pipeline SystEm for *Lotus* genome sequencing project; in prep.), then assignment of the potential protein-encoding genes was carried out by taking both similarity to known genes and computer prediction into consideration.

The RNA-encoding regions were assigned on the basis of sequence similarity to the structural RNAs in the public databases. The prediction by the tRNAscan-SE program (Lowe and Eddy 1997) was also taken into account for tRNA genes.

3.4 Characteristic Features of the Genome and Genes in *L. japonicus*

As of January 2002, a total of 640 seed clones had been selected. Among them, 56 clones, covering 5,473,195 bp genome regions, were annotated and deposited in the public databases (Sato et al. 2001). In this section, features of *L. japonicus* genome found through the analysis of these regions will be described.

3.4.1 Genome Features

In summary, complete structures of 605 potential protein-encoding genes were deduced in the 5,473,195 bp regions of the genome. In addition, partial structures of 69 potential protein-coding genes at the terminal regions of the clones and the contigs, and 172 pseudogenes, which contain either frameshifts or termination codons in the original coding regions, were identified.

If we simply divide the total genome size (472 Mb) by the size of the regions sequenced, 605 potential protein-encoding genes could be estimated to account for approximately 1.1% of the total gene constituents. However, this is probably an underestimation due to the following reasons. First, the clones sequenced so far were likely to have been derived from gene-rich regions of the genome because they were selected based on the ESTs. Second, Ito et al. (2000) reported that significant portions of the prometaphase chromosomes of *L. japonicus* are either heavily or moderately condensed, strongly suggesting the presence of regions rich in repetitive sequences and poor in protein-encoding genes in the genome. Therefore, we speculate that the genes identified so far represent more of the gene constituents than as simply calculated.

The general features of the 605 genes in *L. japonicus*, whose complete structures were deduced, were compared to those of *A. thaliana* (Table 1). The average length of the coding exons (266 bp) and the average number of introns per gene (3.8) were quite similar between the two plant species. However, the average length of genes including introns (2,712 bp) was longer in *L. japonicus* than in *A. thaliana* (1,918 bp), due to a longer average intron length in *L. japonicus.* The average gene density of the sequenced regions of the *L. japonicus* genome was one gene in every 8,120 bp, twice as low as that in *A. thaliana.* It should be noted that this could be overestimated because, as described previously, it is quite likely that the clones containing the genomic regions of higher gene contents were preferentially selected and sequenced.

Conservation of gene content and order (synteny) was occasionally observed between segments of the genome of *A. thaliana* and each sequenced clone of *L. japonicus*, although no clear macrosynteny has been detected so far. In most cases, subsets of conserved genes were identified in multiple seg-

Table 1. Structural features of the assigned protein-encoding genes in *L. japonicus* and *A. thaliana*

Features	*L. japonicus* 605 genes	*A. thaliana* 6,451 genes
Gene length including introns (bp)	168–19,890 (2,712)	78–17,203 (1,918)
Product length (amino acids)	16–1,816 (426)	25–4,706 (427)
Genes with introns	469 (78%)	4,906 (76%)
Number of introns/genes	0–37 (3.8)	0–48 (4.0)
Exon length (bp)	3–5,451 (266)	2–5,966 (256)
Intron length (bp)	30–5,687 (375)	23–2,989 (157)
GC content of exons	45%	44%
GC content of introns	33%	32%
Gene density (kb/gene)	8.1	4.5

Structural features of the 605 potential protein-coding genes assigned in 5.4 Mbp regions of L. *japonicus* genome (Sato et al. 2001) and the 6,451 genes previously assigned in Kazusa *Arabidopsis* genome sequencing project are listed. Average values are shown in parentheses

ments of the *A. thaliana* genome. Therefore, at the level of TAC-clone size (about 100 kb), it may be feasible to align much of the *L. japonicus* genome with segments of the genome of *A. thaliana.*

3.4.2 Functional Assignment and Characteristic Features of the Potential Protein-Encoding Genes

To obtain information on function, BLASTX analysis of the 605 potential protein-encoding genes was carried out with the nr databases, using a *P*-value of less than 1e-20 as a cutoff. The result indicated that 187 (31%) of them were homologues of genes of known function, and 200 (33%) showed similarity to hypothetical genes mostly of those in *A. thaliana* (Fig. 1). The remaining 218 (36%) showed no significant similarity to any registered genes (Fig. 1). This class of genes may carry out legume-specific functions with protein motifs not found in other organisms.

The potential protein-coding genes whose function could be assigned were grouped into 14 categories according to different biological roles, based on the principle of Riley (Fig. 1). Although the available gene number in *L. japonicus* is small for carrying out precise comparison at this stage, the distribution pattern of genes in each category is basically similar to that in *A. thaliana.* The detailed information on each protein-encoding gene is listed in the *L. japonicus* genome database at http://www.kazusa.or.jp/lotus/.

The transcriptional level of each potential protein-encoding gene was roughly monitored by counting the number of matched *L. japonicus* ESTs in-house and in the public DNA databases (described in Sect. 2). Of the 605 genes assigned so far, 213 (35%) carried matched EST sequences. Among the EST-matched genes, 16 (2.6%) were found in 20 or more EST files, suggesting it to be a class of highly transcribed genes.

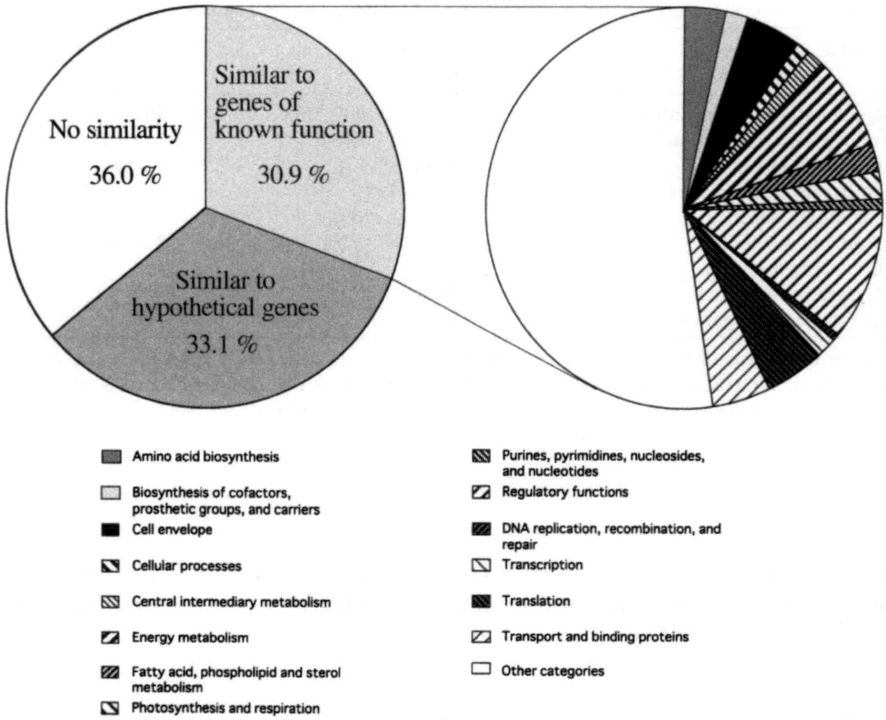

Fig. 1. Functional classification of the assigned protein-encoding genes. The *left pie chart* shows the result of similarity search of the 605 potential protein-encoding genes against the nr database. The *right pie chart* shows functional classification of the genes whose function could be anticipated from the result of similarity search

The genome of *L. japonicus* contains a relatively high number of repetitive sequences. Of the 846 potential protein-encoding genes and gene segments, 161 (31%) were related to retroelements (*gag* and *pol*). The average density of such retroelement-related genes was one in every 21 kb. It is noteworthy that 126 of these 161 were pseudogenes which contain frameshifts or termination codons within the coding regions.

Tandemly repeated gene arrays were often found in the sequenced regions of the *L. japonicus* genome, as is the case in *A. thaliana*. Among these thus far found with tandem repetition, the genes showing sequence similarity to that for anther-specific proline-rich protein (APG) were found on three TAC clones, LjT09C23, LjT07K08, and LjT31L24, as the tandem arrays consist of three, two, and nine genes, respectively. A characteristic feature of these genes is the presence of the "GDSL" family lipase motif, which is found in both prokaryotes and eukaryotes, and is thought to be involved in the regulation of development and morphogenesis in plants (Brick et al. 1995).

4 Linkage Mapping

Genetic linkage maps are essential for genome analysis, including map-based cloning and construction of physical maps. In *L. japonicus*, three genetic linkage maps have been constructed using either an intraspecific or an interspecific F_2 population (details of these genetic linkage maps are described in Chap. 11 by Harada et al.). In order to link the nucleotide sequence information generated from the genome sequencing project to genetic information and phenotypic traits, the sequenced clones have been genetically localized onto the genetic linkage map.

To map the sequenced TAC clones, two types of PCR-based DNA markers, simple sequence repeat length polymorphism (SSLP), and derived cleaved amplified polymorphic sequence (dCAPS), were generated utilizing the sequence information on each clone. For generation of SSLP markers, simple sequence repeats (SSR) such as (AT)n, (GT)n, and (AAT)n that are equal to or longer than 15 bp were searched on the TAC nucleotide sequences determined. Primer pairs were then designed on the flanking sequences of each SSR so that amplified fragment lengths were in the range 100–220 bp. PCR products were resolved on agarose gel, and the primer sets giving a polymorphism between the parents of the mapping population, accessions Miyakojima MG-20 and Gifu B-129 were selected. In the *L. japonicus* genome, the SSRs were estimated to occur every 12 kb. The most abundant class of SSRs were (AT)n repeats, followed by (CT)n and (AAG)n repeats (Fig. 2). Forty-five percent of the primer pairs tested so far were polymorphic, and (AT)n and (CT)n SSRs with large number of repeat units (>12 times) were highly polymorphic.

In the cases where no SSLP was found, single nucleotide polymorphisms (SNPs) between the parental genotypes were searched for. Oligonucleotides were designed based on the sequence information of the TAC clones, mostly from intergenic regions, and the corresponding regions of the genome of the accession Gifu B-129 were amplified by PCR, followed by sequence analysis. If SNPs were identified by comparing the sequences between two parents, they were converted into dCAPS markers. PCR reactions and restriction enzyme digestion were performed and the reaction mixture was analyzed on agarose gel to detect polymorphisms.

Genetic mapping of the generated DNA markers was performed using the 127 F2 mapping population derived from a cross between two accessions of *L. japonicus*, Miyakojima MG-20 and Gifu B-129, and the results were integrated onto the linkage map (Hayashi et al. 2001). As these markers are PCR-based, many of them have also been successfully mapped in another linkage map based on an interspecific F2 mapping population established from a cross between *L. japonicus* Gifu B-129 and *L. filliculis* (Sandal et al., 2002). Therefore, it is possible to integrate these two *Lotus* linkage maps using those markers mapped onto both of them.

Since almost all the markers thus generated are codominant, they are useful for genetic mapping. These PCR-based markers with the surrounding

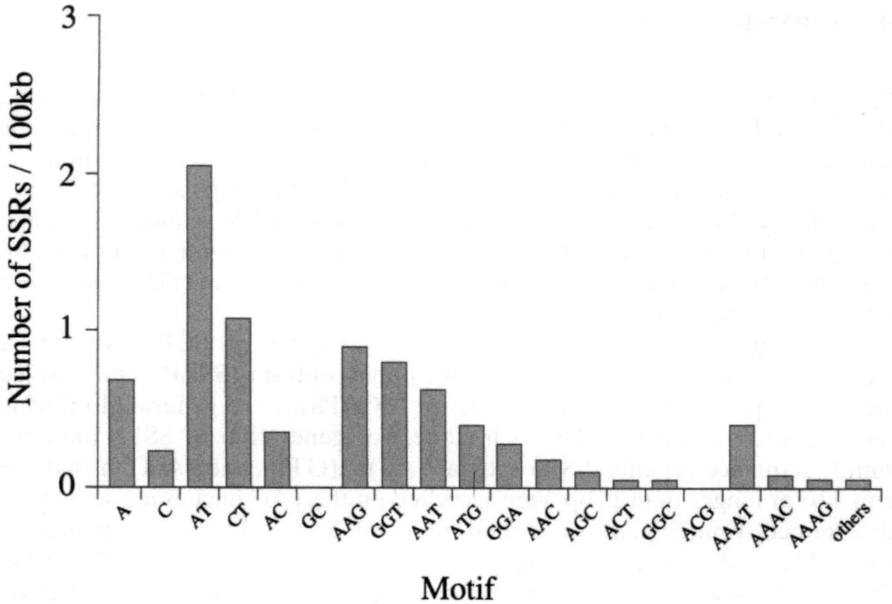

Fig. 2. Frequency of SSRs in *L. japonicus* genome. The number and type of SSRs/100kb of the genome are indicated. Analysis was based on 8.1 Mbp genome regions covered by 83 TAC clones

DNA sequences will facilitate map-based cloning in *L. japonicus*. The detailed information on these markers is available through our web database at http://www.kazusa.or.jp/lotus/.

5 Chloroplast Genome

Along with the progressive accumulation of nuclear genome sequence information, the nucleotide sequence of the entire chloroplast genome (150,519 bp) of *L. japonicus* has been determined (Kato et al. 2000). The circular double-stranded DNA contains a pair of inverted repeats of 25,156 bp which are separated by a small and a large single copy region of 18,271 and 81,936 bp, respectively. A total of 84 predicted protein-coding genes, including 7 genes duplicated in the inverted repeat regions, 4 ribosomal RNA genes, and 37 tRNA genes (30 gene species), representing 20 amino acids species, was assigned on the genome, based on similarity to genes previously identified in other chloroplasts. Most of the genes predicted in the *L. japonicus* chloroplast genome were present in the chloroplast genome of two dicotyledonous plants, tobacco and *A. thaliana*. However, *rpl22*, a gene coding for chloroplast ribosomal protein CL22, was not identified in the chloroplast genome of *L. japonicus*. The absence of *rpl22* gene in a large cluster of ribosomal protein genes was first reported in soybean (Spielmann et al. 1988), then in subclover (*Trifolium subterraneum*) (Milligan et al. 1989), and in other legumes (Gantt et al. 1991). It was reported that a functional copy of the *rpl22* gene was located in the nucleus of pea, strongly suggesting that the missing *rpl22* is present in the nuclear genomes in the legume species, including *L. japonicus*.

Comparison of the genome structures among three dicotyledonous plants, *L. japonicus*, tobacco (Shinozaki et al. 1986), and *A. thaliana* (Sato et al. 1999) revealed inversion of a large segment spanning *rbcL* to *rps16* (positions 5,161–56,176) in the LSC region of the *L. japonicus* chloroplast. This inversion has been reported to be a unique event in the evolution of Leguminosae, providing a synapomorphy for a clade that includes most of the Papilionoideae, one of the three subfamilies (Doyle et al. 1996).

6 Conclusions

An enormous amount of information on the gene and genome structures in *L. japonicus* is accumulating by EST and genome sequencing. Such information, together with the material resources generated during the process of these analyses, including cDNA and genomic libraries and DNA markers, provides a platform for genomics as well as conventional genetic and physiological studies in the future. Gene catalogues generated by the EST information allows one to estimate easily whether orthologues of the genes of interest are present in the genome, which can be followed by immediate isolation of the corresponding cDNA clones from the library. The DNA markers and genome sequences accelerate identification and isolation of genes causing mutants. Furthermore, these data and material resources are extremely useful in the comprehensive study of gene function. One example is transcriptional profiling using array technology. Macroarrays of cDNA were generated by a consortium of the Japanese *L. japonicus* community using the cDNA clones constructed for EST analysis, and identification is in progress of genes whose transcriptional levels vary among organs or during the various biological processes such as nodule formation (H. Kochi, pers. comm.).

Comparison of genetic information carried by various living organisms (comparative genomics) holds promise of identifying genes common or unique to individual organisms, and of studying the process of gene and genome evolution. In plants, a large number of ESTs have been collected from a variety of species (http://www.ncbi.nlm.nih.gov/dbEST/index.html). These include plants of both scientific and agronomic importance such as, to name a few, *Glycine max*, *Medicago truncatula*, *A. thaliana*, *Zea mays*, and *Lycopersicon esculentum*. Genomic sequences that cover the entire and the significant portions of the genomes of *A. thaliana* and rice, respectively, are also available. As another example of progress, comparison of the genome structure of *L. japonicus* with that of *A. thaliana* has revealed microsynteny between the genomes of these two species (S. Sato, in prep.). Such information is extremely useful for isolation of orthologous genes in *L. japonicus*.

The EST and genome sequencing project will continue, and the collected information will be posted on our web database at http://www.kazusa.or.jp/lotus/.

References

Altschul SF, Gish W, Miller W, Myers EW Lipman DJ (1990) Basic local alignment search tool. J Mol Biol 215:403–410

Asamizu E, Nakamura Y, Sato S, Tabata S (2000a) Generation of 7137 non-redundant expressed sequence tags from a legume, *Lotus japonicus*. DNA Res 7:127–130

Asamizu E, Nakamura Y, Sato S, Tabata S (2000b) A large scale analysis of cDNA in *Arabidopsis thaliana*: generation of 12,028 non-redundant expressed sequence tags from normalized and size-selected cDNA libraries. DNA Res 7:175–180

Bonaldo MF, Lennon G, Soares MB (1996) Normalization and subtraction: two approaches to facilitate gene discovery. Genome Res 6:791–806

Brick DJ, Brumlik MJ, Buckley T, Cao JX, Davies PC, Misra S, Tranbarger TJ, Upton C (1995) A new family of lipolytic plant enzymes with members in rice, *Arabidopsis* and maize. FEBS Lett 377:475–480

Burge C, Karlin S (1997) Prediction of complete gene structures in human genomic DNA. J Mol Biol 268:78–94

Cooke R, Raynal M, Laudié M, Grellet F, Delseny M, Morris PC, Guerrier D, Giraudat J, Quigley F, Clabault G, Li YF, Mache R, Krivitzky M, Gy IJ, Kreis M, Lecharny A, Parmentier Y, Marbach J, Fleck J, Clément B, Philipps G, Hervé C, Bardet C, Tremousaygue D, Lescure B, Lacomme C, Roby D, Jourjon MF, Chabrier P, Charpenteau JL, Desprez T, Amselem J, Chiapello H, Höfte H (1996) Further progress towards a catalogue of all *Arabidopsis* genes: analysis of a set of 5000 non-redundant ESTs. Plant J 9:101–124

Covitz PA, Smith LS, Long SR (1998) Expressed sequence tags from a root-hair-enriched *Medicago truncatula* cDNA library. Plant Physiol 117:1325–1332

Doyle JJ, Doyle JL, Ballenger JA, Palmer JD (1996) The distribution and phylogenetic significance of a 50-kb chloroplast DNA inversion in the flowering plant family Leguminosae. Mol Phylogenet Evol 5:429–438

Endo M, Kokubun T, Takahata Y, Higashitani A, Tabata S, Watanabe M (2000) Analysis of expressed sequence tags of flower buds in *Lotus japonicus*. DNA Res 7:213–216

Gantt JS, Baldauf SL, Calie PJ, Weeden NF, Palmer JD (1991) Transfer of rpl22 to the nucleus greatly preceded its loss from the chloroplast and involved the gain of an intron. EMBO J 10:3073–3078

Gyorgyey J, Vaubert D, Jimenez-Zurdo JI, Charon C, Troussard L, Kondorosi A, Kondorosi E (2000) Analysis of *Medicago truncatula* nodule expressed sequence tags. Mol Plant Microbe Interact 13:62–71

Handberg K, Stougaard J (1992) *Lotus japonicus*, an autogamous, diploid legume species for classical and molecular genetics. Plant J 2:487–496

Hayashi M, Miyahara A, Sato S, Kato T, Yoshikawa M, Taketa M, Hayashi M, Pedrosa A, Onda R, Imaizumi-Anraku H, Bachmair A, Sandal N, Stougaard J, Murooka Y, Tabata S, Kawasaki S, Kawaguchi M, Harada K (2001) Construction of a genetic linkage map of the model legume *Lotus japonicus* using an intraspecific F_2 population. DNA Res 8:301-310

Hebsgaard SM, Korning PG, Tolstrup N, Engelbrecht J, Rouze P, Brunak S (1996) Splice site prediction in *Arabidopsis thaliana* DNA by combining local and global sequence information. Nucleic Acids Res 24:3439–3452

Höfte H, Desprez T, Amselem J, Chiapello H, Caboche M, Moisan A, Jourjon MF, Charpenteau JL, Berthomieu P, Guerrier D, Giraudat J, Quigley F, Thomas F, Yu DY, Mache R, Raynal M, Cooke R, Grellet F, Delseny M, Parmentier Y, Marcillac G, Gigot C, Fleck J, Philipps G, Axelos, M, Bardet C, Tremousaygue D, Leacure B (1993) An inventory of 1152 expressed sequence tags obtained by partial sequencing of cDNAs from *Arabidopsis thaliana*. Plant J 4:1051–1061

Imaizumi-Anraku H, Kawaguchi M, Koiwa H, Akao S, Syono K (1997) Two ineffective-nodulating mutants of *Lotus japonicus* – different phenotypes caused by the blockage of endocytotic bacterial release and nodule mutation. Plant Cell Physiol 38:871–881

Ito M, Miyamoto J, Mori Y, Fujimoto S, Uchiumi T, Abe M, Suzuki A, Tabata S, Fukui K (2000) Genome and chromosome dimensions of *Lotus japonicus*. J Plant Res 113:435–442

Kato T, Kaneko T, Sato S, Nakamura Y, Tabata S (2000) Complete structure of the chloroplast genome of a legume *Lotus japonicus*. DNA Res 7:323–330

Kawaguchi M (2000) *Lotus japonicus* "Miyakojima" MG-20: an early flowering accession suitable for indoor handling. J Plant Res 113:507–509

Kawasaki S, Murakami Y (2000) Genome analysis of *Lotus japonicus*. J Plant Res 113:497–506

Liu Y-G, Shirano Y, Fukaki H, Yanai Y, Tasaka M, Tabata S, Shibata D (1999) Complementation of plant mutants with large genomic DNA fragments by a transformation-competent artificial chromosome vector accelerates positional cloning. Proc Natl Acad Sci USA 96:6535–6540

Lowe TM, Eddy SR (1997) tRNAscan-SE: a program for improved detection of transfer RNA genes in genomic sequence. Nucleic Acids Res 25:955–964

Martirani L, Stiller J, Mirabella R, Alfano F, Lamberti A, Rautoiu SE, Iaccarino M, Gresshoff PM (1999) T-DNA tagging of nodulation- and root-related genes in *Lotus japonicus*: expression patterns and potential for promoter trapping and insertional mutagenesis. Mol Plant Microbe Interact 12:275-284

Men AE, Meksem K, Kassem MA, Lohar D, Stiller J, Lightfoot D, Gresshoff PM (2000) A bacterial artificial chromosome (BAC) library of *Lotus japonicus* constructed in an *Agrobacterium*-transformable vector V41. Abstr Mol Genet Model Legumes John Innes Centre:97

Milligan BG, Hampton JN, Palmer JD (1989) Dispersed repeats and structural reorganization in subclover chloroplast DNA. Mol Biol Evol 6:355–368

Newman T, de Bruijn FJ, Green P, Keegstra K, Kende H, McIntosh L, Ohlrogge J, Raikhel N, Somerville S, Thomashow M (1994) Genes galore: a summary of methods for accessing results from large-scale partial sequencing of anonymous *Arabidopsis* cDNA clones. Plant Physiol 106:1241–1255

Sandal N, Krusell L, Ketelsen T, Madsen LH, Stougaard J (2000) Mapping and map based cloning in *Lotus japonicus*. Abstr Mol Genet Model Legumes John Innes Centre:18

Sandal N, Krusell L, Radutoiu S, Olbryt M, Pedrosa A, Stracke S, Sato S, Kato T, Tabata S, Parniske M, Bachmair A, Ketelsen T, Stougaard J (2002) A genetic linkage map of the model legume *Lotus japonicus* and strategies for fast mapping of new loci. Genetics 161:1673–1683

Sato S, Kotani H, Nakamura Y, Kaneko T, Asamizu E, Fukami M, Miyajima N, Tabata S (1997) Structural analysis *of Arabidopsis thaliana* chromosome 5. I. Sequence features of the 1.6 Mb regions covered by twenty physically assigned P1 clones. DNA Res 4:215–230

Sato S, Nakamura Y, Kaneko T, Asamizu E, Tabata S (1999) Complete structure of the chloroplast genome of *Arabidopsis thaliana*. DNA Res 6:283–290

Sato S, Kaneko T, Nakamura Y, Asamizu E, Kato T, Tabata S (2001) Structural analysis of a *Lotus japonicus* genome. I. Sequence features and mapping of fifty-six TAC clones which cover the 5.4 Mb regions of the genome. DNA Res 8:311–318

Schauser L, Handberg K, Sandal N, Stiller J, Thykjflr T, Pajuelo E, Nielsen A, Stougaard J (1998) Symbiotic mutants deficient in nodule establishment identified after T-DNA transformation of *Lotus japonicus*. Mol Gen Genet 4:414–423

Schauser L, Roussis A, Stiller J, Stougaard J (1999) A plant regulator controlling development of symbiotic root nodules. Nature 402:191–195

Shinozaki K, Ohme M, Tanaka M, Wakasugi T, Hayashida N, Matsubayashi T, Zaita N, Chunwongse J, Obokata J, Yamaguchi-Shinozaki K, Ohto C, Torazawa K, Meng BY, Sugita M, Deno H, Kamogashira T, Yamada K, Kusuda J, Takaiwa F, Kato A, Tohdoh N, Shimada H, Sugiura M (1986) The complete nucleotide sequence of the tobacco chloroplast genome: its gene organization and expression. EMBO J 5:2043–2049

Spielmann A, Roux E, von Allmen JM, Stutz E (1988) The soybean chloroplast genome: complete sequence of the rps19 gene, including flanking parts containing exon 2 of rpl2 (upstream), but rpl22 (downstream). Nucleic Acids Res 16:1199

Szczyglowski K, Shaw RS, Wopereis J, Copeland S, Hamburger D, Kasiborski B, Dazzo FB, de Bruijin FJ (1998) Nodule organogenesis and symbiotic mutants of the model legume *Lotus japonicus*. Mol Plant Microbe Interact 11:684–697

The Arabidopsis-Genome Initiative (2000) Analysis of the genome sequence of the flowering plant *Arabidopsis thaliana*. Nature 408:796–815
Uberbacher EC, Mural RJ (1991) Locating protein-coding regions in human DNA sequences by a multiple sensor-neural network approach. Proc Natl Acad Sci USA 88:11261–11265
Yamamoto K, Sasaki T (1997) Large-scale EST sequencing in rice. Plant Mol Biol 35:135–144
Webb KJ, Sk¿t L, Nicholson MN, Jorgensen B, Mizen S (2000) *Mesorhizobium loti* increases root-specific expression of a calcium-binding protein homologue identified by promoter tagging in *Lotus japonicus*. Mol Plant Microbe Interact 13:606–616

10 Introduction of an Early Flowering Accession "Miyakojima" MG-20 to Molecular Genetics in *Lotus japonicus*

M. KAWAGUCHI[1] and R. NISHIMURA[2]

1 Introduction

Leguminosae is the third largest family of the angiosperms. The family includes the remarkable biodiversity ranging over the levels of the morphology, secondary metabolites, seed proteins, and biological interactions with insects and others. The family also includes many important plants closely related to human life. However, genes responsible for the causing generation of leguminous biodiversity and usefulness have not been elucidated at the molecular level. In order to identify and characterize such genes – for example, genes controlling symbiotic nitrogen fixation with rhizobia – certain model legumes that enable progress in molecular genetics are definitely required. In general, transformation of leguminous plants is more difficult than experimentally common plants such as *Arabidopsis* and tobacco. Two herbaceous plants of *Lotus japonicus* and *Medicago truncatula* have emerged from the beginning of 1990s as transformable diploid legumes (Barker et al. 1990; Handberg and Stougaard 1992). *L. japonicus* is known as a determinate-type nodulation plant like soybean and develops slender legumes. On the other hand, *M. truncatula* is known as an indeterminate-type nodulation plant such as pea and vetch and forms spiral legumes. *L. japonicus* has six chromosomes (Kawakami 1930) and a small genome size (442 Mb per haploid of a widely used accession "Gifu" B-129) (Ito et al. 2000). The generation time is 3–4 months and up to 6,000 seeds can be obtained from one plant (Handberg and Stougaard 1992). After the demonstration of such features, the infrastructures of genomics and molecular genetics of *L. japonicus* have been developed rapidly (for review, see Stougaard 2001). The present situation is, however, far different from those of *Arabidopsis* and rice, both of which have long histories. In order to develop the *L. japonicus* as the model system for legumes more efficiently, we began a search for an accession that allows indoor genetics, as has been carried out with *Arabidopsis*. The earliest flowering plant from the southernmost island of Japan met the need and has been accessioned as "Miyakojima" MG-20. Next, we determined that Miyakojima MG-20 is a suitable crossing partner of

[1] Department of Environmental Sciences, Faculty of Science, Niigata University, 8050 Ikarashi, Niigaya City, 950-2181, Japan
[2] Dedpartment of Life Sciences, Graduate School of Arts and Sciences, University of Tokyo, 3-8-1 Komaba, Meguro-ku, Tokyo, 153–8902 Japan

Biotechnology in Agriculture and Forestry, Vol. 52
Nagata/Tabata (Eds.) Brassicas and Legumes

a standard accession of *L. japonicus*, namely, "Gifu" B-129. Finally, from derived cleaved amplified polymorphism (dCAPS) analysis (Michaels and Amasino 1998; Neff et al. 1998) using the single nucleotide polymorphisms (SNPs) found in the Gifu–Miyakojima genome sequences, we showed that a pleiotropic mutant with an early nodulation phenotype named *astray* (*Ljsym77*) is tightly linked with a candidate gene *LjBZF*. Thus, the present status of the *L. japonicus* Miyakojima accession has been established and its significance in legume genetics will be described in this chapter.

2 Search of the Early-Flowering Plant That Enables Indoor Genetics

L. japonicus is distributed in temperate regions of Asia from Japan to Pakistan (Grant and Small 1996). Since Handberg and Stougaard (1992) demonstrated *L. japonicus* as a leguminous model plant, Gifu B-129 (an accession collected on a bank of the Sakai river in Gifu Prefecture, Japan, by Isawo Hirayoshi) has been used widely. *L. japonicus* accessions, including Gifu B-129, require relatively intense light condition for continuous flowering. Therefore, the genetic analysis of *L. japonicus* has been conducted frequently in a greenhouse equipped with supplementary lights. In order to facilitate indoor genetics as with *Arabidopsis*, the earliest flowering plants suitable for indoor genetics is desirable. So we searched for such plants.

In Japan, the island of Hokkaido is at the cold northern limit, and Okinawa in the subtropical zone is at the southern limit of the habitat of *L. japonicus*. Seeds of 15 individual plants that cover this range of habitat (Fig. 1) were collected during my own (M.K.) collecting trip and by kind gifts from research workers. The seeds were scarified by a treatment with sand paper and then sown on "Power Soil" (Kanto Hiryo Kogyo, Tokyo, Japan). Plants were grown in a biotron LH-300 cabinet (Nippon Medical & Chemical Instruments, Tokyo, Japan) under the following conditions: 18/6 h day/night cycle, a light intensity of 150 $\mu E\,s^{-1}\,m^{-2}$, at 25 °C, 70–90% humidity. Flowering time indicates average days from sowing to first flowering in four plants. The results are shown in Fig. 1. Plants growing at a more northern latitude than Gifu (for example, Tono and Towa from the Tohoku region; see Fig. 1) were shown to be later flowering. "Taiki" from Hokkaido and "Oga" and "Aomori" from the Tohoku region never flowered under these conditions. Even if the late-flowering plants were grown under supplementary mercury lights in the greenhouse, Oga hardly flowered and was shown to be the latest-flowering accession among collected plants in Japan. In contrast, plants growing in geographic regions south of Gifu ("Ashizuri" from the Shikoku region and "Kaseda" from the Kyushu region) flowered earlier than Gifu. Judging from the cline in flowering time, it was speculated that the earliest flowering plant might grow in the southernmost region of the habitat of *L. japonicus* under natural conditions. It is reasonable when one considers that photoperiodicity is a variety of basic adaptation of

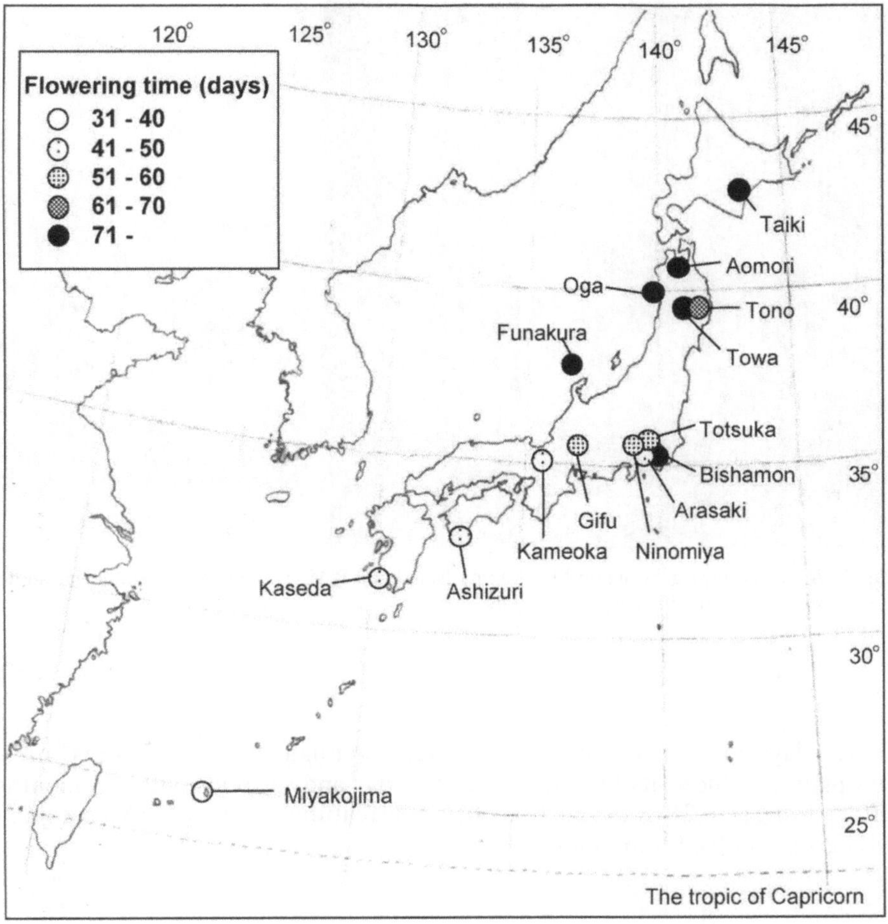

Fig. 1. Distribution and flowering time in 15 Japanese accessions of *L. japonicus*

plants to environments. This speculation was proven by the finding of a plant of *L. japonicus* from Miyakojima Island, Okinawa (Fig. 2A). The seeds of the plants were collected on 25 January 1998. Although *L. japonicus* is a long-day plant, the plants set many flowers and pods even under winter conditions (day length approximately 11 h, temperature approximately 16 °C). The siblings were then seen to flower at the earliest among 15 accessions of *L. japonicus* (Fig. 1). The accession was named "Miyakojima" MG-20.

Miyakojima MG-20 stably produced flowers and pods under illumination of fluorescent lamps. The first flowering of Miyakojima was 35 days prior to forming a tussock under a light intensity of $150 \mu E s^{-1} m^{-2}$ on average. The generation time (from a seed to the next generation of seeds) of Miyakojima

Fig. 2. **A** *L. japonicus* growing on Miyakojima Island and **B** Miyakojima MG-20 in flower under fluorescent light at the intensity $57\,\mu E\,s^{-1}\,m^{-2}$

was 58 days under this regime. Even under lower light intensity ($57\,\mu E\,s^{-1}\,m^{-2}$), the plant produced its first flowers in 52 days and set pods within 3 months after sowing (Fig. 2B; Kawaguchi 2000). These features indicate that Miyakojima MG-20 is suitable for indoor handling.

3 An Inbred Line, "Miyakojima" MG-20-S7 and Its Phenotypic Traits

In order to normalize the genetic background, an inbred line of Miyakojima MG-20 was established following the method of Stougaard and Beuselink (1996). Through the S3 generation, seeds from a maximum of four plants were bulked. Single-seed descent was practiced from the S4 to the S7 generation. Self-pollination was performed in an insect-free biotron. Finally, "Miyakojima" MG-20-S7 germplasm was established. Seeds have been provided worldwidely for experimental studies since August 1999.

Miyakojima is phenotypically different from Gifu in characteristics other than early flowering. The developed stem of Miyakojima is green while that of Gifu is red. There are few trichomes around the calyx tube and on the stem in Miyakojima while Gifu has many trichomes. The seed of Miyakojima is darker in color and larger in size, and its leaflets and cotyledons are wider than those of Gifu. The stem and peduncle are thicker and the hypocotyl and peduncle are longer. These phenotypic traits of Miyakojima are also distinct as compared with other accessions distributed in Honshu, the main island of Japan.

Nitrogen-fixing nodules are efficiently developed in Miyakojima in response to the infection of *Mesorhizobium loti* strains such as MAFF 30-3099 and TONO. Recently, the structures of biologically active Nod factors from *M. loti* MAFF 30-3099 have been identified (Niwa et al. 2001). The rate of formation of nodule primordia by the application of Nod factor has been shown to be higher than that of Gifu (Niwa et al. 2001). These characteristics indicate that this accession Miyakojima is suitable as a model *L. japonicus* for experimental studies.

4 Miyakojima MG-20 as a Suitable Crossing Partner of Gifu

As a crossing partner of Gifu, "Funakura" B-581 (Jiang and Gresshoff 1997) and a different species *L. filicailis* (Sandal et al. 2002) have been proposed. DNA polymorphisms by amplified fragment length polymorphism (AFLP) analyses combined with the high efficiency gel scanning (HEGS) system were evaluated among 15 accessions and *L. filicaulis* (Kawasaki and Murakami 2000; Kawaguchi et al. 2001). It was found that Miyakojima has 4.5% polymorphism to Gifu, which is approximately two times larger than the combination of Gifu and Funakura (Kawaguchi et al. 2001). DNA polymorphism between Gifu and *L. filicaulis* (AFLP 36.7%) was, on the other hand, much higher than the combination with Miyakojima. In general, since around 10% DNA polymorphism can provide efficient results for molecular genetic analysis (S. Kawasaki, pers. comm.), it was thus evaluated that Miyakojima has a moderate potential for DNA polymorphism with Gifu. Then in order to assess the genetic analysis, Miyakojima was reciprocally crossed with Gifu; F1 seeds were easily obtained by the crossing. The F1 plants grew well and set as large a number of pods as the parents showed. A nonshattering pod of the F1 plant contained about 20 seeds, allowing the harvest of a large amount of F2 seeds. Red stem and trichomes observed in Gifu were inherited dominantly when Miyakojima was crossed as a female partner. These features are helpful in assessing a successful cross between them. The early flowering phenotype of Miyakojima was not found in F1 plants crossed with Gifu, indicating that the early flowering gene(s) of Miyakojima may work in a recessive manner. Additionally, the heredity of symbiotic mutant phenotypes was examined. Four different kinds of symbiotic mutant on a Gifu background [*Ljsym70* (nodulation-minus); *Ljsym72* (nodulation-minus and mycorrhization-minus); *alb1-1* (cooperative histogenesis defective); *Ljsym78-2* (hypernodulation and hypermycorrhization); Imaizumi-Anraku et al. 1997, 2000; Solaiman et al. 2000; Kawaguchi et al. 2002] were crossed with Miyakojima. The F2 populations demonstrated that the mutant phenotypes were segregated in a

Mendelian manner. These findings indicate that Miyakojima is suitable as a crossing partner for Gifu.

5 Segregation Analyses of *astray* (*Ljsym77*) Mutant Using SNPs Found in Gifu–Miyakojima Genome Sequences

In order to get an overview of plant factors controlling nodule number and organogenesis, a total of 40,000 M2 seeds of *L. japonicus* Gifu B-129 treated with ethylmethane sulfonate (EMS) were sown under nitrogen-deficient conditions either in vermiculite or on agar plates. From this screening, *astray* (*Ljsym77*) was isolated as a pleiotropic mutant with an increased number of nodules (Fig. 3A). The *astray* mutant developed approximately two times more nodules than the wild type (Kawaguchi et al. 2002). Furthermore, detailed characterization of *astray* demonstrated early initiation of nodule development, which is an unprecedented observation in a symbiotic phenotype (Nishimura et al. 2002a). When grown on agar plate in the absence of bacteria, the hypocotyls and shoots of *astray* were longer than those of wild type (Fig. 3B). The direction of lateral root growth was rather horizontal and the greening of both shoots and roots in response to the light was reduced. Taking these characteristics into consideration, *astray* displayed features very reminiscent of the phenotype of *hy5* mutant in *Arabidopsis thaliana*, as reported in detail by Oyama et al. (1997). The *hy5* mutant is inadequate in photomorphogenesis of young seedlings and enhanced in lateral root initiation.

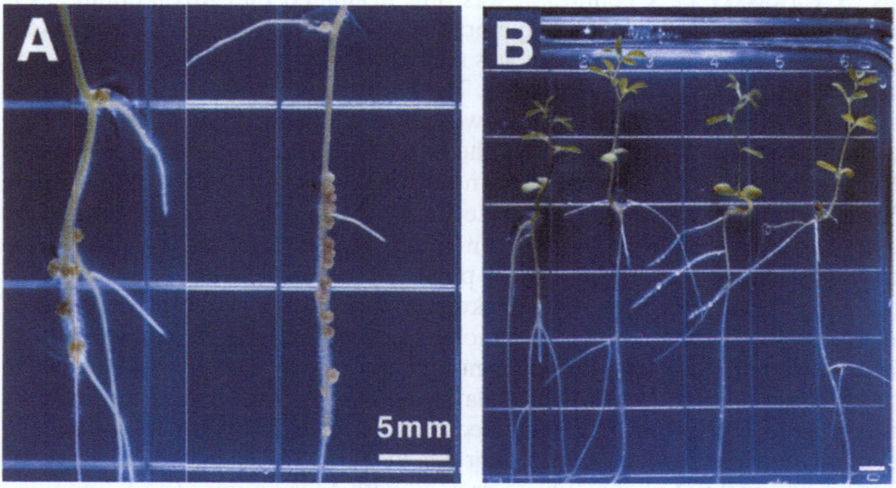

Fig. 3. **A** Root nodules elicited by the inoculation of *M. loti* JRL501. Wild type (*left*) and *astray* (*right*). **B** Seedlings in the absence of *M. loti* strain. Wild type (*left*) and *astray* (*third* to *right*)

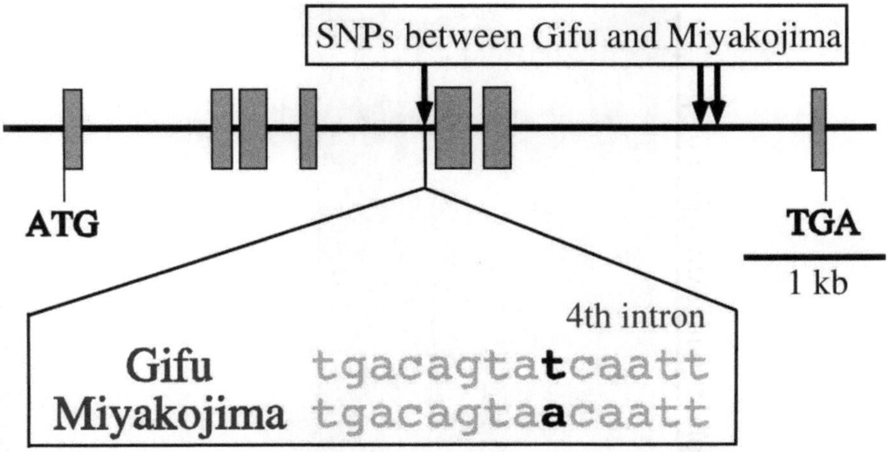

Fig. 4. Gene structure of *LjBZF* and the locations of SNPs detected between Gifu and Miyakojima. *LjBZF* consists of seven exons and six introns

Arabidopsis HY5 encodes the bZIP transcription factor and is known to promote photomorphogenesis as well regulate lateral root development (Oyama et al. 1997; Hardtke et al. 2000).

Deducing from the fact that *astray* phenotype is likely to be the result of the mutation in the *L. japonicus HY5* homologue, we isolated the gene to its full length of 966 bp, based on degenerate polymerase chain reaction (PCR) technique with cDNA, and named it *LjBZF*. The majority of the identical residues in the bZIP domain and the casein kinase II phosphorylation site are well conserved between HY5 and LjBZF. Interestingly, LjBZF includes the zinc finger motif and acidic region in the upstream region, which was not observed in HY5. The closely related genes of *LjBZF* in the database are *STF* in *Glycine max* (Cheong et al. 1998) and CGATG motif-binding protein in *Vicia fava* (GenBank accession No. X97904), which included the zinc finger and acidic regions. The genomic region of *LjBZF* extends about 6.2 kb, and its sequences were analyzed in Gifu and Miyakojima. Although there was no difference in coding regions, three SNPs were identified in introns (Fig. 4). We utilized the SNP in the fourth intron to examine whether *LjBZF* cosegregates with *astray* mutant phenotypes (Nishimura et al. 2002b).

Using the dCAPS method, we detected the SNP between Gifu and Miyakojima for the cosegregation analysis. This method utilizes mismatch PCR primers, in combination with single-nucleotide change, to create a unique restriction site in one of the alleles (Michaels and Amasino 1998; Neff et al. 1998). We designed primers with a single mismatch to amplify about the 180 bp PCR product such that wild type *LjBZF* alone should be digested with the endonuclease *Cla*I (Fig. 5, left). The difference in size between the 180-bp fragment and 160-bp digestion fragment was resolved on 5.0% GTG agarose gel,

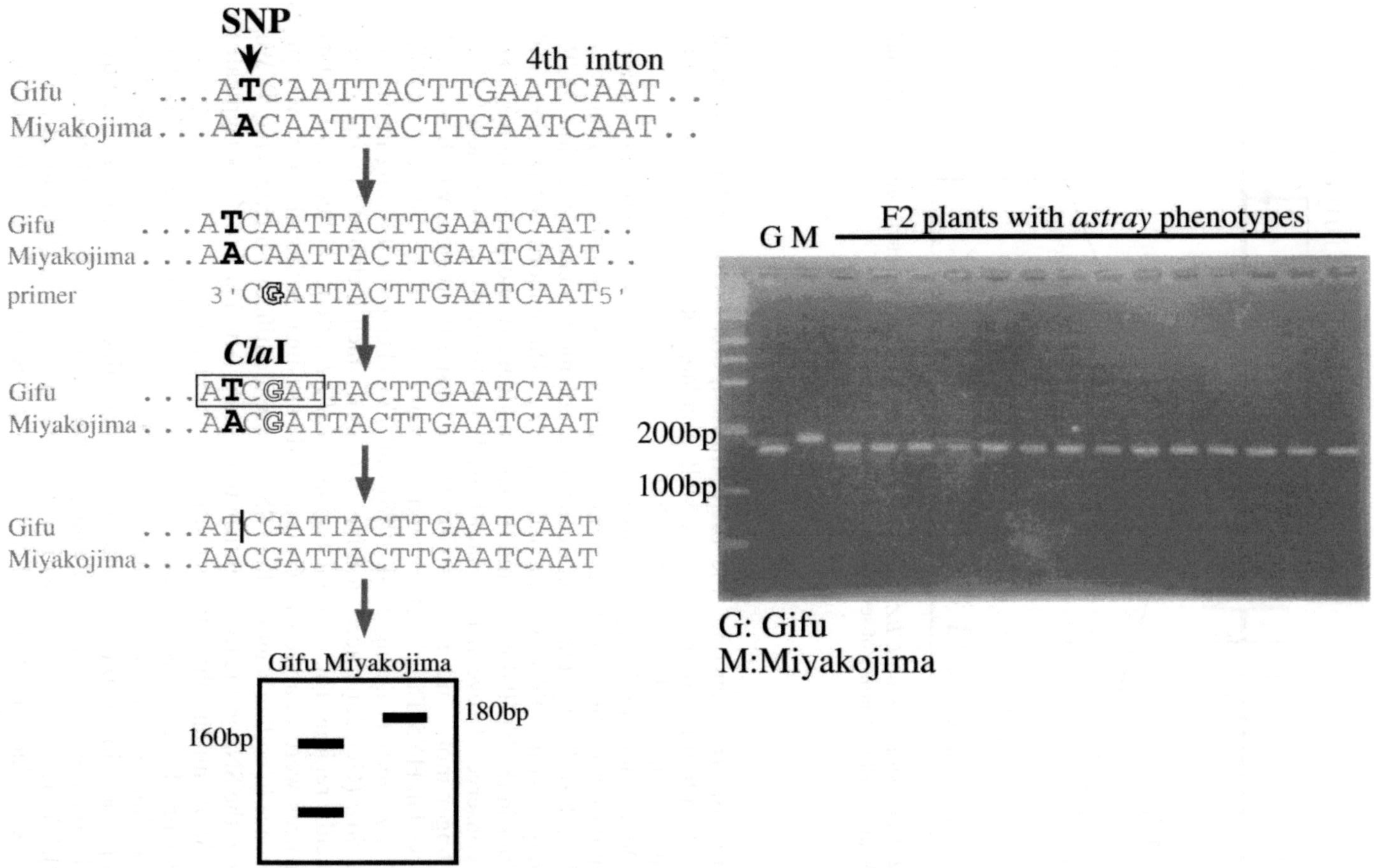

Fig. 5. *Left* Outline of the detection of polymorphism between Gifu and Miyakojima by dCAPS analysis. *Right* Segregation analysis of genomic DNA in F2 progeny. Fragments on the *far left* are DNA markers. *G*, *M*, and *F2* denote digested PCR products derived from Gifu, Miyakojima, and F2 progeny genomic DNA, respectively

and cosegregations of *LjBZF* and *astray* phenotypes were certificated among all F2 plants we tested (Fig. 5, right) (Nishimura et al. 2002b).

6 Current Developments Concerning *L. japonicus* Miyakojima MG-20 in Genomics and Molecular Genetics of Legume

Fundamental research on the genomics and the genetics of *L. japonicus* has recently been started with Miyakojima MG-20. The Kazusa DNA Research Institute, Chiba, Japan, initiated a large-scale expressed sequence tags (EST) analysis from 2-week old seedlings and pods (Asamizu et al. 2000) and constructs of a transformation-competent bacterial artificial chromosome (TAC) library with an insert size of ca. 100 kb on average were produced. Based on the TAC clones, the genome sequencing project was started. The chloroplast genome for this accession was sequenced completely as a legume for the first time (Kato et al. 2000). The sequence data will be helpful for in studying phylogenetic relationship in legumes with remarkable diversity, as well as in photosynthetic research. Transformation procedures by *Agrobacterium* can be applied for Miyakojima" MG-20, but probably due to its higher sensitivity to cytokinin, the addition of lower concentration of cytokinin 6-benzyl amino purine (BAP) 0.2 μg/ml is recommended for the callus formation and plant regeneration medium (Kamisawa, Aoki, Umehara, and Hiroshi Kouchi, pers. comm.).

AFLP analysis combined with HEGS revealed that Miyakojima is the plant most diverged from Gifu among the Japanese accessions (Kawasaki and Murakami 2000; Kawaguchi et al. 2001). Based on the above data, high-density mapping using F2 populations derived from a cross Gifu Miyakojima has been started by the *L. japonicus* consortium. At present, over 500 DNA markers, mainly on AFLP and microsatellites, have been mapped and converged to six linkage groups that correspond to chromosome numbers (Hayashi et al. 2001; Sato et al. 2001). The recombinant inbred line population has also been constructed. Taken together with these infrastructures and the easy handling of Miyakojima, it is fully expected that Miyakojima will facilitate the positional cloning of the genes conferring root symbiosis to legumes.

7 Future Prospects

Although leguminous plants have symbiosis with microorganisms, a remarkable diversity, and are valuable resources for the world population, several barriers have delayed the elucidation of gene function characteristic of legumes. This chapter has highlighted one accession of *L. japonicus* derived from the southern limit of the natural growth, Miyakojima MG-20. Positioning Miyakojima MG-20 with the crossing partner of Gifu B-129 and the detection of DNA polymorphism between them allowed not only the construction of high-density linkage map of *L. japonicus*, but also the segregation analysis between the *astray* mutant phenotype and *LjBZF* having legume-specific motifs. In addition, the utility of AFLP and codominant markers allowed us to identify the first hypernodulating gene, *Har1* (*LjSym78*), which seems to be identical to the gene conferring the soybean hypernodulation phenotype (Nishimura et al. 2002c). Thus Miyakojima MG-20 has contributed greatly to constructing the linkage map and to elucidating the gene function characteristic of legumes.

Today, more than 300,000 ESTs of the model and crop legumes have been registered in public databases. This is the third largest database among all living things. Although the generation of a large scale ESTs has value in and of itself, it is generally difficult to elucidate their biological functions except by expressional analysis using microarrays. Another experimental merit of Miyakojima MG-20 is the early flowering trait that makes it suitable for indoor handling. Once a rapid and stable transformation system of Miyakojima MG-20 is established, the function of legume-specific genes illuminated by EST analysis will be clarified much more efficiently. There is a clear need for improvement in the transformation system of Miyakojima MG-20 in future.

8 Protocols

8.1 Growth Conditions for *L. japonicus* Miyakojima MG-20

1. Seeds were gently scarified with sand paper or immersed in sulfuric acid for 25 min. The seeds were washed in five to seven changes of distilled water.
2. Well-imbibed seeds were sown in pots filled with "Power Soil". The soil surface was covered with wet vermiculite to prevent the proliferation of green algae.
3. Plants were grown in a growth cabinet (Biotron LH-300) under 18-h light/6-h dark cycle at 25 °C, 70–90% humidity at a light intensity of $150\,\mu E\,s^{-1}\,m^{-2}$. The light was provided with white fluorescent tubes. (Miyakojima MG-20 was able to set flowers and pods even under weak light condition, such as $50\,\mu E\,s^{-1}\,m^{-2}$.)
4. Blown pods were harvested and stored in a thin envelope to facilitate shattering the seeds.

8.2 dCAPS Analysis Using SNPs Found in Gifu–Miyakojima Genomic Sequences

1. A dCAPS primer was designed as CGATTACTTGAATCAAT that incorporates a mutation to create a *Cla*I restriction site in the Gifu allele

fragment. The dCAPS primer and a complete match primer AACTAGT-GTTGGTTAGTAGA were used to amplify the 182-bp fragment.

2. Sixteen micrograms of genomic DNA were used as a template for PCR. PCR reaction was performed in 50 μl of the supplemented buffer containing 200 μM dNTPs, 25 pmol of each primer, and 2 units of Taq polymerase (Takara Ltd., Shiga, Japan). Cycling conditions were 30 s at 94 °C, 30 s at 52 °C, 30 s at 72 °C. The cycles were repeated 30 times.
3. Five microliters of the PCR reactions were digested with 2–5 μl of *Cla*I endonuclease in 20 μl of the appropriate restriction buffer.
4. After digestion, samples were separated by electrophoresis for 90 min on 5% GTG agarose gel (Takara) in 1×TAE buffer and stained by ethidium bromide.

References

Asamizu E, Nakamura Y, Sato S, Tabata S (2000) Nucleotide generation of 7137 non-redundant expressed sequence tags from a legume, *Lotus japonicus*. DNA Res 7:127–130

Barker DG, Bianchi S, Blondon F, Dattée Y, Duc G, Essad S, Flament P, Gallusci P, Genier G, Guy P, Muel X, Tourneur J, Dénarié, Huguet T (1990) *Medicago truncatula*, a model plant for studying the molecular genetics of the *Rhizobium*-legume symbiosis. Plant Mol Biol Rep 8:40-49

Cheong YH, Yoo CM, Park JM, Ryu GR, Goekjian VH, Nagao RT, Key JL, Cho MJ, Hong JC (1998) STF1 is a novel TGACG-binding factor with a zinc-finger motif and a bZIP domain which heterodimerizes with GBF proteins. Plant J 15:199–209

Cyranoski D (2001) Japanese legume project may help to fix the nitrogen problem. Nature 409:272

Grant WF, Small E. (1996) The origin of the *Lotus corniculatus* (Fabaceae) complex: a synthesis of diverse evidence. Can J Bot 74:975–989

Handberg K, Stougaard J (1992) *Lotus japonicus*, diploid legume species for classical and molecular genetics. Plant J 2:487–496

Hardtke CS, Gohda K, Osterlund MT, Oyama T, Okada K, Deng XW (2000) HY5 stability and activity in *Arabidopsis* is regulated by phosphorylation in its COP1 binding domain. EMBO J 19:4997–5006

Hayashi M, Miyahara A, Sato S, Kato T, Yoshikawa M, Taketa M, Hayashi M, Pedrosa A, Onda R, Imaizumi-Anraku H, Bachmair A, Sandal N, Stougaard J, Murooka Y, Tabata S, Kawasaki S, Kawaguchi M, Harada K (2001) Construction of a genetic linkage map of the model legume *Lotus japonicus* using an intraspecific F2 population. DNA Res 8:301–310

Imaizumi-Anraku H, Kawaguchi M, Koiwa H, Akao S, Syono K (1997) Two ineffective-nodulating mutants of *Lotus japonicus*: different phenotypes caused by the blockage of endocytotic bacterial release and nodule maturation. Plant Cell Physiol 38:871–881

Imaizumi-Anraku H, Kouchi H, Syono K, Akao S, Kawaguchi M (2000) Analysis of *ENOD40* expression in *alb1*, a symbiotic mutant of *Lotus japonicus* that forms empty nodules with incompletely developed nodule vascular bundles. Mol Gen Genet 264:402–410

Ito M, Miyamoto J, Mori Y, Fujimoto S, Uchiumi T, Abe M, Suzuki A, Tabata S, Fukui K (2000) Genome and chromosome dimensions of *Lotus japonicus*. J Plant Res 113:435–442

Jiang Q, Gresshoff PM (1997) Classical and molecular genetics of the model legume *Lotus japonicus*. Mol Plant Microbe Interact 10:59–68

Kato T, Kaneko T, Sato S, Nakamura Y, Tabata S (2000) Complete structure of the chloroplast genome of a legume, *Lotus japonicus*. DNA Res 7:323–330

Kawaguchi M (2000) *Lotus japonicus* 'Miyakojima' MG-20: an early-flowering accession suitable for indoor handling. J Plant Res 113:507–509

Kawaguchi M, Motomura T, Imaizumi-Anraku H, Akao S, Kawasaki S (2001) Providing the basis of genome research in *Lotus japonicus.* Mol Genet Genomics 266:157–166
Kawaguchi M, Imaizumi-Anraku H, Koiwa H, Niwa S, Ikuta A, Syono K, Akao S (2002) Root, root hair, and symbiotic mutants of the model legume *Lotus japonicus*. Mol Plant Microbe Interact 15:17–26
Kawakami, J (1930) Bot Mag 44:319
Kawasaki S, Murakami Y (2000) Genome analysis of *Lotus japonicus*. J Plant Res 113:497–506
Michaels SD, Amasino RM (1998) A robust method for detecting single-nucleotide changes as polymorphic markers by PCR. Plant J 14:381–385
Neff MM, Neff JD, Chory J, Pepper AE (1998) dCAPS, a simple technique for the genetic analysis of single nucleotide polymorphisms: experimental applications in *Arabidopsis thaliana* genetics. Plant J 14:387–392
Nishimura R, Ohmori M, Kawaguchi M (2002a) The novel symbiotic phenotype of enhanced-nodulating mutant of *Lotus japonicus – astray* mutant is an early nodulating mutant with wider nodulation zone. Plant Cell Physiol 43:853–859
Nishimura R, Ohmori M, Fujita H, Kawaguchi M (2002b) A *Lotus* basic leucine zipper protein with a RING-finger motif negatively regulates the developmental program of nodulation. Proc Natl Acad Sci USA 99:15206–15210
Nishimura R, Hayashi M, Wu G-J, Kouchi H, Imaizumi-Anraku H, Murakami Y, Kawasaki S, Akao S, Ohmori M, Nagasawa M, Harada K, Kawaguchi M (2002c) HAR1 mediates systemic regulation of symbiotic organ development. Nature 420:426–429
Niwa S, Kawaguchi M, Imaizumi-Anraku H, Chechetka SA, Ishizuka M, Ikuta, A, Kouchi H (2001) Responses of a model legume *Lotus japonicus* to lipochitin oligosaccharide nodulation factors purified from *Mesorhizobium loti* JRL501. Mol Plant Microbe Interact 14:848–856
Oyama T, Shimura Y, Okada K (1997) The *Arabidopsis HY5* gene encodes a bZIP protein that regulates stimulus-induced development of root and hypocotyl. Genes Dev 11:2983–2995
Sandal N, Krusell L, Radutoiu S, Olbryt M, Pedrosa A, Stracke S, Sato S, Kato T, Tabata S, Parniske M, Bachmair A, Ketelsen T, Stougaard J (2002) A genetic linkage map of the model legume *Lotus japonicus* and strategies for fast mapping of new loci. Genetics 161:1673–1683
Sato S, Kaneko T, Nakamura Y, Asamizu E, Kato T, Tabata S (2001) Structural analysis of *a Lotus japonicus* genome. I. Sequence features and mapping of fifty-six TAC clones which cover the 5.4 mb regions of the genome. DNA Res 8:311–318
Schauser L, Roussis A, Stiller J, Stougaard J (1999) A plant regulator controlling development of symbiotic root nodules. Nature 402:191–195
Solaiman MZ, Senoo K, Kawaguchi M, Imaizumi-Anraku H, Akao S, Tanaka A, Obata H (2000) Characterization of mycorrhizas formed by *Glomus* sp. on roots of hypernodulating mutants of *Lotus japonicus*. J Plant Res 113:443–448
Stougaard J (2001) Genetics and genomics of root symbiosis. Curr Opin Plant Biol 4:328–335
Stougaard J, Beuselink PR (1996) Registration of GIFU B-129-S9 *Lotus japonicus* germplasm. Crop Sci 36:476

11 Genetic Linkage Map of the Model Legume *Lotus japonicus*

K. Harada[1], M. Hayashi[1], S. Sato[2], and M. Hayashi[3]

1 Introduction

Leguminosae is the third largest family of angiosperms and includes agronomically important crops such as soybean, pea, and bean. However, molecular genetic analyses of agronomically important traits of these crops have been hampered by a large genome size or complexity, and/or low transformation frequencies. Regarding these aspects, a legume *Lotus japonicus* (Regel) Larsen has many appropriate features as a model legume, namely, diploidy (2n = 12), self fertility, small genome size (432 Mb, Pedrosa et al. 2002; 442 Mb, Ito et al. 2000; 494 Mb, Kawasaki and Murakami 2000), short life cycle, and high transformability with *Agrobacterium* (Handberg and Stougaard 1992; Stiller et al. 1997).

Genetic linkage maps are essential for genome analysis, including map-based cloning and construction of physical maps. Jiang and Gresshoff (1997) tried to construct a genetic linkage map of *L. japonicus* with an F_2 population derived from a cross between Gifu B-129-S9 and Funakura B-581 using DNA amplification fingerprinting (DAF). As the level of DNA polymorphism between these two parents was low, a smaller number of linkage groups than the number of chromosome pairs was obtained. On the other hand, Sandal et al. (2002) developed a genetic linkage map of *L. japonicus* using an F_2 population made from an interspecific cross between *L. japonicus* Gifu B-129-S9 and *L. filicaulis*. A high level of DNA polymorphism between the parents was observed by amplified fragment length polymorphism (AFLP), and 524 AFLP markers and some additional sequence-specific polymerase chain reaction (PCR) markers were mapped on six linkage groups. On chromosome 3, a pericentric inversion between the parental species was demonstrated by fluorescence in situ hybridization (FISH) (Pedrosa et al. 2002). The segregation distortions and suppression of recombination in the F_2 population were well described by comparing cytogenetic marker localization (Sandal et al. 2002). In search for useful mapping partners in *L. japonicus* accessions, Kawaguchi et al. (2001) found that Miyakojima MG-20 is the most polymorphic among 15

[1] Faculty of Horticulture, Chiba University, 648 Matsudo, Matsudo, Chiba 271–8510, Japan
[2] Kazusa DNA Research Institute, 2-6-7 Kazusa-kamatari, Kisarazu, Chiba 292-0818, Japan
[3] Department of Biotechnology, Graduate School of Engineering, Osaka University, 2-1 Yamadaoka, Suita, Osaka 565-0871, Japan

Biotechnology in Agriculture and Forestry, Vol. 52
Nagata/Tabata (Eds.) Brassicas and Legumes

accessions of *L. japonicus* when compared with Gifu B-129 by the AFLP technique. Miyakojima MG-20 was, therefore, selected as a suitable partner to establish a mapping population.

Gifu B-129 and Miyakojima MG-20 show noticeable cytological differences with respect to chromosome size and morphology, as reported by Ito et al.(2000). Chromosome 1 of Gifu B-129 is submetacentric and the largest chromosome of the complement, while chromosome 2, the second largest chromosomes, is metacentric and bears the nucleolar organizing region (NOR). In the accession Miyakojima MG-20, however, the homologous chromosome 1 is acrocentric and is the second largest chromosome of the complement, whereas the NOR is located on the largest submetacentric chromosome, homologous to chromosome 2 in Gifu B-129.

Using FISH with BAC clones (Pedrosa et al. 2002) from Gifu B-129 and transformation-competent artificial chromosome (TAC) clones (Sato et al. 2001) from Miyakojima MG-20, a reciprocal translocation was found to be responsible for cytological differences between chromosome 1 and 2 (Hayashi et al. 2001). The borders of the translocations were identified by FISH and by alignment toward the *L. filicaulis* × *L. japonicus* Gifu B-129 linkage map (Hayashi et al. 2001).

Based on the segregation of genotypes of AFLP, simple sequence repeat (SSR), and derived cleaved amplified polymorphic sequence (dCAPS) markers, together with the information of translocated regions, a genetic linkage map of *L. japonicus* was constructed using an F_2 population derived from a cross of Gifu B-129 × Miyakojima MG-20 (Hayashi et al. 2001).

In this chapter the procedure for map construction and information on our map of *L. japonicus* are summarized, and the potential benefit from using this linkage map will be discussed.

2 Construction of the Linkage Map

2.1 Plant Materials and DNA Isolation

The 127 F_2 plants of the mapping population were derived from a single cross between two accessions of *Lotus japonicus,* Gifu B-129 and Miyakojima MG-20. The parents and the F_2 plants were grown in a greenhouse and leaves were collected for DNA isolation. Total DNA from leaves was extracted and purified using the DNeasy plant mini kit (QIAGEN Hilden, Germany).

2.2 AFLP Analysis

The AFLP reaction was performed according to Vos et al. (1995) with a slightly modified AFLP Analysis System I (Invitrogen, Carlsbad, CA, USA) procedure or by using *Hin*dIII/*Taq*I digestion. A total of 100–150 ng of genomic DNA of each parent and F_2 plant were completely digested with *Eco*RI and *Mse*I or *Hin*dIII and *Taq*I. Digested DNAs were subjected to the ligation reaction with adapters which correspond to restriction sites. After ligation, reaction mixtures were diluted ten times with 1 mM Tris-HCl/0.1 mM EDTA solution. For amplification of restricted and ligated fragments, a two-step protocol was followed. The first step included the selective pre-amplification of adapter-ligated DNA with primers having one selective nucleotide. In the second step, selective amplification of pre-amplified DNA was performed with primers having two additional nucleotides (i.e., a total of three nucleotides). All amplification reactions were performed with TaKaRa EX Taq (TaKaRa Shuzo, Kyoto, Japan). Duration of the pre-amplification program was 2 min at 94 °C; then 20 cycles of each 30 s at 94 °C, 30 s at 56 °C and 1 min. at 72 °C. Selective amplification was performed as follows: two cycles each of 30 s at 94 °C, 30 s at 68 °C, and 1 min at 72 °C; then annealing temperature was lowered by 0.7 °C for each of 16 cycles: followed by 23 cycles of each 30 s at 94 °C and 30 s at 56 °C, and 1 min at 72 °C.

Electrophoresis was conducted by high efficiency genome scanning (HEGS) (Kawasaki and Murakami 2000; Kawaguchi et al. 2001) with non-denaturing 11–13% polyacrylamide running gels and 5% stacking gels. The gels were stained by Vistra Geen (Amersham Biosciences, Piscataway, NJ, USA) and electrophoresis patterns were detected with FluorImager 585 (Amersham Biosciences). Typical polymorphic bands of the parents and codominant bands in F_2 plants are seen in Figs. 1 and 2, respectively. E-M- and H-T-represent primer combinations for selective amplification and E-M-G, E-M-M, and E-M-C represent AFLP marker, where the last G or M shows presence in Gifu B-129 or Miyakojima MG-20, respectively, and C shows a codominant marker.

2.3 Generation of SSR and dCAPS Markers

For generation of SSR markers, sequence repeats such as (AT)n, (GT)n, and (AAT)n of ≥15 bp were sought on the TAC nucleotide sequences that were determined. Primer pairs were then designed using the Consed program (Gordon et al. 1998) on the flanking sequences of each SSR so that amplified fragment sizes were in the range 100–220 bp. PCR was performed in a total volume of 10 μl containing 20 ng of *L. japonicus* genomic DNA, 1×PCR buffer (TaKaRa Shuzo), 1 unit of TaKaRa Taq (TaKaRa Shuzo), 0.2 mM dNTPs and 0.5 μM of each designed primer. Reactions were run as follows: 2 min at 94 °C; then 30 cycles each of 45 s at 94 °C, 45 s at 55 °C, and 2 min at 72 °C; followed by a final 10-min extension at 72 °C. PCR products were resolved on a 3% MetaPhor agarose gel (BMA, Rockland, IL, USA), and the primer sets giving polymorphisms between Miyakojima MG-20 and Gifu B-129 were selected and used for scoring 127 F2 plants.

In the cases where no simple sequence repeat length polymorphism was found, single nucleotide polymorphisms (SNPs) between the parents were searched for. Oligonucleotides were designed based on the sequence information of the TAC clones, mostly from intergenic regions, and the corre-

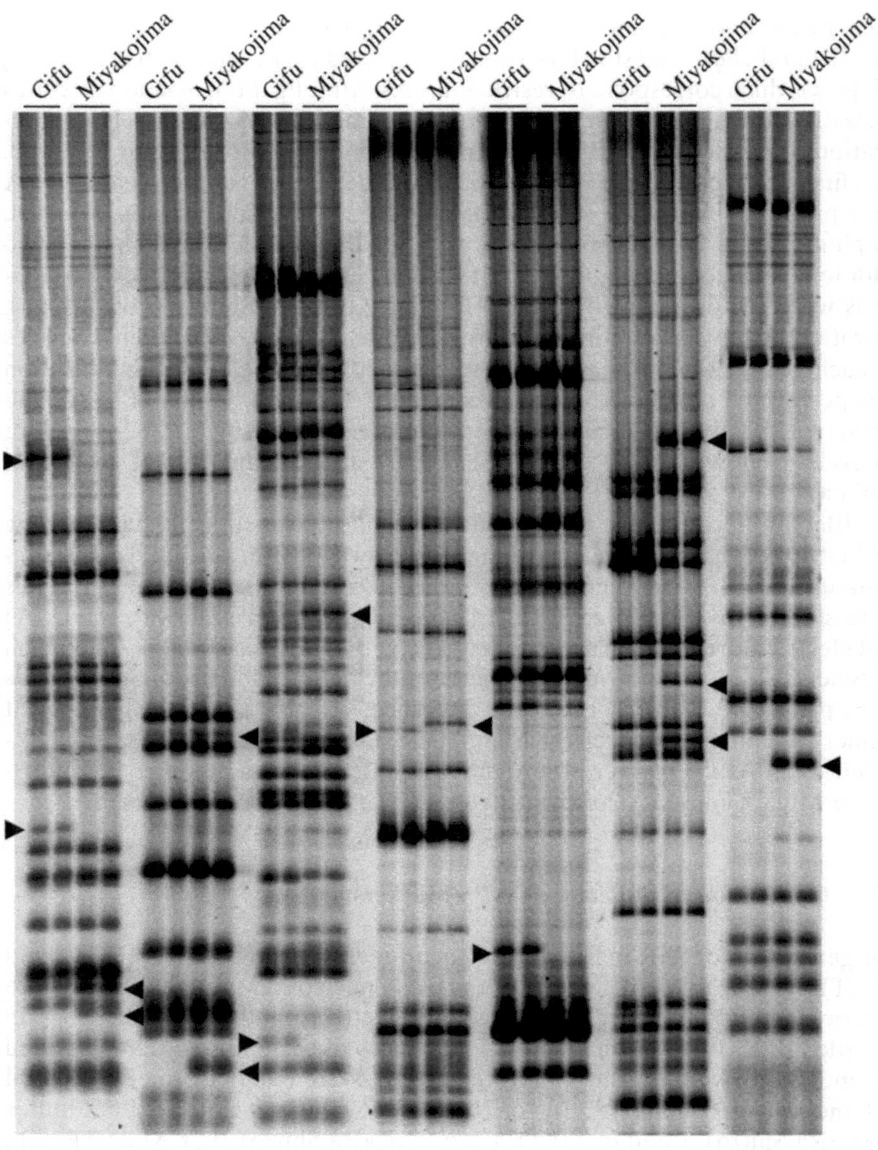

Fig. 1. Detection of AFLPs between the parents, Gifu B-129 and Miyakojima MG-20. Only a part of the analysis is shown. Polymorphic bands are indicated by *arrowheads*

sponding regions of the genome of the accession Gifu B-129 were amplified by PCR, followed by sequence analysis. If SNPs were identified by comparing the sequences between the parents, they were converted into dCAPS markers, as facilitated by the dCAPS finder program (Neff et al. 1998). PCR reactions were performed using the same conditions used for amplification of SSR markers. Aliquots of the PCR product(10 μl) were digested for 2 h in 15 μl

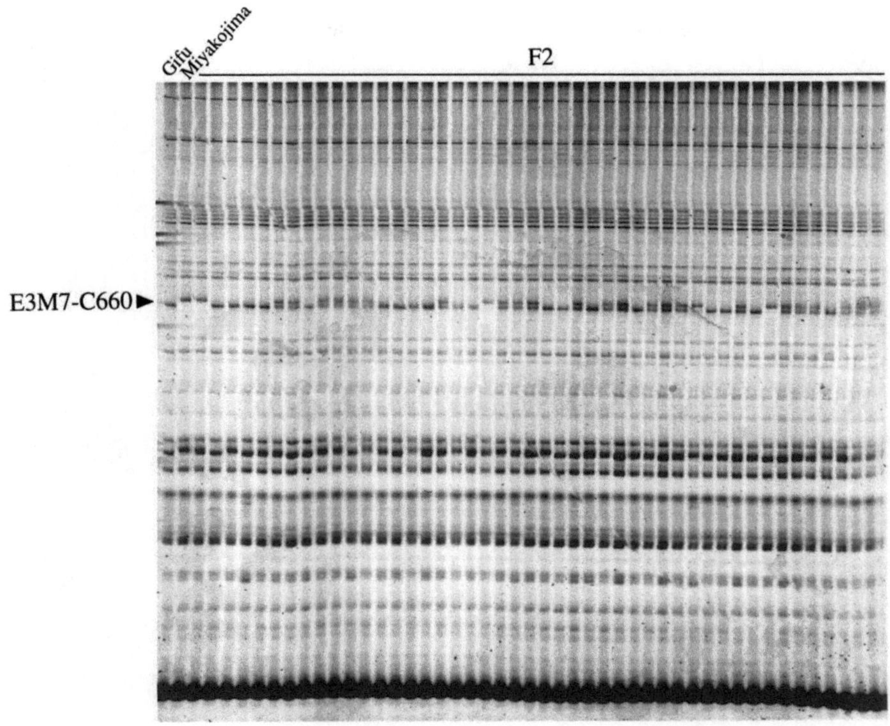

Fig. 2. Segregation of polymorphic bands in the F_2 population by AFLP analysis. An example of codominant PCR products is shown by *arrowheads*

(total volume) with 2–5 units of the appropriate restriction endonuclease, and the reaction mixture was analyzed on a 3% MetaPhor agarose gel to detect polymorphisms.

Examples of segregation of PCR products of SSR and dCAPS markers in the F_2 population are presented in Fig. 3.

2.4 Other PCR-Based Markers

Specific primer sets were developed for the Miyakojima MG-20 cDNA clones that are highly homologous to the nitrate transporter of *Arabidopsis*, the transcription factor IAA16 of *Arabidopsis*, and the thioredoxin of soybean. The PCR fragments showed length polymorphism between parents, and these primer sets were used as DNA markers named LjNIT-1, LjIAA16–1, and Lj096, respectively. When primer pairs for Sctt008 and Satt567, which were originally SSR markers of soybean, were used for PCR, products of PCR were observed in Gifu B-129 but were absent in Miyakojima MG-20. These were also included in the map as DNA markers.

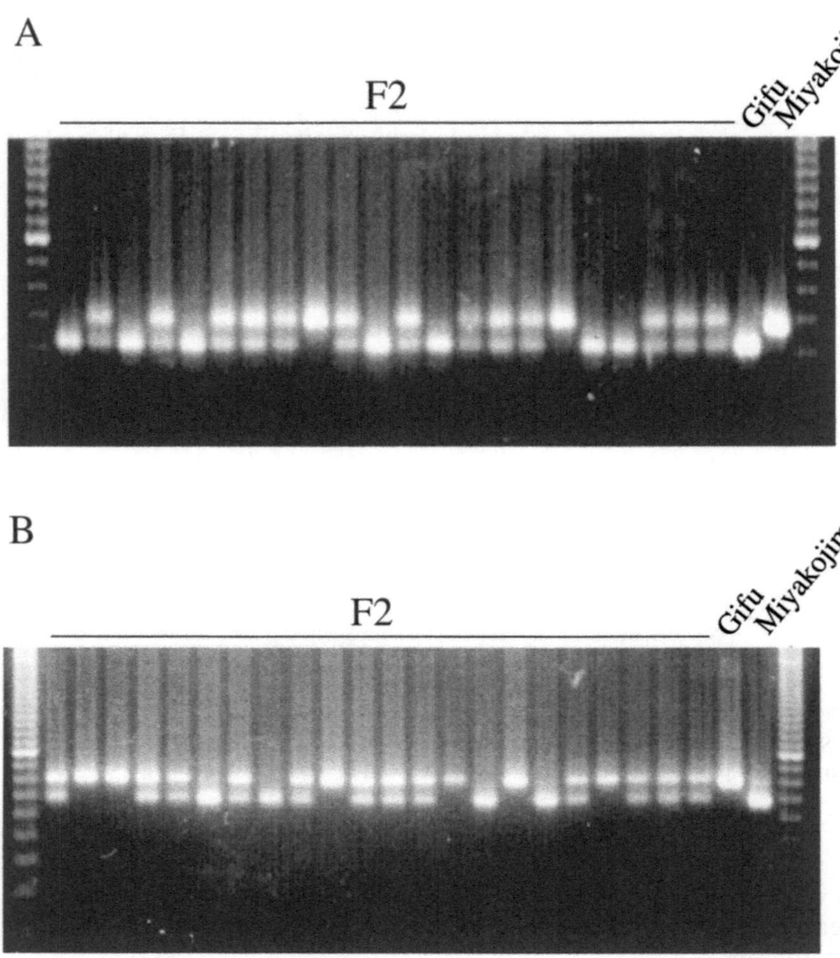

Fig. 3. Segregation of PCR products of **A** SSR and **B** dCAPS markers in the F_2 population

2.5 Linkage Map Construction

The test for segregation ratios of 3:1 or 1:2:1 was performed by χ^2 statistics. Codominant AFLP markers were identified and a framework map was constructed using these markers and other codominant markers, including SSR, dCAPS, and EST markers. The MAPMAKER/EXP, Version 3.0b program (Lander et al. 1987; Lincoln et al. 1992) was used for map construction. In this calculation, the *compare* and the *try* commands were used to determine the most likely order. Following these steps, the *three-point* command was used to confirm the marker order on the framework map. Finally, the positions for dominant markers derived from respective parents (maternal Gifu B-129 and paternal Miyakojima MG-20) were merged into this map using the *try* command. We constructed a separate linkage map for each parent. Recombination frequencies were converted into map distance in cM using the Kosambi mapping function. The linkage criteria of the markers was a LOD score >3.0 and a maximum distance of 37.2 cM.

2.6 Characteristics of AFLP Markers on the Map

Compared with other DNA marker systems, AFLP is an efficient technique to generate large numbers of markers for construction of a genetic map (Mackill et al.1996; Keim et al. 1997) While some researchers have reported relatively uniform distributions of AFLP markers (Keim et al. 1997; Becker et al. 1995), others have observed a strong clustering of them, often in association with telomeric or centromeric regions (Alonso-Blanco et al.1998; Qi et al. 1998) .To avoid clustering of markers, we used two enzyme sets, *Eco*RI/*Mse*I and *Hin*dIII/*Taq*I in our AFLP analysis. Even with this precaution we still observed some clusters, one of which was associated with the breakpoint of the translocation, and a few gaps (Fig. 4). Approximately 1–1.5 polymorphic bands per primer pair were observed between Gifu B-129 and Miyakojima MG-20. One hundred and four (52 sets) of AFLP markers were codominantly scored. In total, 224 (Gifu B-129) and 216 (Miyakojima MG-20) AFLP markers were positioned on the linkage maps.

2.7 PCR-Based Markers Derived from TAC Clones

In addition to the AFLP markers, 45 SSR and 13 dCAPS markers (TM markers), developed on the basis of the sequence information from Miyakojima MG-20 TAC clones (Sato et al. 2001), were included in the linkage analysis. All of these markers except one (TM0011) were codominant. Primer sequences for PCR, product sizes for the parents, restriction enzymes for digestion, and expected fragment sizes are listed in Tables 1 and 2.

2.8 Development of a Linkage Map

The primary map that was compiled by the MAPMAKER program had five linkage groups, including an unusually long one, and a group of excluded markers. Because there are six pairs of chromosomes in *L. japonicus*, the number of linkage groups was smaller by one than expected. By the alignment of the two linkage maps derived from Gifu B-129 × Miyakojima MG-20 and *L. filicaulis* × *L. japonicus* Gifu B-129, and FISH with BAC clones from Miyakojima MG-20, it became clear that the unusually long linkage group from our primary data corresponded to chromosomes 1 and 2. The linkage groups were numbered based on chromosome size of Gifu B-129, and the top of each linkage group corresponded to the short arm of the chromosome

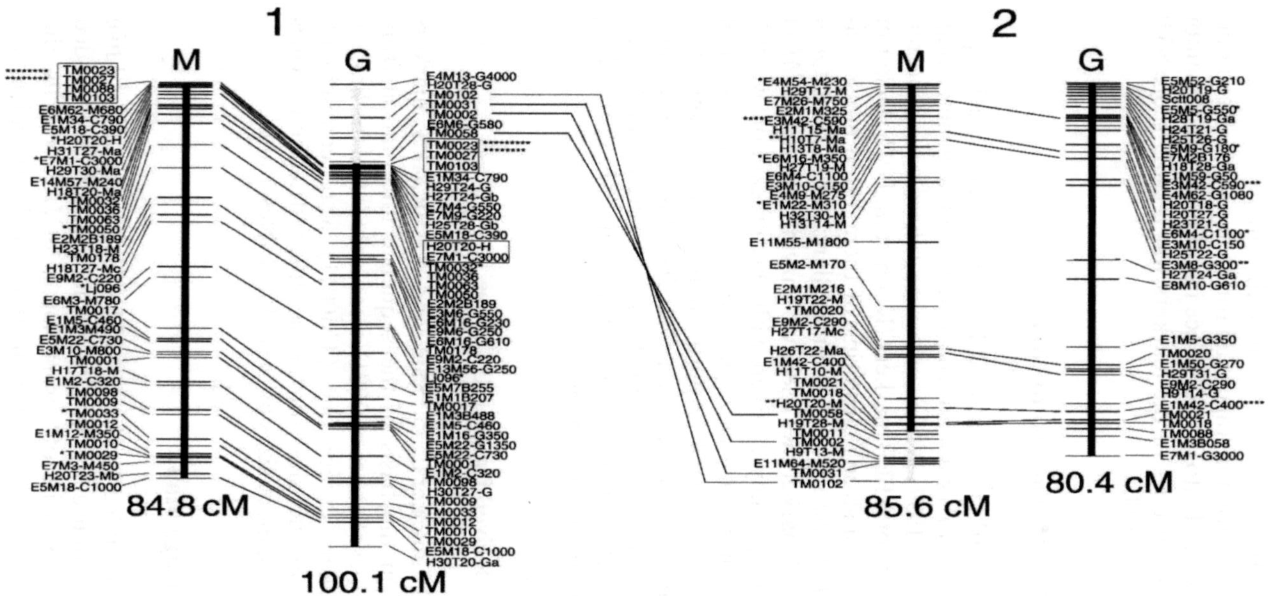

Fig. 4. Genetic linkage map of *Lotus japonicus*; maternal Gifu B-129 (G) map on the *right* and paternal Miyakojima MG-20 (M) map on the *left*. *Asterisks* and *boxed letters* indicate the markers mapped on the same locus. *Gray bars* and *slanted bars* show the main translocated regions and the regions of distorted segregation, respectively. Codominant Gifu B-129 and Miyakojima MG-20 markers are connected by a *line*, except for the TM0088 marker mapped on Gifu B-129 linkage group 2, and on Miyakojima MG-20 linkage group 1. *Bar* = 10 cM. (Hayashi et al. 2001)

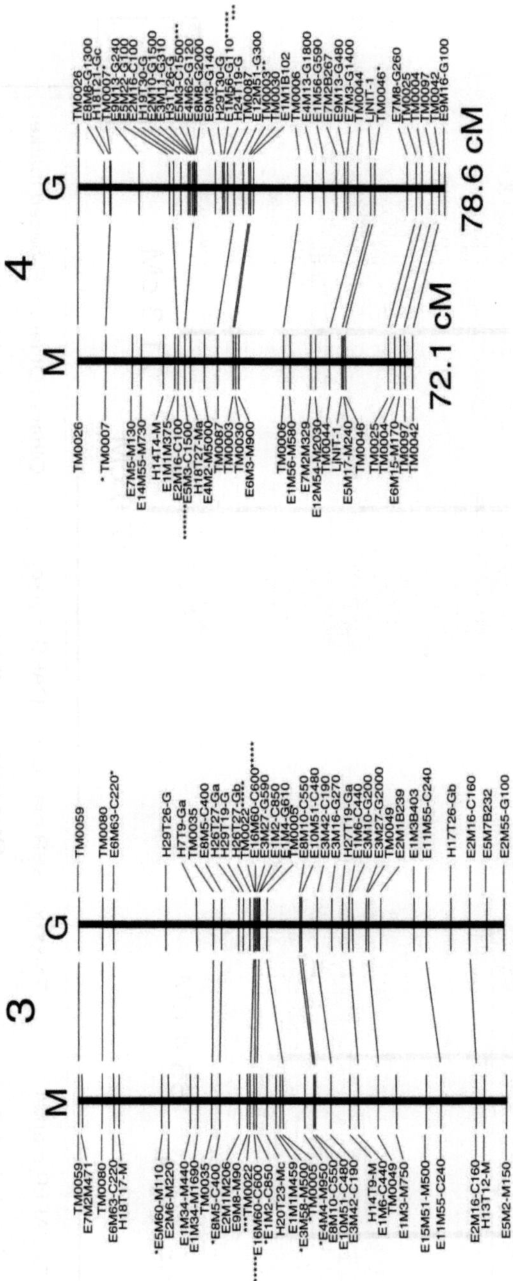

Fig. 4. *Continued*

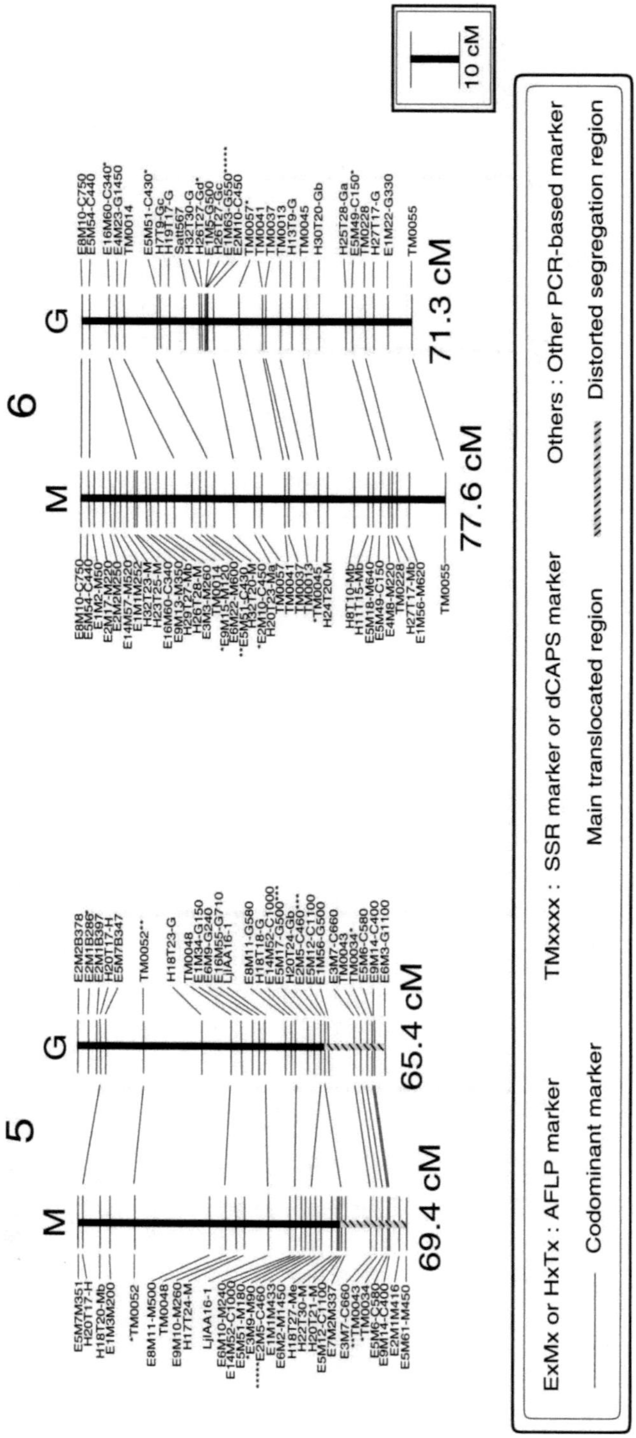

Fig. 4. *Continued*

Table 1. List of SSR markers (Sato et al. 2001)

Clone name	Marker name	Modif	Number of repeats	Product size		Fw primer (5′ to 3′)	Rv primer (5′ to 3′)	Chr
				Miyakojima	Gifu			
LjT09C23	TM0001	AT	10	136	144	TCTGTTGTATCCTCATTATCC	CTCCTCATTTATATATAATTGTCA	1
LjT13B22	TM0009	CT	10	149	105	CCTGTCTACCAAACGCTAC	ACTTTCAATTAACATAAAGCAG	1
LjT01I13	TM0010	CT	12	187	205	AAACCAGTGTTTCAGCTTGT	TTATATATTCTCTCTCTGCCC	1
LjT03HI3	TM0012	CT	20	144	154	GAGAGGAATGTCGTAGGAAG	GCTTCTTCTTCAATCTCTGC	1
LjT13004	TM0016	AT	16	186	220	TTTAGTGCCTTGTAGGTGAG	AATTCCTACATAAACTTCAGTG	1
LjT6I17	TM0017	AT	15	149	141	CGAATTTGATGGTATTGTATG	AGAATGCTCAAAATAACAATC	1
Lj17G08	TM0023	AT	12	185	179	CATAAGCACAACAATTCATAG	GTTGTGGTTCAAAGTTAGGG	1
L1T08B24	TM0027	CT	33	178	158	AGGATAATTACATTCCACCTC	TCTTGCAATATCTATATGACTGG	1
LjT11C13	TM0029	CT	15	152	148	CCTATATAACCTTATTCAAATTGG	ACGAAAACAAAACCCTGCTG	1
LjT05P21	TM0032	AT	17	168	160	CTTACCACTTAAGCTTGCTG	GTTTAATTTTCTCCCATTTC	1
LjT07E11	TM0033	AAT	18	189	159	AGTGTACTCGGATTAACACC	TTAAACATTTGAAATAGATGTAAC	1
LjT03K03	TM0036	ATC	9	171	180	GATGTGACGGTGAGTTATTG	AGAGAGAAGTGGAGCTTACG	1
LjT03J05	TM0050	GT	11	149	165	ATTTGTTGGGATACATTGAC	GCAGGTATCCATCAATTCTC	1
LjT09L22	TM0063	AT	14	163	155	AAATTGAAAAGTTGGGACGG	GCATGAGAGAATGAGACCTATAAG	1
LjT15K21	TM0088	AT	30	161	151	AGCTATAGTTCATAAAATCACCA	ATTACCACCATCTCTCTGGC	1
LjT04C07	TM0098	AAT	18	181	190	CTCCACCTTTAGAGGGTATG	AAAAGAGTGAGTTTGAAAGC	1
LjT26E16	TM0103	AT	23	192	182	TGACAAGAGCTTCATAAGAG	GATGAAGTACAGACACCGAC	1
LjT06Kll	TM0002	CT	15	159	179	AGCGATCTACATTCAAGAG	AGCGTTCTCTCAGTGTTG	1, 2[a]
LjT12K07	TM00ll[b]	AT	44	222	250	TGAACCGATATCCATATCTAT	TAATTTTGAAGTTTGGGGAC	1, 2[a]
LjT08D14	TM0018	ATG	11	153	147	GTTTGAGCAAGTTAGAGGTG	CGGATAAGAAAGGTAGAAGAG	2
LjT02F05	TM0020	CA	16	180	190	GCAGGCTGTGTTA AAGCATC	TTCTCATGACAGTCATCA AAC	2
LjT04I02	TM0021	CT	16	157	151	GCTTCATCTTTGTGATACTTAAGTAA	CTGTTC,TATCA AGCCACA AG	2
LjT15H09	TM0005	GT	16	159	171	ACAAAACACAAOAACTTTTGG	CACTACTCATTTACCiCGCAC	3
LjT11G09	TM0022	AT	24	168	148	CATTACCTAGTCTATGTTTTCC	TAAAGGTCCATTCATATrGC	3
LjTI0E18	TM0035	AACr	17	110	95	TACATAAATCAGCTCTGCATGCT	CCACTTCCACTGTGCTTCTG	3
LjT17K09	TM0049	AT	14	141	159	TGGGTTAGCTTACCTGTTTC	ATGTCCTGATCAAATCTTTTC	3
LjT13Ml4	TM0059	AT/CT	7/5	160	166	TCCTTCATTCATTCATAACC	TCTACTAAGAGAATCTAAAACTCG	3
LjT14GT02	TM0080	AT	14	137	131	AACAAAATACTAAACTATAGCAAAG	CGTCCCACAACTCTCTTTAC	3
LjT06A20	TM0030	AAT	18	142	121	GTCCAACACGGGAGAATGAG	CATAGAAACCTAAGCATGAGTC	4
LjT04F23	TM0044	ATO	9	139	147.	TCTCTATGATCACCrTTCTTA AG	TCCAACCTTTATCTTTATTAGC	4
LjT07K08	TM0046	CT	16	155	143	ATCTAACCAAAACGTGCTTC	TTCTTCTCCCTTTCTCTCTTGG	4
LjT14P20	TM0087	AAAT	7	141	129	AGCTGTCCTATGATCAGAACT	AAAAGGGTTCAAAATAGAATAG	4
LjT15N19	TM0097	AT	11	183	179	TTGTGTTTCTGATCTATGTAGC	CTTAACTTTAAAGTTGAA AGTTGG	4
LjT02M03	TM0034	AAT	12	183	192	TCACTATCTCCTTAGATCACAC	GGTTGGTTTGTATTGCGTGC	5
LjT01O22	TM0043	AT	12	164	154	AAGAAAAGTGAGTGTGTGCT	GGCCAATAATAAGATTGAGC	5
LjT05P01	TM0048	AT	12	162	146	TTA ATTACTATTCTCTCGACTGTTCT	CGTAAATAATAGCATTTCTCC	5
LjT08018	TM0052	CT	21	183	177	CTATATCACCTCTAGTCACTCTG	CCACATATGATGATCATTTTC	6
LjT14007	TM0013	AT	20	206	216	GTTGTTACACTCAATATGTCCC	CCATTATCATTATCATTATCATATC	6
LjT17C05	TM0014	AAC	10	159	153	AACCACAGACAAGATTATCG	TCATCATCTATAATCATTCAAC	6
LjT17H19	TM0041	AT	8	173	183	GCAGAGAAGAAACGGCTTCG	TCÕTTTGATCATATAATCACAACCÕ	6
LjT14B06	TM00045	CT	13	149	141	CTCTTCATGTTCTTTCAAGC	AGCACAATCACATACCTACC	6
LjT05E07	TM0055	AAG	7	147	153	TATAACCGATGCTCACACAC	GATTAACGAACACGCAGTAG	6
LJT03B03	TM0057	CT	14	138	132	CAAAGATAAATGCAGATGGC	CTTTCTATAAACAGTGAAACTGG	6
LjT31L24	TM0228	AT	14	134	144	TCTTTTGTTAATTCAAGTTGAC	AACTATTCCATTAATGTTTCTTC	6

[a] Markers on the region of translocation between Gifu B-129 and Miyakojima MG-20.
[b] Mapped as a Miyakojima MG-20 dominant marker.

Table 2. List of dCAPS markers (Sato et al. 2001)

Clone name	Marker name	Enzyme	Product size (bp)		Fw primer (5′ to 3′)	Rv primer C5′ to 3′)	Chr
			Miyakojima	Gifu			
LjT43N05	TM0178	*Hind*III	145	117 + 28	TCATTCACACTTCCTTTTATAAAGCT	AGGATTCGAACCCTGAGGAG	1
LjT16N13	TM0031	*Pst*I	122 + 27	149	AATGGAAGGATCCAAGCTCCAACTGCA	AAGTGAAGTTGTTTTATCTGAGG	1, 2[a]
LjT01K12	TM0058	*Hid*III	171	142 + 29	ACTAAGTTGCTTGTAACITATTGTCAAG	GTCCCATTATAATGCCTTCC	4
LjT08G20	TM0102	*Nde*I	134 + 29	163	CAGCCTGCAAAACATTAATTGGATACATA	AACTACAATGTCTCCAATGC	4
LjT09A24	TM0003	*Ssp*I	170	145 + 25	GAGGCTGAGGGCAGACAGAGAAATA	GCATCAATACTTGAGTTCCTTC	4
LjT15D01	TM0004	*Bsp*1407I	137	108 + 29	AAAAAGATTACATAAAATTGTTTTGTAC	ACATGAATGCTGTCCTGTCC	4
LjT17P02	TM0006	*Hha*I	188 + 29	217	CAACTTTTGCCCCTGTCTGAACAATAGC	AGCACTGCTCAATCAAAGCC	4
LjT10J15	TM0007	*Kpn*I	187	159 + 28	TCGCAATTTCGATTTTGAGCCGGGTGGT	TAAATAGCGGCCGAAATAGC	4
LjT08D16	TM0025	*Hinf*I	128 + 29	157	GAAGTTAGTTAGTTAGTTAGTTAGTGACr	TGACAGCAAAGAAAGCATCG	4
LjT09I17	TM0026	*Hind*III	223	194 + 29	TTTTTATAAATATTACCATATTTAAAAGC	GGAGAGTTTGTTCCCAAGAC	4
LjT10L16	TM0042	*Dra*I	111 + 27	138	TAAGAGTTGGGACATATGGATTGTTTAA	GTAAATTTATGTGTGTATATTGCC	4
LjT05B20	TM0019	SphI	137 + 25	162	TCTGCACCCTTCTCCCAGCATCGCA	AAGAGCAGTGATTATCATCG	5
Lj16K17	TM0037	*Acl*I	126 + 25	154	GGTCATATTTTCAATTGTAATTTAACGT	CAAGCAAACTCCATAACCTG	6

[a] Markers on the region of translocation between Gifu B-129 and Miyakojima MG-20.

Table 3. Descriptive data of Gifu B-129 and Miyakojima MG-20 map (Hayashi et al. 2001)

	Gifu B-129	Miyakojima MG-20
Number of markers	287	277
Number of genetically separated loci	206	201
Total length of linkage groups, cM	487.3	481.6
Average distance between markers, cM	1.70	1.74
Average between genetically separated loci, cM	2.38	2.41

(Pedrosa et al. 2002; Sandal et al. 2002). We divided the TM markers on the connected linkage group and those in the excluded marker group into two groups, putative linkage groups 1 and 2. Then we constructed framework of the linkage groups 1 and 2 for the two accessions independently, using the separated TM markers and other codominant markers (some of the AFLP markers and other PCR-based markers). The common framework of linkage groups 3 to 6 was also developed with codominant markers. Then dominant markers (mostly AFLP markers) were integrated in each linkage group, and two separate linkage maps corresponding to Gifu B-129 and Miyakojima MG-20 were obtained. The order of TM markers in linkage groups 1 and 2 thus obtained was consistent with that in the *L. filicaulis* × *L. japonicus* Gifu B-129 map (Sandal et al. 2002), and that in ideograms of Miyakojima MG-20 and Gifu B-129 (Hayashi et al. 2001). Each linkage map that consisted of 6 linkage groups contained 113 codominant markers that were used for construction of framework map. The total number of markers was 287 and 278 in the Gifu B-129 and the Miyakojima MG-20 linkage maps, respectively (Fig. 4, Table 3). The total lengths of the linkage maps are 487.3 cM for Gifu B-129 and 481.6 cM for Miyakojima MG-20; the average distance between markers (and genetically separated loci in parentheses) was 1.70 cM (2.38 cM) for Gifu B-129 and 1.74 cM (2.41 cM) for Miyakojima MG-20 (Table 3). Markers used for map construction are shown in Table 4. The map and marker information are available over the Internet at http://www.kazusa.or.jp/lotus/.

On the linkage group 1, characteristic clustering of markers was observed in the region corresponding to the short arm of chromosome 1, where 20 markers of Gifu B-129 and 20 markers of Miyakojima MG-20 were mapped at the identical position (Fig. 4). This feature suggests reduced rate of meiotic recombination and can in part be explained by the presence of the small-translocated segment in this region. At present, however, we cannot exclude that an additional rearrangement, such as an inversion, contributes to this suppression. Remarkable clustering was also observed on the top region of linkage group 2, and relatively long gaps were detected in linkage groups 2 and 5 (Fig. 4). A region of distorted segregation was identified on the bottom part in linkage group 5 (Fig. 4). The segregation ratios of all the markers mapped on this region were significantly different from the theoretically deduced ones. These markers showed high genotypic frequencies for maternal Gifu B-129, suggesting that loci involved in gametogenesis and sterility might be located in this region.

Table 4. Number of markers along linkage group (Hayashi et al. 2001)

Linkage group	AFLP		SSR	dCAS	Other PCR-based marker	Total
	EcoR I/*Mse* I	*Hin* d III/*Taq* I				
1 Gifu	38 (13)	9 (1)	18 (18)	4 (4)	1 (1)	70 (37)
Miyakojima	35 (13)	11 (1)	18 (18)	1 (1)	1 (1)	66 (34)
2 Gifu	29 (8)	12	4 (4)	0	1	46 (12)
Miyakojima	23 (8)	18	5 (4)	3 (3)	0	49 (15)
3 Gifu	28 (10)	10	6 (6)	0	0	44 (16)
Miyakojima	31 (10)	10	6 (6)	0	0	47 (16)
4 Gifu	31 (3)	8	5 (5)	8 (8)	1 (1)	53 (17)
Miyakojima	18 (3)	3	5 (5)	8 (8)	1 (1)	35 (17)
5 Gifu	24 (9)	6 (1)	4 (4)	1 (1)	1 (1)	36 (16)
Miyakojima	25 (9)	8 (1)	4 (4)	1 (1)	1 (1)	39 (16)
6 Gifu	16 (7)	13	7 (7)	1 (1)	1	38 (15)
Miyakojima	21 (7)	12	7 (7)	1 (1)	0	41 (15)

Numbers in parentheses indicate number of codominant markers.

Thus we succeeded in constructing a linkage map of *L. japonicus* based on an intraspecific cross. This map will provide the basis for molecular genetic analysis of the traits in *L. japonicus*, especially nitrogen-fixing symbiosis, mycorrhization, host-pathogen interaction, secondary metabolism, seed components, and morphogenesis. If a trait is identified as being controlled by a single gene, the gene can be mapped on the linkage map using the codominant markers first, and bulked segregant analysis with HEGS (Kawasaki and Murakami 2000; Kawaguchi et al. 2001) using AFLP markers around the mapped locus could efficiently locate the responsible gene. By analysis of genome synteny with other leguminous crops, this map will provide clues for the identification and isolation of agronomically important genes.

3 Conclusion and Perspectives

We have developed a genetic linkage map of the model legume *Lotus japonicus* (Regel) Larsen based on amplified fragment length polymorphism (AFLP), simple sequence repeat polymorphism (SSRP) and derived cleaved amplified polymorphic sequence (dCAPS). The F_2 mapping population was derived from a cross between two *L. japonicus* accessions, Gifu B-129 and Miyakojima MG-20. Because a reciprocal translocation has been found to be responsible for the cytological differences between chromosome 1 and 2 of the two accessions, and the borders of the translocation have been identified, the markers from the main translocated region were located on linkage groups 1 and 2 of the two accessions, respectively. The framework of linkage groups 1 and 2 for the two accessions was constructed independently with these markers and other codominant markers. The common framework of linkage groups 3 to 6 was also developed with codominant markers. Then dominant markers were integrated separately in each linkage group of the accessions. The resulting linkage group correspond to the six pairs of chromosomes of *L. japonicus* and consist of 287 markers with 487.3 cM length in Gifu B-129 and 277 markers with 481.6 cM length in Miyakojima MG-20.

In order to make a saturated high density map of *Lotus japonicus*, we will integrate more SSR and cDNA markers onto the current map. As SSR markers are generally codominant and robust, they make it easy to map genes for qualitative traits. Recently, there has been a dramatic increase in the number of cDNAs in *Lotus japonicus* whose partial sequences have been determined, so we will be able to develop many cDNA markers based on PCR. These markers could be used for synteny analyses between *Lotus japonicus* and leguminous crops, which will help the identification and isolation of agronomically important genes. Development of recombinant inbred lines (RILs) derived from a cross between Gifu B-129 and Miyakojima MG-20, which provides a permanent mapping population, is being finished at Kazusa DNA Research Institute. RILs are ideal for mapping of quantitative trait loci (QTL), and a new linkage map using RILs could be constructed based on the DNA markers on the F_2 map.

Acknowledgements. This study was partly supported by a grant from Program for Promotion of Basic Research Activities for Innovative Biosciences, Special Coordination Funds for Promoting Science and Technology from Ministry of Education, Culture, Sports, Science, and Technology, the Japanese Government and the Kazusa DNA Research Foundation.

References

Alonso-Blanco C, Peeters AJ, Koornneef M, Lister C, Dean C, van den Bosch N, Pot J, Kuiper MTR (1998) Development of an AFLP based linkage map of *Ler*, Col and Cvi *Arabidopsis thaliana* ecotypes and construction of a *Ler*/Cvi recombinant inbred line population. Plant J 14:259–271

Becker J, Vos P, Kuiper M, Salamini F, Heun M (1995) Combined mapping of AFLP and RFLP markers in barley. Mol Gen Genet 249:65–73

Gordon D, Abajian C, Green P (1998) Consed: a graphical tool for sequence finishing. Genome Res 8:195–202

Handberg K, Stougaard J (1992) *Lotus japonicus*, an autogamous, diploid legume species for classical and molecular genetics. Plant J 2:487–496

Hayashi M, Miyahara A, Sato S, Kato T, Yoshikawa M, Taketa M, Hayashi M, Pedrosa A, Onda R, Imaizumi-Anraku H, Bachmair A, Sandal N, Stougaard J, Murooka Y, Tabata S, Kawaski S, Kawaguchi M, Harada K (2001) Construction of a genetic linkage map of the model legume *Lotus japonicus* using an intraspecific F2 population. DNA Res 8:301–310

Ito M, Miyamoto J, Mori Y, Fujimoto S, Uchiumi T, Abe M, Suzuki A, Tabata S, Fukui K (2000) Genome and chromosome dimensions of *Lotus japonicus*. J Plant Res 113:435–442

Jiang Q, Gresshoff PM (1997) Classical and molecular genetics of the model legume *Lotus japonicus*. Mol Plant Microbe Interact 10:59–68

Kawaguchi M, Motomura T, Imaizumi-Anraku H, Akao S, Kawasaki S (2001) Providing the basis for genomics in *Lotus japonicus*: the accessions Miyakojima and Gifu are appropriate crossing partners for genetic analyses. Mol Genet Genomics 266:157–166

Kawasaki S, Murakami Y (2000) Genome analysis of *Lotus japonicus*. J Plant Res 113:497–506
Keim P, Schupp JM, Travis SE, Clayton K, Zhu T, Shi L, Ferreia A, Webb DM (1997) A high-density soybean genetic map based on AFLP markers. Crop Sci 37:537–543
Lander ES, Green P, Abrahamson J, Barlow A, Daly MJ, Lincoln SE, Newberg L (1987) MAPMAKER: an interactive computer package for constructing primary genetic linkage maps of experimental and natural populations. Genomics 1:174–181
Lincoln SE, Daly MJ, Lander ES (1992) Construction genetic maps with MAPMAKER/EXP 3.0. In: Whitehead Institute Technical Report, 3rd edn. Whitehead Institute, Cambridge, MA
Mackill DJ, Zhang Z, Redona ED, Colowit PM (1996) Level of polymorphism and genetic mapping of AFLP markers in rice. Genome 39:969–977
Neff MM, Neff JD, Chory J, Pepper AE (1998) dCAPS, a simple technique for the genetic analysis of single nucleotide polymorphisms: experimental applications in *Arabidopsis thaliana* genetics. Plant J 114:387–392
Pedrosa, A, Sandal N, Stougaard J, Schweizer D, Bachmair A (2002) Chromosome map of the model legume *Lotus japonicus*. Genetics 161:1661–1672
Qi X, Stam P, Lindhout P (1998) Use of locus-specific AFLP markers to construct a high density molecular map in barley. Theor Appl Genet 96:376–384
Sandal N, Krusell L, Radutoiu S, Olbryt M, Pedrosa A, Stracke S, Sato S, Kato T, Tabata S, Parniske M, Bachmair A, Ketelsen T, Stougaard J (2002) A genetic linkage map of the model legume *Lotus japonicus* and strategies for fast mapping of new loci. Genetics 161:1673–1683
Sato S, Kaneko T, Nakamura Y, Asamizu E, Kato T, Tabata S (2001) Structural analysis of a *Lotus japonicus* Genome. I. Sequence features and mapping of fifty-six TAC clones which cover the 5.4 Mbp regions of the genome. DNA Res 8:311–318
Schauser L, Roussis A, Stiller J, Stougaard J (1999) A plant regulator controlling development of symbiotic root nodules. Nature 402:191–195
Stiller J, Martirani L, Tuppale S, Chian RJ, Chiurazzi M, Gresshoff PM (1997) High frequency transformation and regeneration of transgenic plant in the model legume *Lotus japonicus*. J Exp Bot 48:1357–1365
Vos P, Hogers R, Bleeker M, Reijans M, van de Lee T, Hornes M, Fritjers A, Pot J, Kuiper M, Zabeau M (1995) AFLP: a new technique for DNA fingerprinting. Nucleic Acids Res 23: 4407–4414

12 Construction of High-Density Map, Genome Library, and Saturation Mapping of Nodulation Genes

S. Kawasaki, Y. Murakami, H. Imaizumi-Anraku, A. Shimizu, and I. Mikami

1 Introduction

Legumes have a unique characteristic in that they fix atmospheric nitrogen by symbiosis with *Rhizobium*, vital in maintaining sustainable world agriculture. Legumes are also important crops in their own right, with a high content of seed proteins with a good balance of amino acids among them (Sarwar et al. 1978). *Lotus japonicus* as a model legume is attracting the attention of researchers because of its high transformation susceptibility (Thykajaer et al. 1995), and a relatively small genome size (475 MB, Kawasaki and Murakami 2000; Murakami and Kawasaki, submitted). From the standard accession Gifu (G129) several mutant libraries have been constructed for use in functional genomics (Szczyglowski et al. 1998; Kawaguchi et al. 2002). The most interesting topic in its regard is no doubt the genetic dissection of the nodulation process, including that of the related mycorrhizal development. If essential components of the process can be clarified, transferring the ability of the nitrogen fixation to other important crops, including the graminaceous family, would be a great blessing for world agriculture and for environment protection. However, the current status of infrastructures for genomics in *Lotus japonicus* is still far behind those of *Arabidopsis* and rice, whose genomes have been sequenced completely (The Arabidopsis Genome Initiative 2000; Goff et al. 2002; Yu et al. 2002). The first linkage map of *Lotus japonicus* has only recently been published (Hayashi et al. 2001) since the appropriate crossing partner to Gifu for genetic analyses, accession Miyakojima, has been reported only very recently (Kawaguchi et al. 2001). In addition, a high quality genome library competent for physical analysis of the full genome has also been recently constructed (Kawasaki and Murakami 2000; Murakami and Kawasaki, submitted), as were some other genome libraries of binary vectors (Men et al. 2001; Sato et al. 2001).

Our high quality genome library of *Lotus japonicus*, with an average insert size of 138 kb and covering eight genome equivalents, has been constructed recently (Murakami and Kawasaki, submitted), and arrangement into a physical map covering the total genome is being planned. For that purpose, for a

Plant Physiology Department, National Institute of Agrobiological Sciences, Kannon-dai 2-1-2, Tsukuba, Ibaraki 305-8602 Japan

Biotechnology in Agriculture and Forestry, Vol. 52
Nagata/Tabata (Eds.) Brassicas and Legumes

high-density map with more than 2,000 markers and a system for efficiently correlating the markers and the library clones that are also needed, we have developed a highly efficient mapping system, HEGS (high efficiency genome scanning: Shimizu et al., submitted). This prototype was used for analyzing the phylogeny of Japanese accessions (Kawaguchi et al. 2001) and for constructing the first genome map of *Lotus japonicus* derived from the cross between accessions Gifu (B129) and Miyakojima (MG20) (Hayashi et al. 2001). However, this was far from providing a basis for physical mapping and was also not sufficient for effective positional cloning. The HEGS can provide a still higher-density map or markers closer to target genes, as is much needed for many purposes. In the cases of application to the genomes of rice (Shimizu et al., submitted) and barley (Hori et al., submitted), maps with more than 1,000 markers were constructed in 6 months, each by a single researcher with inexpensive equipment. Also, for *Lotus japonicus*, a map with about 1,000 markers was constructed (Wang and Kawasaki, in preparation) and an increase to 2,000 markers within a year is planned. As for preparing the close markers to the target genes in positional cloning, HEGS is very powerful, and practically unlimited numbers of markers can be provided, as needed, even for the recombination-poor and repetitive-sequences crammed centromeric regions, or for plants with much larger genome sizes, such as peas (4.0–4.4 GB; Arumuganathan and Earle 1991). Saturating maps with about 40–50 markers were constructed in the region around 1–2 cM distant from the earliest steps of nodulation-controlling genes *LjSym 70* and *72* by Murakami et al. (submitted) and Imaizumi et al. (submitted), respectively, and also in 6 months. The very high density of markers in these regions seems to be due to their location close to or within centromeres. With other systems it might have been more difficult to provide markers to cover the centromere regions with the density to enable physical map construction. This system is also useful for efficient phylogeny analysis (Kawaguchi et al. 2001). The HEGS system is flexible and can be combined with several marker systems, such as AFLP (amplified fragment length polymorphism), SSRP (simple sequence repeats polymorphism), RAPD (random amplified polymorphic DNAs), CAPS, heteroduplex and so on, and is useful for efficient high-resolution QTL analyses. By correlating the BAC clones to the corresponding HEGS map markers, we plan to apply this system to the arrangement of the total genome library into a physical map.

In this report, we describe the methodology for constructing the bases of genome analyses, namely, the genome library, and the efficient marker providing system HEGS. Especially the latter system and its applications to *Lotus japonicus* and some other crops will be reported, with some examples. Our descriptions of applications to the marker tagging of genes for systematic positional cloning and for constructing high-density marker maps may be informative for those intending to pursue functional genomics. Finally, we will review our plan to integrate the genome library and the marker systems as a physical map covering the total genome. RLGS (restriction landmark genome scanning), developed by Hatada et al. (1991), has inspired us greatly regarding more efficient genome scanning that does not use isotopes and that can be carried out in a day. HEGS is an answer to the need to quickly process more data while utilizing the possibility of PCR to the fullest.

2 Construction of a High-Quality BAC Library of *Lotus japonicus*

In combination with a high-resolution mapping system, a genome library of high quality is the basis of genome analysis. Accordingly, we have developed a highly efficient mapping system, as described in the following section, and a long insert BAC genome library of such a quality as to permit arrangement into a physical map covering the total genome as a base for genome analysis of *Lotus japonicus*. This map can also be used for systematic positional cloning of regulative genes, for example, of the nodulation process. In the past, we experienced difficulties in constructing genome libraries of wild plants (wild buckwheat: Nagano et al.; wild mungbean: Kaga et al.; rice blast fungus: Nishimura et al. 1998) compared with those of cultivated crops such as rice (Nakamura et al. 1997). Our problems were probably due to the presence of some inhibitory materials influencing pure DNA preparation and to the high activity of persistent nucleases. To remove these inhibitory materials, we believe that the protoplast method is effective in separating the DNA source from the heavy polysaccharides that should be excluded from preparation of plugs of long-genome DNAs. The basic procedure used to construct the rice BAC library, as described by Nakamura et al. (1997), is valid even now. Although wild plants are also recalcitrant for preparing protoplasts, the use of a high concentration (1%) of pectolyase Y23 (Kyowa Kasei, Osaka, Japan) could overcome this low digestibility of wild cell walls. As for the materials of protoplasts, young soft leaves grown in shaded sites are preferable. Taking these precautions, we were able to construct a genome library of 138 kB of average insert size (Fig. 1), consisting of 30,000 clones (Murakami and Kawasaki, submitted). The vector pBACLAC, developed by Asakawa (Nakamura et al. 1997), was superior in its clearer blue color development than the other BAC vectors. We preferred this rather simple vector to those complicated ones with transformability, such as BIBAC, because the former can accommodate larger inserts than the latter (94 kb with PCLD04541, Men et al. 2001; 100 kb with TAC, Sato et al. 2001), which are longer by about 20 kb than the simple BAC vectors such as pBACLAC. We also found that the BAC vector of the single copy is apparently more stable than the multicopied vectors such as cosmids and PCLD04541. We also believe that for large-scale genome analysis in the >100 kb range, which corresponds to about >0.3 cM in *Lotus japonicus*, the genetic analysis strategy is simpler and more reliable than the long-insert complementation strategy, which does not always guarantee efficient transformation. Moreover, if one uses the HEGS system, no paucity may be in of close markers around the target genes. The longer the insert size of the library, the more efficient is the process of editing the genome library into physical contigs. In order to cover 95% of the

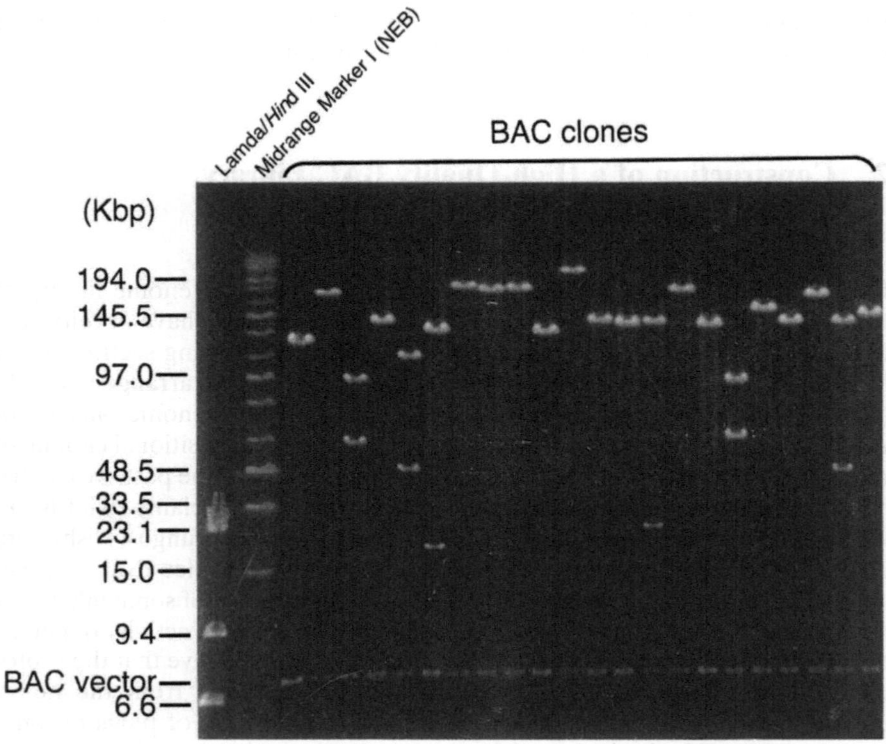

Fig. 1. CHEF electrophoresis pattern of the *Lotus japonicus* BAC library. Average insert size 138 kb. 30,000 clones covers about 8 genome equivalents of the 475 MB genome (by Y. Murakami). Random clones were digested by *Not* I and CHEF electrophoresed at 10-s pulse time. Note that the paucity of the cutting sites of *Not* I in the *Lotus* genome, the enzyme recognizes 8 bases sequence of GC only, and average fragment length of 64 kb is expected for the genome of 50% of GC content.

Lotus genome, the 140- and 100-kb insert libraries need about 10,000 and 14,000 clones, respectively. To remove contaminating short fragments, which have much higher transformation efficiency than the longer fragments, double electrophoretic exclusion from the sliced plug has proved effective.

The genome size of the *Lotus japonicus* was estimated to be around 475 MB for accession Gifu (B129), and 490 MB for accession Miyakojima (MG-20) (Kawasaki and Murakami 2000, Murakami and Kawasaki, submitted) by cell sorter analysis with DAPI as a the fluorescence reagent and using *Arabidopsis thaliana* Columbia (125 MB) as the size standard. Although the GC content affects the specific signal intensity, its similarity between the sample *Lotus* (39.2%) and standard *Arabidopsis* (35%; The Arabidopsis Genome Initiative 2000) genomes may have alleviated its effect. The GC contents of *Lotus japonicus* could be calculated from the frequency of the *Not* I sites in the BAC library. This genome size means that our BAC library covers about eight genome equivalents of *Lotus japonicus*. If the clones are distributed randomly, the library accounts for 99.98% of the whole genome and will provide a reasonable framework for constructing an assorted genome library covering the total genome. This BAC library also provides a good foundation for the positional cloning of the *Lotus* genes, especially for the nodulation-controlling genes. The rich mutant libraries of *Lotus japonicus*, by providing chemical mutagens (Szczyglowski et al. 1998; Senoo et al. 2000; Kawaguchi et al. 2002), are expected to provide materials for the application of various physiological procedures. The libraries developed by tagging with transposons or T-DNA (Thykajaer et al. 1995; Schauser et al. 1998) would provide the materials for positional cloning as well, because most of the mutants created by these procedures are thought to be caused by culture mutations and certain footprints (Stougaard, pers. comm.).

Membranes or specific clones from this BAC library are available upon request (kawasa@nias.affrc.go.jp).

3 HEGS and Its Application for *Lotus japonicus* Genome Analysis

3.1 HEGS: High-Efficiency Genome Scanning

The rather small genome size of *Lotus japonicus*, which is comparable to that of rice, makes the positional cloning of genes easy relative to the crop legumes. However, this task can be done only with the aid of a convenient marker system. As *Lotus japonicus* is still far from being furnished with maps for genome analysis, we had to develop a system that can provide needed markers for many purposes.

An important characteristic of the HEGS system is its rather thick (1-mm) multilane polyacrylamide gel electrophoresis system (see the Protocols, Sect. 6). This denaturing gel makes handling easy for post-electrophoresis staining, with silver or any fluorescence dyes. The HEGS system is very flexible, while the standard system may consist of a discontinuous non-denaturing gel of 1-mm thickness with the stacking gel to concentrate samples. As for the polymorphism detection system, AFLP (Vos et al. 1995) may be the most convenient, while other systems like SSRP, RAPD, heteroduplex, or systems for detecting SNPs (single nucleotide polymorphism), PCR-SSCP (single-strand conformational polymorphism; Suzuki et al. 1990) and DGGE (denaturing gradient gel electrophoresis; Sheffield et al. 1989), can be combined with this HEGS system, and three to five bands can be simultaneously analyzed using the denaturing gel system, others than RAPD.

For recording fluorescence images of HEGS, we normally used a fluorescence scanner (FluoroImager 575, Amersham Biosciences, Piscataway, USA) with a resolution of 0.1 mm for 25 × 20 cm gel with a 16-bit depth of gradation. This means 5 M pixels for the entire gel. Recently, as digital cameras of this resolution, with a depth of gradation of ≧12 bits, become available on the market at prices of less than one tenth of the scanners, they can be conveniently used for image recording. The important specification is not the picture size or resolution, but the depth of gradation of ≧12 bits, and conventional 8-bit cameras are insufficient. If this condition is fulfilled, the digital camera is more efficient in recording images than scanners. When one uses UV illuminators as the excitation light, an appropriate IR cut filter may be another prerequisite for utilizing CCD cameras, which have a higher sensitivity for longer wavelength light than the naked eye.

Data-processing is often easier with the HEGS system than with AFLP fluorescence sequencer systems because the signals are clearer and higher in S/N ratio. Software for transporting the data into the Excel spreadsheet was developed, which further simplified data processing (HEGS/F2, Shimidzu et al., submitted). The superiority of HEGS in efficiency of processing data is apparent upon comparing the systems, apart from the more obvious advantage of cost. This becomes apparent when one rechecks the original data in a later phase of the study.

3.2 Saturation Tagging of Target Genes

An example of applying HEGS is the marker tagging of target genes in positional cloning. Using bulked segregant analysis (Michelmore et al. 1991), and allotting four lanes for a primer pair, two lanes for parents and another two lanes for bulked samples of about ten each of dominant or recessive homos, the HEGS/AFLP system can provide dozens of linked markers to the target genes (An example of application to a disease resistance gene is seen in Fig. 2.) This procedure was used for saturation tagging of nodule development-controlling genes *LjSym 70* (nod–, myc+) and *LjSym 72* (nod–, myc–). (For mutant description see Kawaguchi 2002.) In *LjSym70*, 20 runs of electrophoresis were done with 8 gels to cover all the combinations of the primer pairs with +3 selective nucleotides; 64 primer pairs for each restriction site of 6- and 4-base cutters, *Eco* RI, and *Mse* I; making $64 \times 64 = 4096$ combinations of primer pairs in total. As the average number of polymorphic bands generated per primer pair was about 2 or 3 out of about 50 bands, this is equivalent to the screening of about 8,000–12,000 markers to find those close to the target gene. By increasing the F_2 population of recessive mutant-typed homozygotes to 814 for *LjSym70*, 42 bands were found within 0.25–1.0 cM to the target gene. By further increasing the F_2 population, these markers can be resolved into separate loci. In the case of *LjSym 72*, from the 2,048 primer pair combinations, 54 markers were selected with 260 F_2 recessive homozygotes in the region of 0.4–1.8 cM from the target (Fig. 3). As both *LjSym 70* and *72* are located in the marker-dense, probably centromeric regions of chromosomes 1 and 3 (Hayashi et al. 2001), respectively, the marker density in these regions is much higher (34–24 markers per cM) than the average (4–8 markers/cM). Because the centromeric regions are low in recombination frequency, and stuffed with a high density of repetitive sequences, only this high density of markers will make it possible to construct a physical map in the target regions. As the markers' density is higher than the resolution available in the conventional size of the F_2 population, it may be called as "saturation tagging".

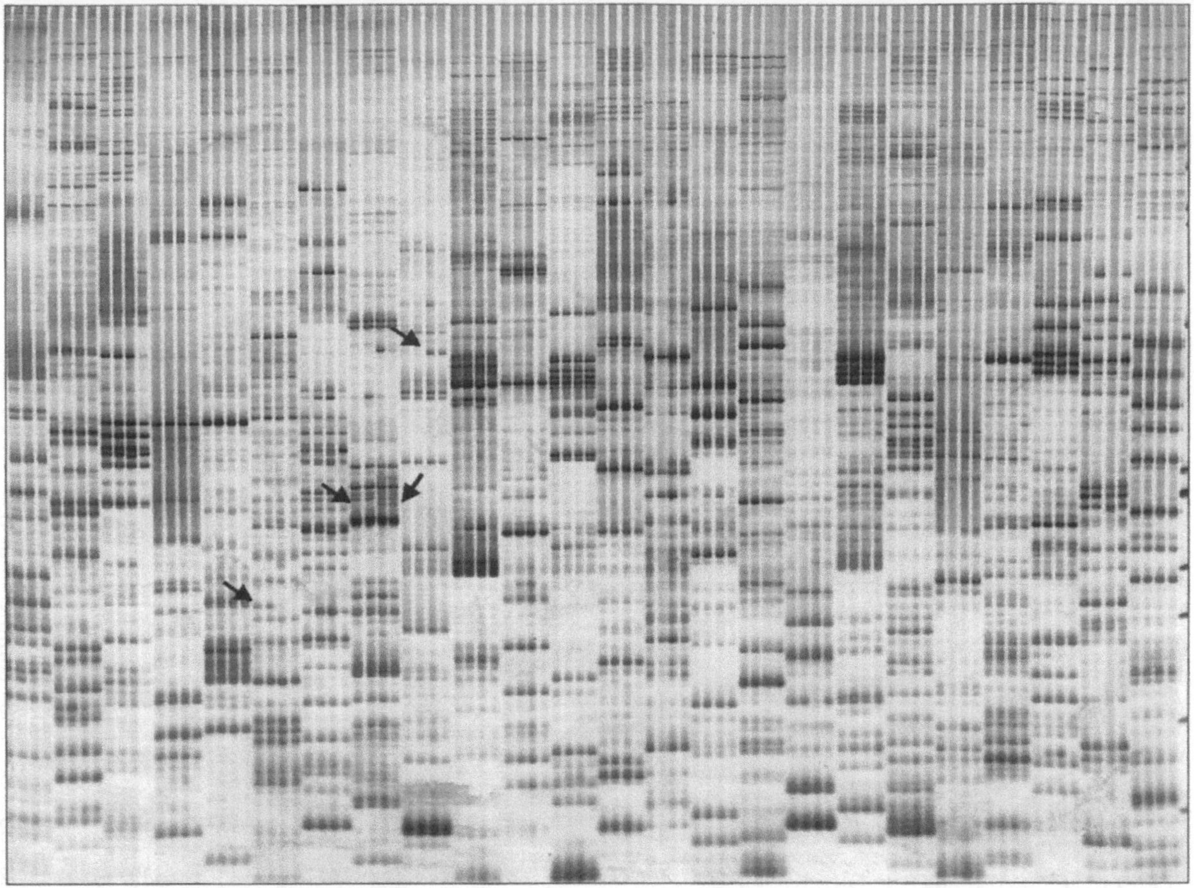

Fig. 2. An example of HEGS/AFLP bulked segregant analysis for the screening of the blast resistance gene *Pi-ta*^{2} linked HEGS/AFLP markers of rice (cv. Pi No4 with *Pi-ta*^{2}×cv. Nipponbare: susceptible). 25 primer pairs/gel × 8 gels = 200 primer pairs can be screened in a day. About 50 bands are detected by the fluorescence staining with VistraGreen (Amersham-Pharmacia) and the candidates of polymorphic bands linked to the target gene are indicated by *arrows* (by C. Wang)

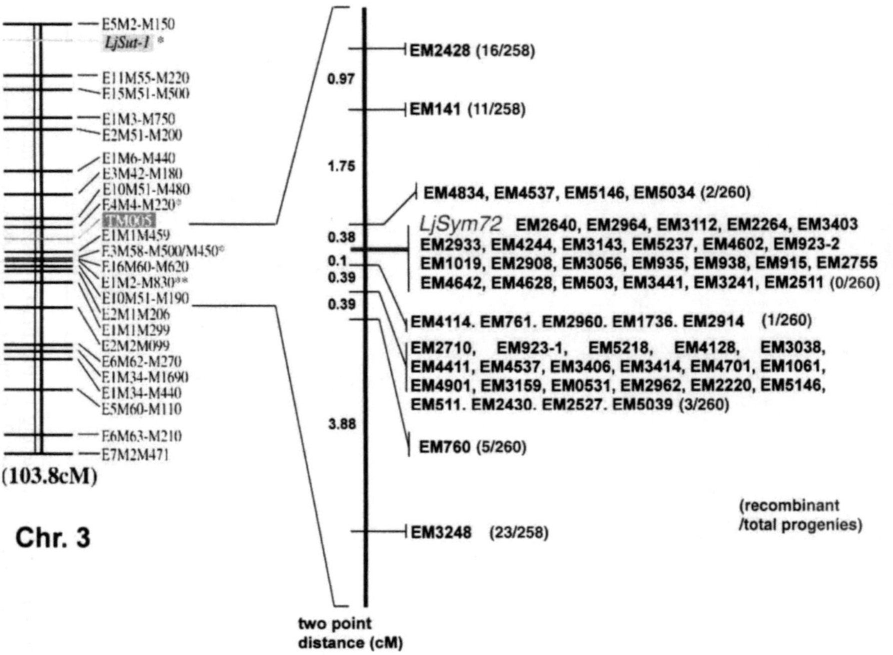

Fig. 3. Saturation mapping of *LjSym72* with the markers in a 2-cM region around the gene. Some 54 markers were mapped from about 6,000 polymorphic band candidates derived from 2,048 primer-pair combinations with the 260 recessive homo F_2 population. The higher marker density (about 24 markers/cM) than the average (4–8 markers/cM) may be due to the locus of the gene in the centromeric region (by H. Imaizumi-Anraku). The localization of the marker-dense region around *Sym 72* in chromosome 3 is indicated in correlation with a tentative map by K. Harada et al.

AFLP electrophoreses are done conventionally with large sequencing gels or fluorescence sequencers. However, the efficiency for detecting clear bands is higher in HEGS, making a scan of all 4,096 primer pairs and screening of the close markers possible in about 2 months by a single researcher. As the primers are not needed for fluorescence or radioactive labeling, and the apparatus is not expensive, the total cost of HEGS may be only some thousand dollars. As each lab worker can use their own apparatus, several thousands of samples can be electrophoresed every day in a laboratory. A PCR amplifier, ABI 9600 dual 384 (Applied Biosystems, Foster City, USA), can process $768 \times 3 = 2304$ samples a day.

3.3 Phylogeny Analysis of Accessions of *Lotus japonicus* and *Lotus filicaulis*

As a preliminary application of the HEGS/AFLP with silver staining, phylogeny analysis of 15 Japanese accessions of *Lotus japonicus* and an accession of *Lotus filicaulis* was done with 16 primer pairs (Fig. 4). Silver staining

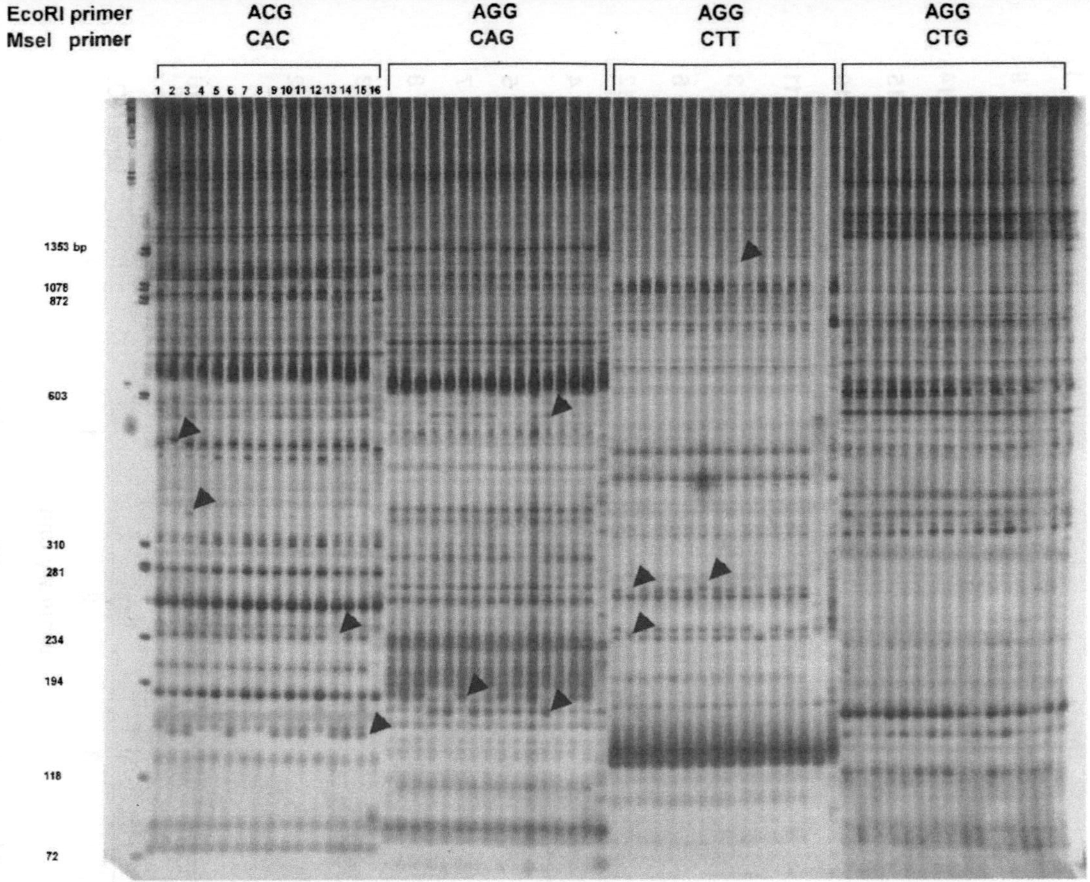

Fig. 4. Phylogeny analysis of 15 Japanese *Lotus japonicus* accessions and *L. filicaulis* with HEGS/AFLP – silver staining (by T. Motomura, in Kawaguchi et al. 2001). *Arrowheads* indicate the polymorphic bands that were summarized into a half matrix to calculate the dendrogram (Fig. 5). A small number of samples can be processed with silver staining without special apparatus. *Lane numbers 1–16* indicate the accessions as indicated in Fig. 5

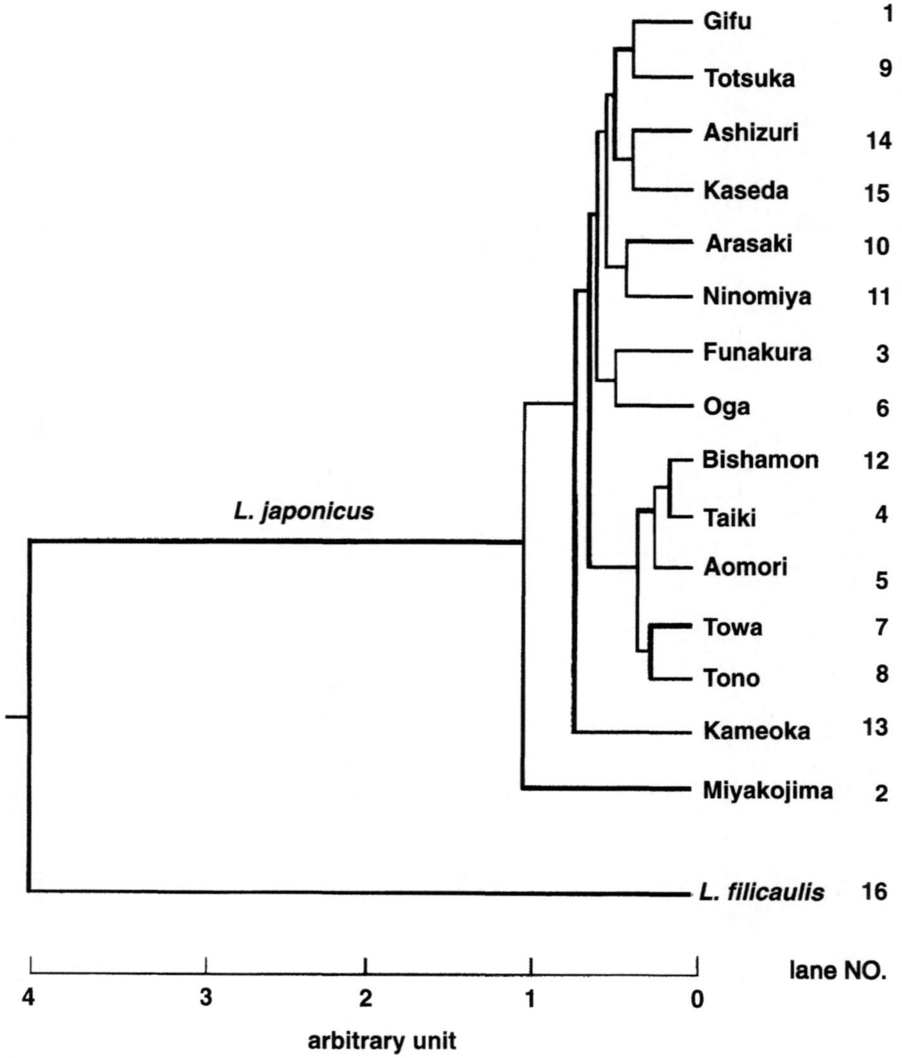

Fig. 5. Dendrogram of the *Lotus japonicus* accessions and a line of *Lotus filicaulis*. The Japanese accessions are named after their sampled locations. Miyakojima was found to be especially divergent from the other Japanese accessions (Kawaguchi et al. 2001). The *numbers* on the accessions indicate the corresponding lanes in Fig. 4

was sufficient and needed no special apparatus other than the gel drier for recording the small number of gels run in the analysis. The half-matrix table of polymorphic bands between the accessions was analyzed with the NEIGHBOR program (UPGM method) of PHYLIP 3.5c (http://evolution.genetics.washington.edu/phylip.html). The deduced dendrogram (Fig. 5) revealed that there are two major subgroups of Japanese accessions, one growing in northeastern and the other in southwestern Japan, and that Miyakojima is the most genetically diverged among the Japanese accessions (Kawaguchi et al. 2001). This analysis could be done in 4 days using HEGS/AFLP electrophoresis. The accessions Miyakojima and Gifu were found to be the best partners for genetic analysis of the Gifu mutant libraries because of their high fertility and smooth and non-distorted recombination between chromosomes. Although the polymorphic rate of 4–6% with HEGS/AFLP is rather low compared to that of more than 40% of polymorphism between *Lotus japonicus* and *Lotus filicaulis*, the high efficiency of HEGS made this low rate sufficiently practical for genetic analyses.

Although this analysis was done for only 16 samples, the high efficiency of HEGS can be verified by using more samples. Analysis of about 100 samples can be done in a week or so with the fluorescence staining system using the large gel.

3.4 Rapid Construction of High-Density Maps and Their Application to QTL Analyses

The high efficiency of the HEGS system made it possible for a single researcher to construct a high-density map from the siblings of an appropriate crossing in a year. From 96 F_9 RILs of Gimbozu (*japonica*)×Kasalath (*indica*), the high-density maps of rice with 956 AFLP and 64 SSRP markers was first constructed with this system by a single researcher in 6 months (Shimizu et al., submitted; Fig. 6), and more than 1,000 RGA markers are being added. Similarly, a map of barley with 1,000 markers was constructed in 6 months by a graduate student (Hori et al., accepted). The presence of an appropriate population of RIL (recombinant inbred line) makes map construction very easy than the case of F_2 population, because HEGS/AFLP mapping consists of high proportion of dominant markers.

A map of *Lotus japonicus* from the cross of Gifu and Miyakojima was first developed by Harada's group (Hayashi et al. 2001) with about 220 codominant AFLP markers; in total, about 280 markers were counted for each accession. However, since this marker density was apparently not sufficient for construction of a physical map, we have now developed a high-density map with more than 1,200 markers, constructed by a researcher in 6 months, and will have about 1,000 more markers of cDNAs and RGAs (resistance gene analogue; Wang et al., in prep.). This will serve as the framework of the physical map, and will help in the efficient cloning of genes.

Probably, the most effective use of the HEGS high-density map would be the efficient QTL analysis of quantitative traits. If the map should be derived from an RIL, all the quantitative traits between the parents can be mapped with unprecedented resolution. As a considerable portion of the markers can be utilized for the other crossing populations, most of the QTL analyses could be done on the HEGS system with high efficiency. For instance, rice *indica/japonica* 233 markers could be processed from 82 primer pairs,

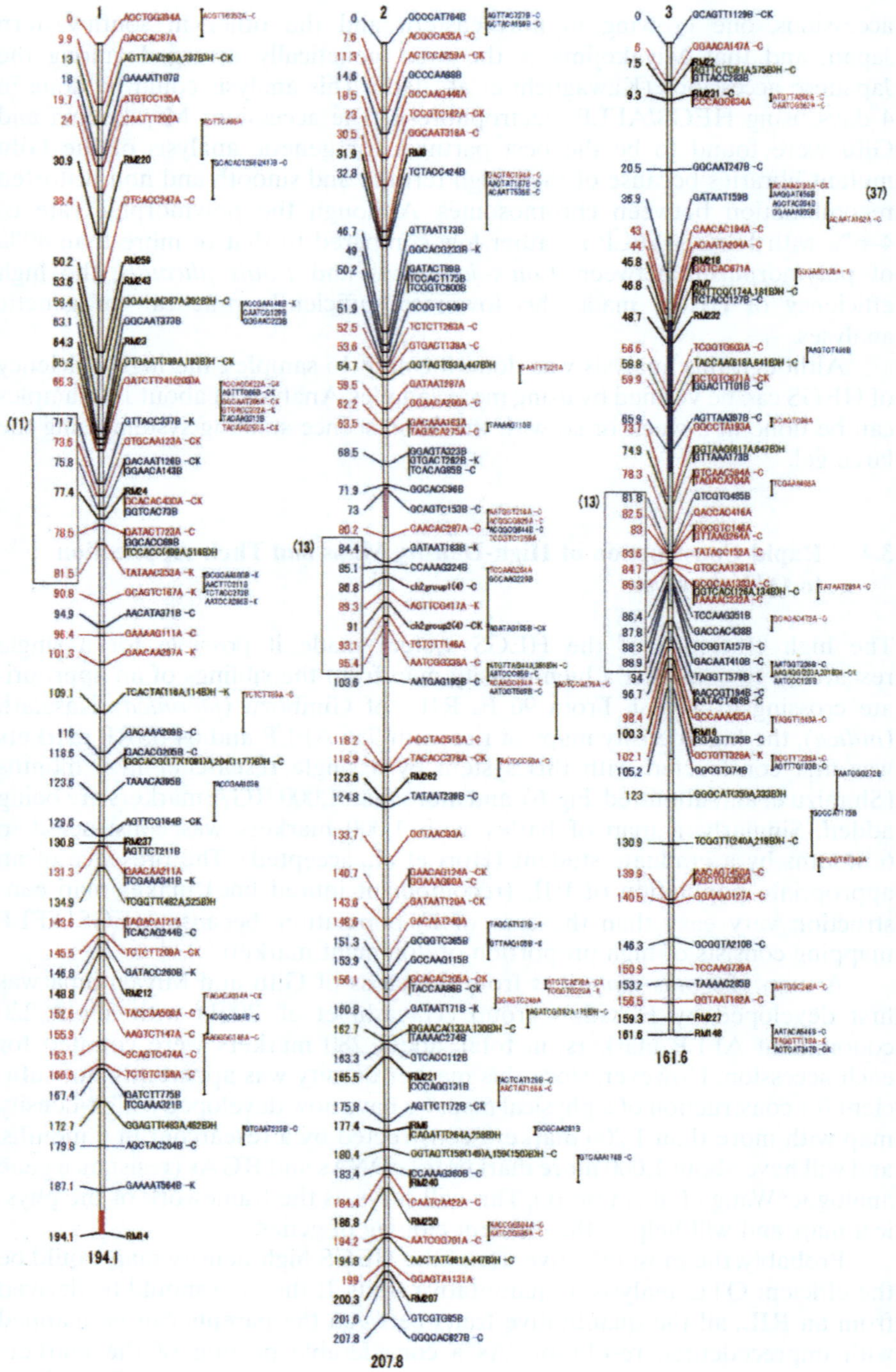

Fig. 6.

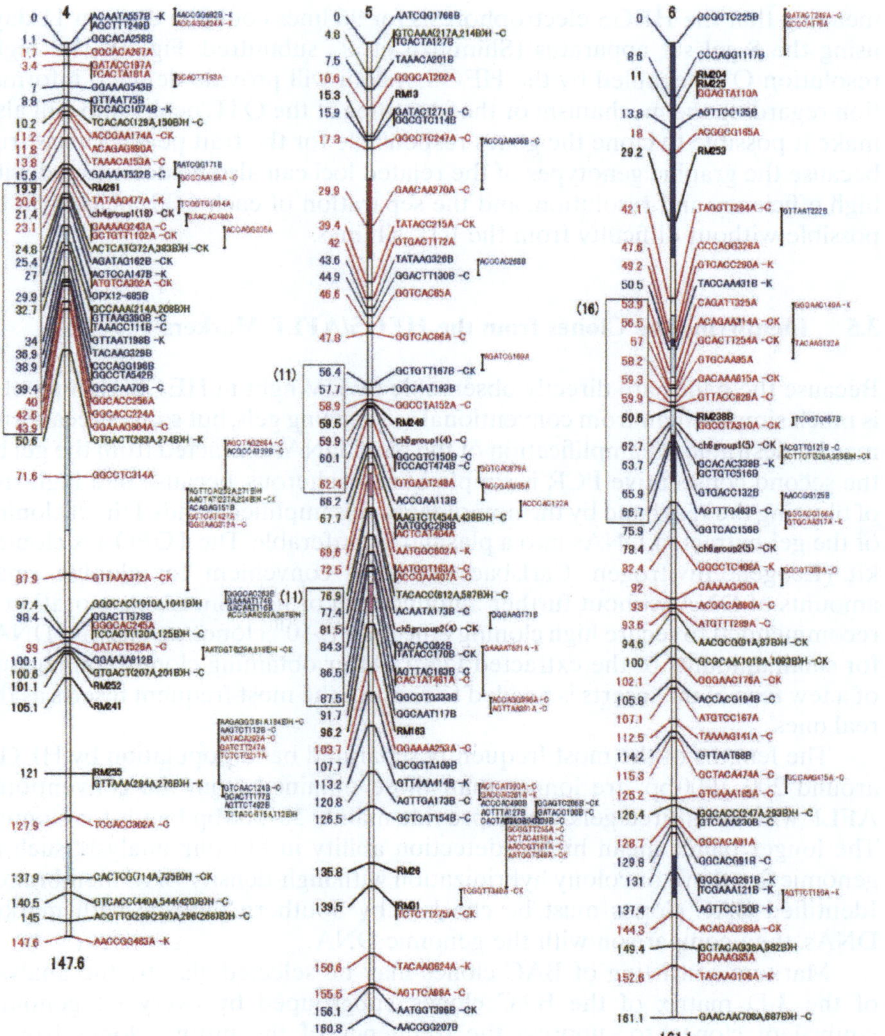

Fig. 6. High-density HEGS map of rice with 956 AFLP/64 SSR markers constructed from Gimbozu × Kasalath RILs. Only numbers 1–6 of 12 chromosomes are presented. AFLP markers are shown with the selective sequences (EcoRI+3/MseI+3), their fragment sizes (bp), and their band origins as: *A* (*red*), Gimbozu associate markers; *B* (*blue*), Kasalath associate markers; *H* (*green*), codominant markers. Availability of the mapped AFLP marker are also indicated at their ends as: *C* common polymorphic markers among *indica* and *japonica* cultivars; *K* polymorphic markers between Gimbozu (lowland *japonica*) and Kahei (upland *japonica*). *Colored* chromosome regions showed a distorted genotype ($P < 0.01$): *blue*; Gimbozu dominant, *red*; Kasalath dominant

meaning that the HEGS electrophoresis for 96 lines could be done in 11 days using the 8-gel set apparatus (Shimidzu et al., submitted; Fig. 7). The high-resolution QTL enabled by the HEGS system will provide detailed information regarding the mechanism or the identities of the QTL peaks, and will also make it possible to clone the genes responsible for the trait peaks. This is true because the graphic genotypes of the related loci can also be determined with high efficiency and resolution, and the separation of each QTL peaks will be possible without difficulty from the RIL siblings.

3.5 Identifying the Clones from the HEGS/AFLP Markers

Because the bands are directly observable by UV light in HEGS, cloning of it is much simpler than from conventional sequencing gels, but some precautions may be desirable. Reamplification of the band DNAs extracted from the gel by the second consecutive PCR is simple but treacherous, because of a high risk of blurring the real band by the homologous misamplified bands. Direct cloning of the gel-extracted DNAs into a plasmid is preferable. The TOPO TA cloning kit (Ecogene/Invitrogen, Carlsbad, USA) is convenient for cloning small amounts of DNA without further amplification or ligation. Electroporation is recommended to secure high cloning efficiency (>10^{10} clones/μg plasmid DNA) for small amounts of the extracted DNA. After obtaining clones, comparison of a few tens of the inserts is needed to confirm the most frequent inserts as the real ones.

The lengths of the most frequently separated band population by HEGS, around 200–1000 bp, are longer than those obtained from the conventional AFLP with denatured gels, which provide mainly 50–300 bp bands for cloning. The longer bands mean higher detection ability in blotting analysis, such as genomic Southern or colony hybridization with high-density BAC membranes. Identified BAC clones must be checked by Southern blotting with marker DNAs, then comparison with the genomic DNA.

Markers consisting of BAC clones may be selected also by the analysis of the 3-D matrix of the BAC clones, subgrouped by every $\leq$1 genome-equivalent clones, to suppress the frequency of the hitting clones from a subgroup to no more than 1, and to make clone identification rather simple. Some genome regions, such as the centromere, are dispersed with a high proportion of repetitive sequences, making the conventional colony hybridization selection of BAC clones almost impossible. In such cases, 3-D PCR selection of the clones through HEGS/AFLP may be the only reliable procedure (Fig. 8).

The selected clones should be compared by their restriction patterns with rather rare cutting enzymes like *Xho* I (+*Not* I) with CHEF (Contour-dumped Hexagonal Electric Field) electrophoresis or by *Hind* III with normal, but long-range electrophoresis. By this comparison, mutual overlapping of the BAC clones will become apparent. Then the most protruding ends of both sides will be amplified for selection for the next steps of walking.

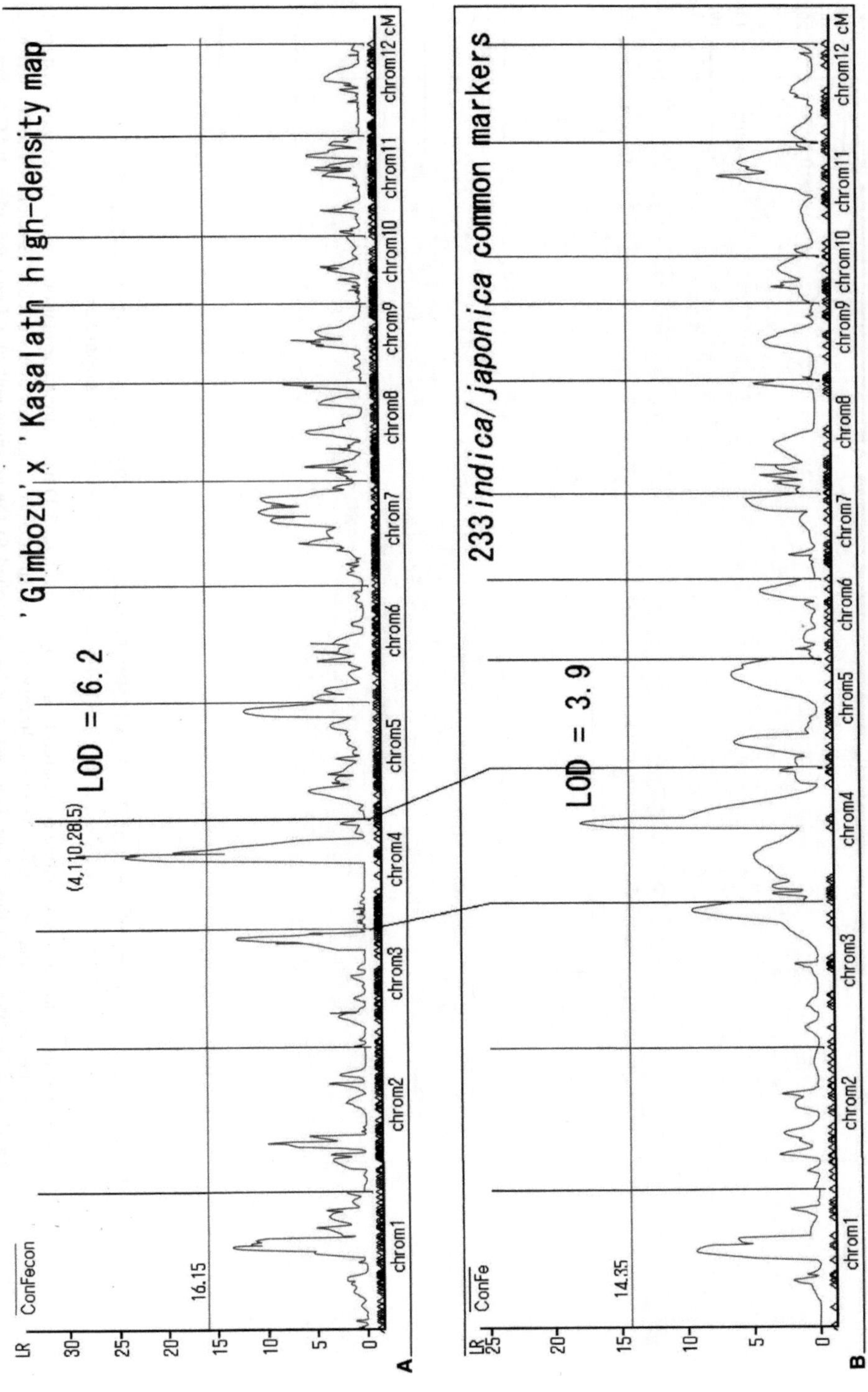

Fig. 7. An example of high-resolution HEGS/QTL analysis in rice. QTL affecting Fe accumulation in the shoot were analyzed with the high-density map of rice with 1,000 HEGS/AFLP markers (by A. Shimizu). The high-density QTL pattern with more than 1,000 markers (A) is represented in a resolution sufficient for practical purposes, with fewer markers seen for 233 with 82 primer pairs (B) (by A. Shimizu)

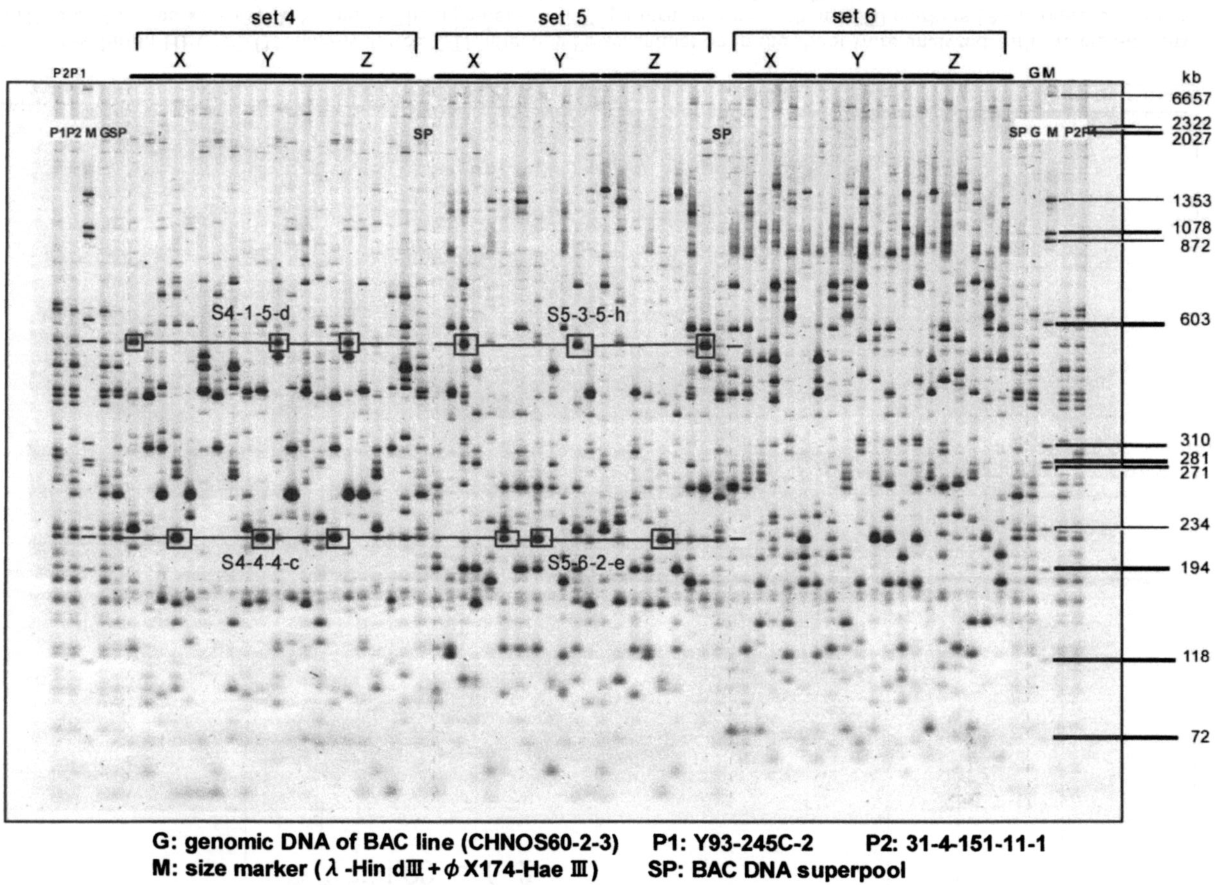

Fig. 8. Identification of a BAC clone correlated to a marker band by a 3-D band clone correlation system. As a model system, the rice blast fungus (*Magnaporthe grisea*) genome library was divided into six subgroups of 6 × 6 × 8 = 288 clones. The clone from the bands was identified as in this figure. This system is being applied to the arrangement of a *Lotus japonicus* BAC library (by I. Mikami)

4 Toward Construction of a Physical Map Covering the Total Genome of *Lotus japonicus*

The 3-D-HEGS electrophoresis system, correlating the clones to genomic marker bands, allowed systematic construction of a database of the corresponding BAC clones and the marker bands of the high-density genome map. After completion of such a database correlating the HEGS map and the genome library (including the nonpolymorphic bands), the network of the plural bands correlating to single clones and vice versa, plural bands to single clone), can be analyzed by a special program. Thus, automated walking to construct contigs around the landmark marker bands is possible. Fingerprinting of 800–4,000 clones at a time is also made possible with the HEGS or multicapillary system. Consequently, 30,000 clones of the *Lotus japonicus* BAC library will able to be scanned for fingerprinting in 38 days by a conventional two-gel set of apparatuses, or in 7.5 days by a special multicolor large gel system. Repeated scanning of the library with various enzymes increases the reliability of the data. The data of 3-D genome electrophoresis and those of fingerprinting can be crosschecked as well. As a result, the full genome-covering physical map with high reliability through several crosschecks can be constructed from the landmarks of the high-density HEGS map within a short period, and at a very reasonable cost. A similar system was proposed for construction of a physical map of the sorghum genome by Klein et al. (2000) using a conventional slab gel sequencer, but it has not yet been completed. We are afraid that the throughput of the conventional sequencing gels may be insufficient for handling the larger, total genome of sorghum (750 Mb).

When completed, such a physical map system will make the positional cloning of target genes very easy, even though the total sequence is not known, and will contribute greatly to the advances of functional genomics. *Lotus japonicus* may work as a model for constructing similar genome analyzing systems for all crops.

5 Conclusion and Future Perspectives

HEGS was applied successfully for saturation mapping of nodulation genes in the centromere regions and for construction of high-density genome maps of rice, barley, and *Lotus japonicus*. Positional cloning has now become an economical procedure with high efficiency by using the HEGS. Cloning of the genes in the centromeric regions was probably made practical with HEGS, and construction of high-density maps can be a provisional procedure for further QTL analyses. Although AFLP is highly effective, detecting polymorphism by SNPs and in RGAs will probably become important in the near future for mapping the functional genes, and for screening the candidates or cluster regions of the resistance genes, and even of the nodulation-controlling genes.

The recently identified nodulation controlling genes seem to have some homologous characters as disease resistance genes (Nishimura et al. 2002, Endre et al. 2002), and LRR (leucine rich repeat)/protein kinase may have the versatility needed for detecting target molecules and transmitting signals required for the next steps.

The still unidentified nodulation mutations can be categorized with high efficiency by systematically analyzing F_2 populations with the HEGS. When the genome library is totally arranged into a physical map, then positional cloning will become a much easier task. Probably all the serial components specific to legume nodulation will be cloned within the next 10 years, thus the cassette of nodulation genes conferring symbiotic nitrogen fixation ability to other crops may be realized. This will fundamentally change agricultural practice, and will contribute greatly to protecting the earth environment.

6 Protocols

The basic gel system follows that of the discontinuous protein gel electrophoresis system of Tris-glycine buffer (Davis 1964). For the non-denaturing system, 13% acrylamide gel is convenient for resolution and handling. About 1 cm of stacking gel is sufficient to compress a 1-cm-thick sample of DNA solution. Large 24 × 27 cm gel plates can accommodate 100 lanes of 1.25-mm-wide wells separated by 1-mm walls, thus providing four pitches of wells against one pitch of the 96-well microplates, or a smaller gel plate set (18 × 18 cm) with 66 lanes of the same pitch (produced by Nihon-Eido, Tokyo, Japan). Both the larger and smaller gel-set apparatuses accommodate four gel sets. The special 8- or 16-parallel syringes with two or four wells pitches, respectively, made sample loading efficient, and 400–800 samples can be applied to 4–8 gel sets in 1–2 apparatuses in 30 min. A single run of 7–8 h can be processed every day for eight gels in two sets of apparatuses without stress. During electrophoresis, the next day's samples can be prepared, or electrophoresis can be run during the night. Normally, the non-denaturing gel system is preferable to the denaturing ones, because the stronger gel makes gel handling much easier, and the bands can be resolved over the main area of the gel. In the case of SSRP, a denaturing gel composition is preferred for normal sequencing in order to avoid some repetitive bands, and the minimum gel concentration is about 8% for easy handling. However, the separation area is shifted to an upper region of the gel, and the denatured DNA dispersed in this region slightly obscures the resolution.

Fluorescence staining is much simpler than silver staining and can preferentially detect the non-denatured DNAs, contributing to clearer images. Since high-resolution digital cameras (>3 M pixels) are available at reasonable prices, expensive fluorescence scanners can be replaced. If fluorescence-labeled nucleotides become affordable as primers, the procedure will be much easier and multicolor detection multiplying the efficiency becomes possible – but they are expensive today, and not selective for double-stranded DNA.

Acknowledgements. This work was supported mainly by funds from Program for Promotion of Basic Research Activities for Innovative Biosciences from BRAIN (Bio-oriented Technology Research Advancement Institution). The work with mutants of *LjSym 70* and *72* is being done in collaboration with Dr. M. Kawaguchi of Niigata University. The work on the high-density map of *Lotus japonicus* is being done in collaboration with Dr. S. Sato and S. Tabata of Kazusa DNA Institute.

References

Arumuganathan K, Earle ED (1991) Nuclear DNA content of some important plant species. Plant Molecular Biology Reporter. Plant Mol Biol 9:208–218

Davis BJ (1964) Disc electrophoresis. 2. Method and application to human serum proteins. Ann NY Acad Sci 121:404–427

Goff SA, Ricke D, Lan TH, Presting G, Wang R, Dunn M, Glazebrook J, Sessions A, Oeller P, Varma H, Hadley D, Hutchison D, Martin C, Katagiri F, Lange BM, Moughamer T, Xia Y, Budworth P, Zhong J, Miguel T, Paszkowski U, Zhang S, Colbert M, Sun WL, Chen L, Cooper B, Park S, Wood TC, Mao L, Quail P, Wing R, Dean R, Yu Y, Zharkikh A, Shen R, Sahasrabudhe S, Thomas A, Cannings R, Gutin A, Pruss D, Reid J, Tavtigian S, Mitchell J, Eldredge G, Scholl T, Miller RM, Bhatnagar S, Adey N, Rubano T, Tusneem N, Robinson R, Feldhaus J, Macalma T, Oliphant A, Briggs S (2002) A draft sequence of the rice genome (*Oryza sativa* L. ssp. *japonica*). Science 296:92–100

Hatada I, Hayashizaki Y, Hirotsune S, Komatsubara H, Mukai T (1991) A genomic scanning method for higher organisms using restriction sites as landmarks. Proc Natl Acad Sci USA 88:9523–9527

Hayashi M, Miyahara A, Sato S, Kato T, Yoshikawa M, Taketa M, Hayashi M, Pedrosa A, Onda R, Imaizumi-Anraku H, Bachmair A, Sandal N, Stougaard J, Murooka Y, Tabata S, Kawasaki S, Kawaguchi M, Harada K (2001) Construction of a genetic linkage map of the model legume *Lotus japonicus* using an intraspecific F2 population. DNA Res 8:301–310

Kawaguchi M, Motomura T, Imaizumi-Anraku H, Akao S, Kawasaki S (2001) Providing the basis for genomics in *Lotus japonicus*: the accessions Miyakojima and Gifu are appropriate crossing partners for genetic analyses. Mol Genet Genomics 266:157–166

Kawaguchi M, Imaizumi-Anraku H, Koiwa H, Niwa S, Ikuta A, Syono K, Akao S (2002) Root, root hair, and symbiotic mutants of the model legume *Lotus japonicus*. Mol Plant Microbe Interact 15:17–26

Kawasaki S, Murakami Y (2000) Genome analysis of *Lotus japonicus*. J Plant Res 113:497–506

Klein PE, Klein RR, Cartinhour SW, Ulanch PE, Dong J, Obert JA, Morishige DT, Schlueter SD, Childs KL, Ale M, Mullet JE (2000) A high-throughput AFLP-based method for constructing integrated genetic and physical maps: progress toward a sorghum genome map. Genome Res 10:789–807

Men AE, Meksem K, Kassem MA, Lohar D, Stiller J, Lightfoot D, Gresshoff PM (2001) A bacterial artificial chromosome library of *Lotus japonicus* constructed in an *Agrobacterium tumefaciens*-transformable vector. Mol Plant Microbe Interact 14:422–425

Michelmore RW, Paran I, Kesseli RV (1991) Identification of markers linked to disease-resistance genes by bulked segregant analysis a rapid method to detect markers in specific genomic regions by using segregating populations. Proc Natl Acad Sci USA 88:9828–9832

Nakamura S, Asakawa S, Ohmido N, Fukui K, Shimizu N, Kawasaki S (1997) Construction of an 800-kb contig in the near-centromeric region of the rice blast resistance gene Pi-ta2 using a highly representative rice BAC library. Mol Gen Genet 254:611–620

Nishhimura M, Nakamura S, Hayashi N, Asakawa S, Shimizu N, Kaku H, Hasebe A, Kawasaki S (1998) Construction of a BAC library of the rice blast fungus *Magnaporthe grisea* and finding specific genome regions in which its transposons tend to cluster. Biosci Biotechnol Biochem 62:1515–1521

Sarwar G, Sosulski FW, Bell JM, Bowland JP (1978) Nutritional evaluation of oilseeds and legumes as protein supplements to cereals. Adv Exp Med Biol 105:415–441

Sato S, Kaneko T, Nakamura Y, Asamizu E, Kato T, Tabata S (2001) Structural analysis of a *Lotus japonicus* genome. I. Sequence features and mapping of fifty-six TAC clones which cover the 5.4 mb regions of the genome. DNA Res 31 8:311–318

Schauser L, Handberg K, Sandal N, Stiller J, Thykjaer T, Pajuelo E, Nielsen A, Stougaard J (1998) Symbiotic mutants deficient in nodule establishment identified after T-DNA transformation of *Lotus japonicus*. Mol Gen Genet 259:414–423

Senoo K, Solaiman MZ, Kawaguchi M, Imaizumi-Anraku H, Akao S, Tanaka A, Obata H (2000) Isolation of two different phenotypes of mycorrhizal mutants in the model legume plant *Lotus japonicus* after EMS-treatment. Plant Cell Physiol 41:726–732

Sheffield VC, Cox DR, Lerman LS, Myers RM (1989) Attachment of a 40-base-pair G + C-rich sequence (GC-clamp) to genomic DNA fragments by the polymerase chain reaction results in improved detection of single-base changes. Proc Natl Acad Sci USA 86:232–236

Suzuki Y, Orita M, Shiraishi M, Hayashi K, Sekiya T (1990) Detection of ras gene mutations in human lung cancers by single-strand conformation polymorphism analysis of polymerase chain reaction products. Oncogene 5:1037–1043

Stougaard J (2001) Genetics and genomics of root symbiosis. Curr Opin Plant Biol 4(4):328–335

Szczyglowski K, Hamburger D, Kapranov P, de Bruijn FJ (1997) Construction of a *Lotus japonicus* late nodulin expressed sequence tag library and identification of novel nodule-specific genes. Plant Physiol 114:1335–1346

Szczyglowski K, Shaw RS, Wopereis J, Copeland S, Hamburger D, Kasiborski B, Dazzo FB, de Bruijn FJ (1998) Nodule organogenesis and symbiotic mutants of the model legume *Lotus japonicus*. Mol Plant Microbe Interact 11:684–697

The Arabidopsis Genome Initiative (2000) Analysis of the genome sequence of the flowering plant *Arabidopsis thaliana*. Nature 408:796–815

Thykjaer T, Stiller J, Handberg K, Jones J, Stougaard J (1995) The maize transposable element Ac is mobile in the legume *Lotus japonicus*. Plant Mol Biol 27:981–993

Vos P, Hogers R, Bleeker M, Reijans M, van de Lee T, Hornes M, Frijters A, Pot J, Peleman J, Kuiper M et al. (1995) AFLP: a new technique for DNA fingerprinting. Nucleic Acids Res 23:4407–4414

Yu J, Hu S, Wang J, Wong GK, Li S, Liu B, Deng Y, Dai L, Zhou Y, Zhang X, Cao M, Liu J, Sun J, Tang J, Chen Y, Huang X, Lin W, Ye C, Tong W, Cong L, Geng J, Han Y, Li L, Li W, Hu G, Huang X, Li W, Li J, Liu Z, Li L, Liu J, Qi Q, Liu J, Li L, Li T, Wang X, Lu H, Wu T, Zhu M, Ni P, Han H, Dong W, Ren X, Feng X, Cui P, Li X, Wang H, Xu X, Zhai W, Xu Z, Zhang J, He S, Zhang J, Xu J, Zhang K, Zheng X, Dong J, Zeng W, Tao L, Ye J, Tan J, Ren X, Chen X, He J, Liu D, Tian W, Tian C, Xia H, Bao Q, Li G, Gao H, Cao T, Wang J, Zhao W, Li P, Chen W, Wang X, Zhang Y, Hu J, Wang J, Liu S, Yang J, Zhang G, Xiong Y, Li Z, Mao L, Zhou C, Zhu Z, Chen R, Hao B, Zheng W, Chen S, Guo W, Li G, Liu S, Tao M, Wang J, Zhu L, Yuan L, Yang H (2002) A draft sequence of the rice genome (*Oryza sativa* L. ssp. *indica*). Science 296:79–92

13 Genome Analysis of *Mesorhizobium loti*: A Symbiotic Partner to *Lotus japonicus*

T. Kaneko[1], K. Saeki[2], and K. Minamisawa[3]

1 Introduction

By the symbiotic nitrogen fixation process, bacteria of the family Rhizobiaceae convert atmospheric dinitrogen (N_2) to ammonia (NH_3), which can be effectively utilized by host legume plants. Symbiotic nitrogen fixation contributes significantly to global nitrogen cycles and agricultural practice. The establishment of rhizobia-legume symbiosis requires the induction of new developmental programs in respective partners. The symbiotic interaction begins with signal exchanges of flavonoids and bacterial nodulation signals, Nod factors, between the two partners (Denarie et al 1996). However, the establishment of nitrogen-fixing symbiosis probably requires more complex steps triggered by reciprocal signal exchanges leading to the organogenesis of nodules and differentiation of the microsymbionts (Niner and Hirsch 1998).

Lotus japonicus is a promising model legume for studying molecular interactions between the partners because of its characteristics, such as small genome size, a short generation cycle, self-compatibility, transposon tagging, and genetic transformation (Handberg and Stougaard 1992; Stiller et al. 1997). Indeed, using the Ac transposon tagging approach, Schauser et al. (1999) first identified a plant regulatory *nin* gene that is responsible for the nodule organogenesis program. Recent progress in large-scale analyses of cDNAs and the genome of *Lotus japonicus* (Asamizu et al. 2000) has also increased the significance and usefulness of *Lotus japonicus* as a model legume.

Mesorhizobium loti (*M. loti*) is a nitrogen-fixing endosymbiont associated with *Lotus japonicus*. The bacterially-encoded *nod*, *fix/nif*, and *exo* genes are apparently responsible for the establishment of functional symbiosis (Denarie et al 1996; Niner and Hirsch 1998). There is, however, increasing evidence that other housekeeping genes influence either nodulation or nitrogen fixation in other bacterial habitats, such as the rhizosphere, infection thread niche, and symbiosome (Niner and Hirsch 1998). The aim of this chapter is to introduce

[1] Kazusa DNA Research Institute, 2-6-7 Kazusa-kamatari, Kisarazu, Chiba 292-0818, Japan
[2] Department of Biology, Graduate School of Science, Osaka University, Toyonaka, Osaka 560–0043, Japan
[3] Graduate School of Life Sciences, Tohoku University, Katahira, Aoba-ku, Sendai 980-8577, Japan

Biotechnology in Agriculture and Forestry, Vol. 52
Nagata/Tabata (Eds.) Brassicas and Legumes

genome analysis and a post-genome approach to *M. loti* in the furtherance of our global understanding of symbiotic interactions.

2 Biology of *Mesorhizobium loti*

M. loti is a gram-negative, rod-shaped, soil bacteria. The free-living cells represent the membrane structure of typical gram-negative bacteria (inner and outer membranes with periplasm), and often possess poly-β-hydroxybutylate granules in the cells (Fig. 1A). The cells grown in YM medium are seen to be approximately 0.5 μm wide and 1.5–2.0 μm long. The taxonomy within the family Rhizobiaceae is still elusive, although all taxa are within the α-subdivision of *Proteobacteria* (Young 1996; van Berkum and Eardy 1998). The family Rhizobiaceae currently includes five genera: *Rhizobium*, *Sinorhizobium*, *Mesorhizobium*, *Azorhizobium*, and *Bradyrhizobium*, collectively called rhizobia. The genus *Bradyrhizobium* is a slow-growing member on laboratory media, whereas the other four genera including *Mesorhizobium* display rapid growth (van Berkum and Eardy 1998; Minamisawa and Mitsui 2000).

Mesorhizobium loti invades the host plant from the tip of the root hair via infection threads, as do other rhizobia (Fig. 2C,D). After the infection thread carrying the bacteria penetrates with ramification to the cortical cells of the *L. japonicus* roots, the bacteria are released into nodule primordial cells and differentiated to bacteroids (Hayashi et al. 2000). The resultant root nodules have a globular shape 1–2 mm in diameter – typical determinate-type legume root nodules (Fig. 2A,B). The bacteroids are enclosed by a plant-derived membrane structure, which is known as the peribacteroid membrane (PBM) (Fig. 1C). The term symbiosome includes the bacteroid, the symbiosome space, and the PBM as a functional unit, where signal exchanges and gene expression take place vigorously between both partners for nitrogen fixation and bacteroid maintenance (Tajima et al. 2000). In the nodule tissues of *L. japonicus* (Fig. 1B), uninfected plant cells have been observed around the infected plant cells harboring bacteroids. Because this cell configuration is very similar to that of soybean nodule tissue, the uninfected cells probably support the carbon supply (dicarboxylic acid production) and nitrogen assimilation (amide and ureide production) for the infected plant cells (Tajima et al. 2000).

During the onset of rhizobia-legume symbiosis, flavonoids secreted by host plant roots cause the induction of rhizobial nodulation genes. These genes are involved in the synthesis and excretion of Nod factors (Denarie et al.1996). However, the flavonoids that induce the expression of *nod* genes in *M. loti* have not been determined yet. Although aldonic acids have been reported to induce *M. loti nod* genes (Gagnon and Ibrahim 1998), it is unlikely that aldonic acids are natural inducers, because high concentrations of the compounds were required for the induction. However, the production of Nod factors by *M. loti* has been achieved by introducing pMP2112, which carries naringenin-inducible *nodD* from *R. leguminosarum* bv. *trifolii* (Lopez-Lara et al. 1995;

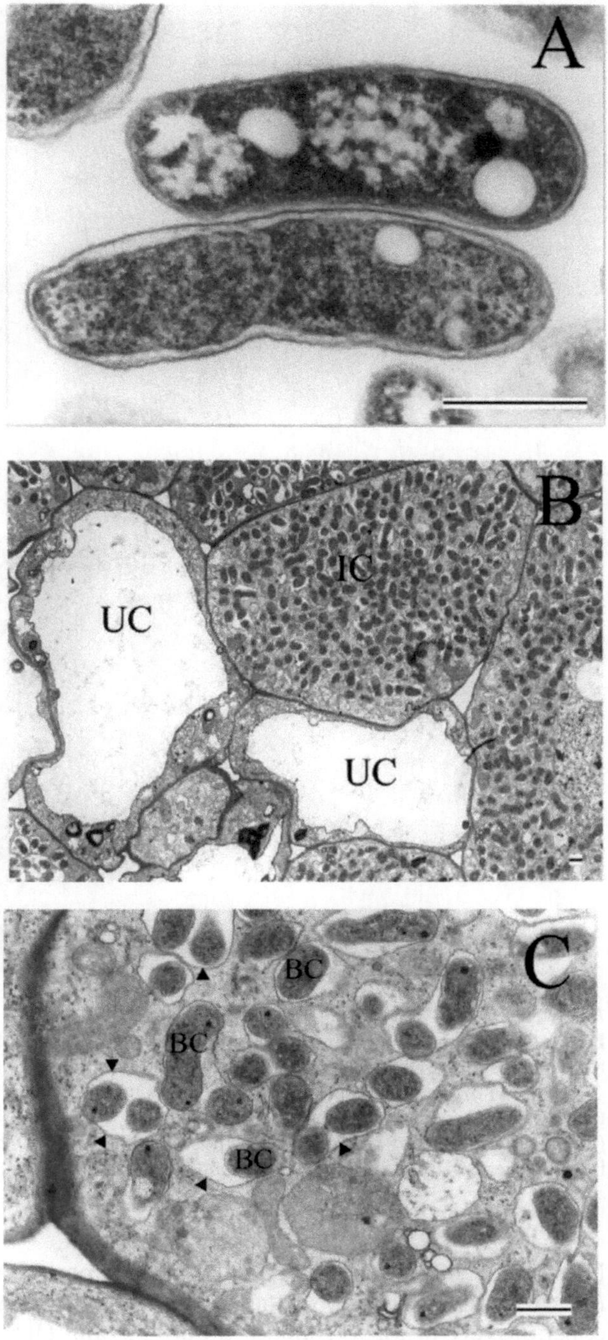

Fig. 1. Transmission electron micrographs of free-living cells of *M. loti* MAFF303099 (**A**) and nodule sections of *Lotus japonicus* "Gifu" inoculated with *M. loti* MAFF303099 (**B** and **C**). **A** Bacterial cells of *M. loti* strain MAFF303099 grown in yeast-extract mannitol broth medium. *Bar* = 0.5 μm. **B** Infected plant cell (*IC*) containing a large number of bacteroids, and uninfected cell (*UC*). *Bar* = 1 μm. **C** High-magnification view of panel B showing bacteroids (*BC*) and peribacteroid membrane (*arrowhead*).

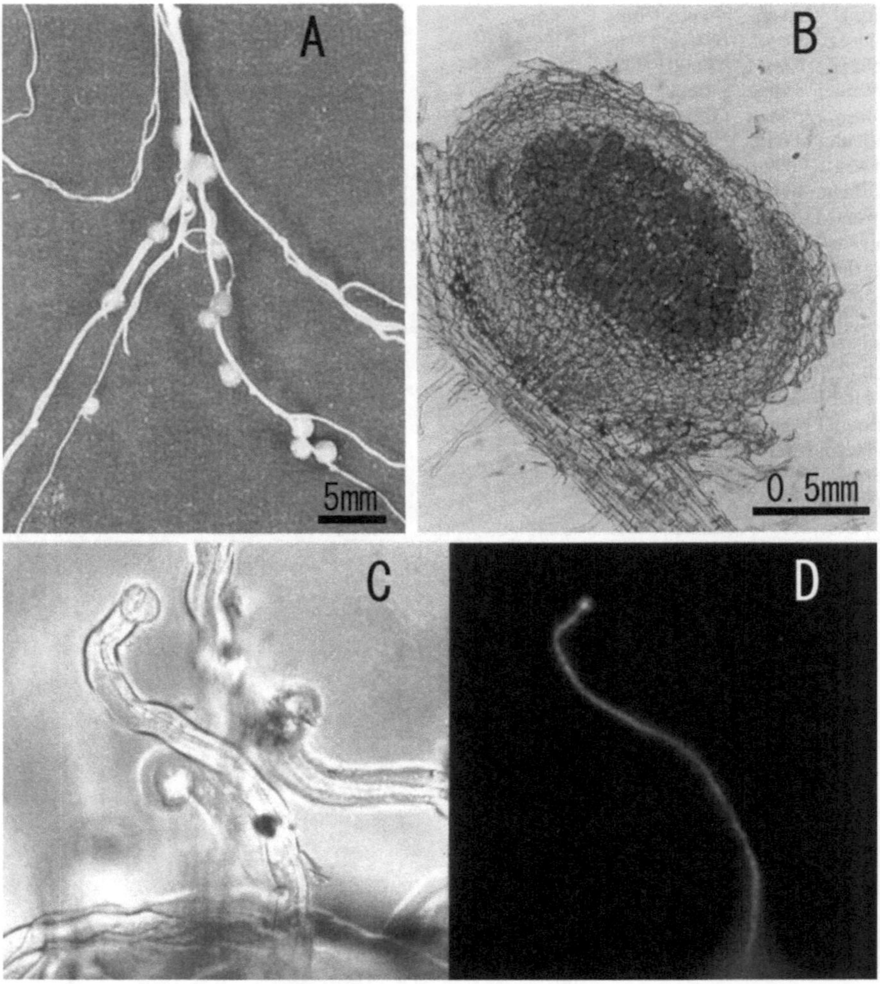

Bar = 1 μm

Fig. 2. Determinate nodules of *Lotus japonicus* "Gifu" inoculated with *Mesorhizobium loti* MAFF303099. **A** Determinate globular nodules (1–2 mm in size) formed on roots of *L. japonicus* 40 days after inoculation. **B** Mature nodule shows infected region (*dark area*). **C** and **D**. Infection thread in root hair *of L. japonicus* "Gifu" inoculated with *M. loti* BN02, a green fluorescent protein (GFP)-tagged strain of JRL501, a derivative of MAFF303099. The same root hair was observed by fluorescent microscopy under a light field (**C**) and fluorescence field with a GFP filter unit (**D**).

Niwa et al. 2001). Recently, co-cultivation experiments of *L. japonicus* and *M. loti* carrying *nodA-lacZ* fusion suggest that *nodD* genes were not expressed constitutively in *M.* loti unlike other rhizobia (Itakura et al., unpubl. data). These facts suggest that an alternative regulation system is involved in *nod* gene induction in *M. loti* as well as flavonoid induction. The structural component of Nod factors was identified using the *M. loti* strain carrying pMP2112

(Lopez-Lara et al. 1995; Niwa et al. 2001). The major component was identified as NodMl-V(C18:1, Me, Cb, AcFuc), an *N*-acetyl-glucosamine pentamer decorated with a $C_{18:1}$ acyl moiety, methyl group, carbamoyl group, and acetylfucose. These Nod factors can elicit root hair deformation, nodule primordia formation, nodulin gene expression, and preinfection thread formation in *L. japonicus* (Niwa et al. 2001).

3 Genome Structure of *Mesorhizobium loti* MAFF303099

The nucleotide sequence of the entire genome of *M. loti* MAFF303099 was determined in 2000 as the first example of a soil bacterium that is a member of *α-Proteobacteria* (Kaneko et al. 2000a,b). To determine the entire genomic sequence, the whole-genome shotgun strategy was adopted using four shotgun libraries with different lengths of DNA fragments that were generated from the total cellular DNA of *M. loti* MAFF303099. Computational assembly of the end-sequence files from these libraries, corresponding to approximately eight times the genome equivalent, constituted three circular units, a single chromosome, and two plasmids designated pMLa and pMLb. The total size of the genome was 7,596,297 bp, consisting of a chromosome of 7,036,071 bp, and pMLa and pMLb of 351,911 bp and 208,315 bp, respectively. The average GC content of the genome was 62.5%.

A total of 7,281 potential protein-encoding genes, two sets of rRNA genes, 50 tRNA genes, and an RNase P RNA gene were identified on the genome. None of the genes contained either an intron or intein. Of the potential protein genes, 3,910 (53.7%) were homologues of genes of known function in the database. The potential protein-encoding genes whose function could be anticipated were grouped into 14 categories with respect to different biological roles. Genes encoding transport systems constituted the largest class of genes (764 genes; 10.5% of the total) in this genome. The assignment of each gene on *M. loti* genome is presented in the Web database, RhizoBase, at http://www.kazusa.or.jp/rhizobase.

One of the notable features of the *M. loti* chromosome is the integration of a horizontally transferred DNA, a symbiosis island. It was first reported that the symbiotic genes of *M. loti* strain ICMP3153 are assigned to be clustered in a 500 kbp DNA region on the chromosome, and that this element is transmittable to nonsymbiotic *Mesorhizobium* species with the ability to form nodules (Sullivan and Ronson 1998). This DNA segment is inserted into a phenylalanine-tRNA gene with 17 bp duplication of the 3′ terminal portion of the gene. A putative symbiosis island in the *M. loti* strain MAFF303099 is flanked by the 17-bp sequence on both sides on the chromosome at coordinates 4,644,792 to 5,255,766. One of the flanking 17 bp sequences of 610,975 bp of DNA segment is a part of the intact phenylalanine-tRNA gene. Moreover, the symbiosis island of MAFF303099 was 20% longer than that of ICMP3153, therefore, the structures of the symbiosis islands are very proba-

Table 1. Comparison of GC contents on *M. loti* genome

	Chromosome		Plasmid	
	non-sym	sym	pMLa	pMLb
Length (bp)	6,425,096	610,975	351,911	208,315
Entire sequence (%)	63.0	59.8	59.3	59.9
Coding region (%)	63.7	60.3	59.8	60.6
Third codon (%)	80.5	70.2	69.6	71.2

bly divergent between the two *M. loti* strains. Indeed, the two islands share 248 kb of DNA, with multiple deletions and insertions up to 168 kb (Sullivan et al. 2002). When GC contents of both coding regions and the third letter of the codons were calculated separately inside and outside of the symbiosis island (Table 1), the GC content of the symbiosis island was lower than outside, and similar to that of the plasmids. Most notably, these units have originated and have been transmitted from other genetic systems.

On the symbiosis island of MAFF303099, 580 protein-coding genes were assigned, in which the following genes were related to nitrogen fixation and symbiosis: *nifABDEHKNQSUWXZ*, *fdxBN*, *fixABCGHINOPQSU*, *nodABCDEFIJMSZ*, *nolBKLOTUVWX*, *noeJKL,* and *dctABD*. Other related genes such as *fixKLJ*, *nodNPQ*, *nolR*, *noeCK,* and *exoPNOMALTKUCYFQIZB* were not identified in the symbiosis island, but were located outside. The consensus sequences for *nod*-box promoter (Rostas et al. 1986) were found upstream to the coding regions of nine genes (*nodA*, *nodB*, *nodM*, *nodS*, *nodZ*, *noeK*, *nolL*, mlr6144, mlr6334) located exclusively within this island. In addition to the characteristics as a cassette for *nod* and *nif* genes in the symbiosis island, other remarkable features noted so far are as follows: (1) out of 580 genes (19.6%) assigned in the symbiosis island, 111 were those for transposon-related functions such as transposase, integrase, recombinase, and resolvase. The high content of the genes in this category may indicate that the island is possibly the hot spot for the insertion of transposable elements. (2) Conjugal transfer protein homologues, *traG-traB-trbC-trbE-trbJ-trbL-trbF-trbG-trbI*, formed a large gene cluster. These genes may be involved in transmission of the symbiosis island. (3) There are two gene clusters that synthesize vitamins, each comprised of four genes for biotin synthesis and a cluster of six genes for that of thiamine. Although nonsymbiotic *Mesorhizobia* require thiamine and biotin, the symbionts are prototrophic for these vitamins (Sullivan and Ronson 1998). These gene clusters may be involved in supplementation of thiamine and biotin for the symbiont. (4) There is a cluster of genes for type III secretion systems, which may influence nodule formation, involving the secretion of proteins into host cells (Marie et al. 2001).

We predicted 510 genes from the symbiosis island, compared with 6,172 outside of the chromosome and 529 on the plasmids, in order to find the paralogous genes. Comparisons were conducted by taking three independent factors of a bit score, the BLAST2 E-value, and the ratio of alignment lengths, into consideration. We first used the BLAST2 algorithm (Altschul et al. 1997) to identify BLASTP alignments that were determined to have an E-value not greater than E-10. A lower threshold of acceptability was set at one-half of the bit score reported by self-comparison of the translated amino acid sequences. Only amino acid sequences whose alignments extended over at least 0.6 of the length of the query sequence were considered similar. We identified 131 paralogs dispersed on the genome (75 genes on the chromosome outside the symbiosis island, 41 on pMLa and 15 on pMLb; Fig. 3A). Notably, paralogs noted so far are as follows: (1) outside of the symbiosis island, second copies of the symbiosis genes: *nodE* (at 3.93 Mb on the chromosome), *nodG* (6.50 Mb), *noeK* (6.25 Mb), *dctABD* cluster (5.98 Mb), and *fixNOQPGHIS* cluster (5.42–5.43 Mb) were identified. (2) Two copies for the sigma-54 factor (*rpoN*), which is involved in transcriptional control of major nitrogen-fixation genes such as *nifHDK* (Kullik et al. 1991), are present on the chromosome, one inside and the other outside the island (2.56 Mb). (3) Five pairs of chaperonin *groES-groEL* genes were present on the genome, one inside and three outside the symbiosis island (1.80, 1.95, and 6.73 Mb), and one on plasmid pMLa (0.30 Mb). (4) pMLa contained a total of 40 genes for the ABC-transporter system, of which eight genes (0.12–0.13 and 0.24–0.25 Mb) were paralogs of those encoding on the island. (5) A gene cluster comprised of ten genes containing four biotin-synthesis genes (mll5836–mll5826) was assigned within the island. Another gene cluster with the same gene set (mll9106–mll9093) was found in 0.07–0.08 Mb on plasmid pMLa. The above-mentioned three-gene clusters assigned on pMLa are of interest in that the symbiosis island and pMLa originated from the same genetic systems.

4 Similarity to *Sinorhizobium melilotii* Genes

Sinorhizobium meliloti strain 1021 (the microsymbiont of alfalfa) genome was completely sequenced in 2001 as an alternative genus of rhizobia (Galibert et al. 2001). The genome was also larger in size (6,691,694 bp) than many other complete bacterial genomes, while the replication component carrying megaplasmids considerably differ from them (Fig. 1B). A total of 6,204 potential protein-encoding genes were identified on the genome of *S. meliloti*. Rhizobial genomes of *M. loti* and *S. meliloti* have over 6,000 protein-encoding genes. Because of their similarity, the expectation is to find a genetic backbone on the genomes conserved as necessary for the rhizobial lifestyle of the undersoil environment and for symbiosis.

The genetic complements of *M. loti* and *S. meliloti* were compared with higher (bit-score/self≥0.5, E-value ≤ E-10, hit length≥0.6 query) and lower (E-value ≤ E-4) stringencies. Comparison with the higher stringency indicated that 2,157 (29.6%) of *M. loti* protein genes had counterparts in *S. meliloti*. With the lower stringency, 5,245 (72.0%) of *M. loti* genes showed similarity to *S. meliloti* genes. The higher-stringency comparison indicated that significant

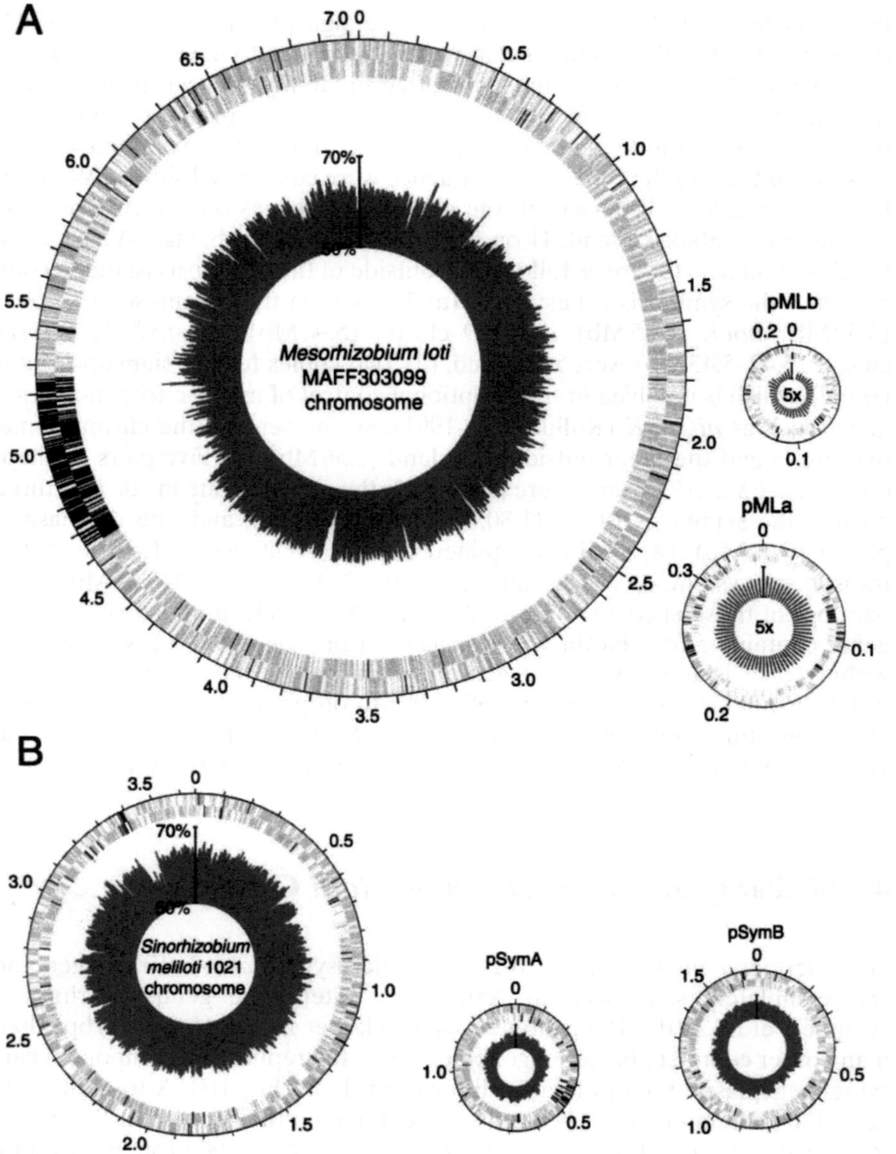

Note that the tip of the root hair is curled

Fig. 3. Circular representation of genomes of two rhizobia. **A** Three replicons of *M. loti* MAFF303099. Both pMLa and pMLb are drawn at 5× magnification. The symbiotic island locates between 4.64–5.26 Mb on the chromosome. **B** Three replicons of *S. meliloti* 1021. The *scale* indicates the location in Mb starting from the *Pac*I recognition site. The *bars* in the outermost and the second circles show the positions of the putative protein-encoding genes in the clockwise and counterclockwise directions, respectively. Genes indicating similarity to ones on the symbiosis island are depicted in *black,* and those indicating lower than the threshold are in *halftone*. The *innermost circle* with a *scale* shows the average GC% calculated by a window size of 10 kbp

common genes are conserved between the genomes. That with lower stringency meant that most members of a given gene family could be grouped into a cluster. When a reverse comparison was made, 2,184 (35.2%) and 4,956 (79.9%) of *S. meliloti* genes showed sequence similarity to *M. loti* genes under the defined higher and lower stringency conditions, respectively. Twenty-eight percent of *M. loti* genes had no similarity to *S. meliloti* genes with lower stringency. When the reverse comparison was made, 20% of *S. meliloti* genes showed no similarity. These latter sets of results likely reflect the genus specificity.

For investigating distribution of certain protein domains, each gene was analyzed to make alignments with the known protein domains in the Pfam database using the HMM algorithm (Sonnhammer et al. 1998). The result indicated that 3,471 out of 7,281 (47.7%) of potential protein-encoding genes in *M. loti* had matches in 996 protein domain species, and 3,306 (53.3%) *S. meliloti* genes showed significant similarities to 1,002 domains. Some of gene families related to the transporter (ABC transporter family) and the regulator (two-component response regulator, LysR family, AraC family, GntR family) existed in similar numbers in both the *M. loti* and *S. meliloti* genomes. These may show evidence for genetic backbones on the genome.

The symbiosis island on the *M. loti* chromosome has 510 genes, of which 108 showed significant similarity to 131 genes dispersed in the *S. meliloti* genome (60 genes on the chromosome, 62 genes on pSymA, 29 genes on pSymB) (Fig. 1B). The *S. meliloti* genome possesses two interesting regions that show significant similarity to regions on the symbiosis island of *M. loti*. First, the symbiotic genes (*nod*, *nif*, and *fix*) exist as some clusters inside a region with lower GC content (58.2%) between 0.40 and 0.50 Mb on pSymA (Barnett et al. 2001). The other region with lower GC content (58.2%) between 3.36 and 3.45 Mb on *S. meliloti* chromosome seemed to have been transferred horizontally (Capela et al. 2001). This region includes 13 genes with implied significant similarity to three transposases, four components of ABC transporter, one proline hydroxylase, one transcriptional regulator, two dipeptidases, and two hypothetical proteins. The notable features on the two regions suggest a significant relationship with the symbiosis island on *M. loti* chromosome.

5 An Organized Cosmid Library/Database of *Mesorhizobium loti*

M. loti has historically been less intensively studied than other major species, such as *Sinorhizobium meliloti* and *Bradyrhizobium japonicum*. Studies of *M. loti* have been performed mostly on five representative strains, NZP2037, NZP2213, NZP2235, NZP2238, and MAFF303099 (for review, Saeki and Kouchi 2000); however, molecular genetic resources are very limited. The former four NZP strains were isolated from *Lotus corniculatus* or *Lotus tenuis*

(Jarvis et al. 1982), MAFF303099 being the only strain originally isolated from the nodule of *Lotus japonicus*. Therefore, a well organized DNA segment bank of the *M. loti* MAFF303099 genome is anticipated to assist study in many areas of genetics, molecular biology, and biotechnology – in a manner similar to the way help has come from the Kohara library on an *E. coli* chromosome (Kohara et al. 1987). Such a bank would consist of a complete set of mutually overlapping plasmid clones that covers the entire genome of *M. loti* and an electronic data set of detailed characteristics of these clones. The bank would enable researchers to easily obtain a desired gene or operon of interest for functional analysis. The construction of an ordered cosmid bank and database to meet the above prerequisites is described below (Hattori et al. 2002).

The vector chosen for the clone bank was a cosmid vector pKS800 with an IncP replication origin, which is derived from the vector pLAFR1 (Friedman et al. 1982) to have cloning sites of *Nhe*I/*Pac*I/*Bam*HI/*Pme*I/*Xho*I, replacing the original *Eco*RI site. Although the size of pKS800 is more than 20 kbp and can contain an insert of about 25–30 kbp DNA, this vector has advantages over conventional cosmid vectors with a ColE1 replication origin, because it is conjugatively transferable to rhizobia and stably maintained as a low copy plasmid in both *Escherichia coli* and rhizobia. In fact, it has scarcely accumulated deletions in cloned DNA fragments. In addition, the eight-base rare cutters, *Pac*I and *Pme*I, in the cloning site have only three and six restriction sites in entire genome of *M. loti* MAFF303099, respectively, and are thus suitable for mapping and further manipulations.

Initial library construction was performed as a step in whole genome sequencing to confirm the overlaps of contigs. Partial *Sau*3AI digests of total DNA from *M. loti* MAFF303099 were size-fractionated and ligated to pKS800 digested with *Bam*HI to make up more than 3,000 clones. *E. coli* strains harboring individual cosmid clones were maintained in 96-well microtiter plates, out of which glycerol stocks in a total of 41 plates were stored at –80 °C. After "miniprep" of each clone, sequences of both ends of the insert were determined, and were subjected to BLAST search against the whole genome sequence of *M. loti* MAFF303099 to estimate the insert size and to choose clones to cover the entire genome. Since the assignment of open reading frames (ORFs) had been done, efforts were made not to separate an ORF and/or an apparent operon in different cosmid clones. Rather, efforts were made to cover every ORF and/or every set of tightly contiguous ORFs entirely by at least one particular ordered clone. About 400 clones were chosen as candidates for a mutually overlapping ordered library. The integrity and stability of each clone were assessed by *Bam*HI and *Eco*RI restriction analysis of DNA preparations from multiple independent colonies obtained after streaking *E. coli* glycerol stock on LB agar, because usefulness of a genomic library might be impaired by rearrangement of cloned segments that can occur both in vivo and in vitro. In cases of in vivo rearrangement, agarose gel electrophoresis patterns of digested DNA preparation disagrees with the calculated patterns based on the whole genome nucleotide sequence. Less than 1% of the candidate clones gave different patterns, indicating that in vivo segment rearrangement was a very minor event in the collection. Detailed inspection of end sequences revealed about 1% of the clones containing very short *Sau*3AI fragment of 10–200 bp, together with expected long fragments of more than 20 kbp. The short inserts were probably the result of in vitro rearrangement derived from co-ligation of more than one insert DNA to the vector DNA.

Approximately 360 clones were established to cover about 99.6% of the chromosome (7,036,071 bp) with eight gaps: the average overlap size was about 6.3 kbp. The larger plasmid pMLa (351,911 bp) was entirely covered by 17 clones, whereas about 98.8% of the smaller plasmid pMLb (208,315 bp) was covered by 11 clones with 2 gaps. The sizes of the 8 gaps were 0.9–8.9 kbp, and more than half of them corresponded to the region containing genes probably controlling DNA replication, recombination, and repair. It may be noted that 40 clones with an average overlap of 6.4 kbp were compiled to encompass the "symbiosis island" on the chromosome at coordinates 4,644,792–5,255,766 (Kaneko et al. 2000a, b), which corresponds but is not identical to the 500 kbp symbiosis island in the *M. loti* strain ICMP3153 (or R7A) that was derived from NZP2238 following plant passages (Sullivan and Ronson 1998; Sullivan et al. 2000).

Ordered overlapping cosmid libraries of symbiosis-specific plasmid and chromosomal regions had been put together for *Rhizobium* sp. NGR234 (Perret et al. 1991) and *B. japonicum* USDA110 (Göttfert et al. 2001), respectively, and an ordered BAC library of the entire genome of *S. meliloti* 1021 had been constructed (Capela et al. 1999). However, the ordered cosmid library of *M. loti* MAFF303099 described above is the most easily transferred and maintained in the authentic rhizobial host (Hattori et al. 2002). The *M. loti* MAFF303099 library clones can be used for gene disruption by homologous recombination using a derivative of a particular clone with another incompatible IncP plasmid with different sequences. Then the original clones may be used for complementation studies.

Currently, a database is being constructed to relate detailed characteristics of respective clones. The data to be included will be (1) positional parameters, start and end positions of the inserts; (2) sequence files with an annotated gene table; and (3) results of restriction analysis and other integrity data. This database will be operated as a Web service by 2003 and will be a link from the RhizoBase (http://www.kazusa.or.jp/rhizobase/) and other *Lotus-Mesorhizobium*-related Web pages.

6 Concluding Remarks

In terms of biology and genetics, *M. loti* has historically been less intensively studied than other major rhizobial species such as *S. meliloti* and *B. japonicum*. The establishment of genome sequences and an organized cosmid library will significantly facilitate the design of future experiments. Moreover, post-genome approaches such as micro(macro)array and proteome analyses hold promise of strongly supporting various aspects of experimental design.

Comparative genomics among rhizobia and the phylogenetically close relatives would throw a new light upon the principle of rhizobial genome structures and their evolution as plant-associated and nitrogen-fixing microbes.

All protein functions of the *M. loti* genome have been deduced from their similarity to known proteins of other organisms. Many ORFs do not have such similarities. Therefore, systematic mutational analyses of *M. loti* by an ordered cosmid library and/or minitransposons mutagenesis are ultimately required. *M. loti* mutants showing delay nodulation and cancellation of symbiotic nitrogen fixation may be useful phenotypes for a full understanding of the symbiotic process of *M. loti* as combined with molecular genomic approaches of *L. japonicus*.

Rhizobia such as *M. loti* have survived in soil and inoculant environments outside host plants. Therefore, genomic approaches directed toward understanding environmental adaptation of rhizobia will also be important for agricultural maintenance and environmental conservation (Minamisawa and Mitsui 2000).

Acknowledgements. We thank Bin Ye (Tohoku University) and Noriyuki Nukui (Tohoku University) for taking electron micrographs and photographs of *M. loti* cells and *L. japonicus* nodules, respectively. Appreciation is expressed to PROBRAIN (Japan) for support of researches by K. Saeki and K. Minamisawa.

References

Altschul SF, Madden TL, Schaffer AA, Zhang J, Zhang Z, Miller W, Lipman DJ (1997) Gapped BLAST and PSI-BLAST: a new generation of protein database search programs. Nucleic Acids Res 25:3389–3402

Asamizu E, Watanabe M, Tabata S (2000) Large scale structural analysis of cDNAs in the model legume, *Lotus japonicus*. J Plant Res 113:451–455

Barnett MJ, Fisher RF, Jones T, Komp C, Abola AP, Barloy-Hubler F, Bowser L, Capela D, Galibert F, Gouzy J, Gurjal M, Hong A, Huizar L, Hyman RW, Kahn D, Kahn ML, Kalman S, Keating DH, Palm C, Peck MC, Surzycki R, Wells DH, Yeh KC, Davis RW, Federspiel NA, Long SR (2001) Nucleotide sequence and predicted functions of the entire *Sinorhizobium meliloti* pSymA megaplasmid. Proc Natl Acad Sci USA 98:9883–9888

Capela D, Barloy-Hubler F, Gatius MT, Gouzy J, Galibert F (1999) A high-density physical map of *Sinorhizobium meliloti* 1021 chromosome derived from bacterial artificial chromosome library. Proc Natl Acad Sci USA 96:9357–9362

Capela D, Barloy-Hubler F, Gouzy J, Bothe G, Ampe F, Batut J, Boistard P, Becker A, Boutry M, Cadieu E, Dreano S, Gloux S, Godrie T, Goffeau A, Kahn D, Kiss E, Lelaure V, Masuy D, Pohl T, Portetelle D, Puhler A, Purnelle B, Ramsperger U, Renard C, Thebault P, Vandenbol M, Weidner S, Galibert F (2001) Analysis of the chromosome sequence of the legume symbiont *Sinorhizobium melilo*ti strain 1021. Proc Natl Acad Sci USA 98:9877–9882

Denarie J, Debeelle F, Prome JC (1996) *Rhizobium* lipo-chitooligosaccharide nodulation factors: signaling molecules mediating recognition and morphogenesis. Annu Rev Biochem 65:503–535

Friedman AM, Long SR, Brown SE, Buikema WJ, Ausubel FM (1982) Construction of a broad host range cosmid cloning vector and its use in the genetic analysis of *Rhizobium* mutants. Gene 18:289–296

Gagnon H, Ibrahim RK (1998) Aldonic acids: a novel family of nod gene inducers of *Mesorhizobium loti*, *Rhizobium lupini*, and *Sinorhizobium meliloti*. Mol Plant Microbe Interact 11:988–998

Galibert F, Finan TM, Long SR, Puhler A, Abola P, Ampe F, Barloy-Hubler F, Barnett MJ, Becker A, Boistard P, Bothe G, Boutry M, Bowser L, Buhrmester J, Cadieu E, Capela D, Chain P, Cowie A, Davis RW, Dreano S, Federspiel NA, Fisher RF, Gloux S, Godrie T, Goffeau A, Golding B, Gouzy J, Gurjal M, Hernandez-Lucas I, Hong A, Huizar L, Hyman RW, Jones T, Kahn D, Kahn ML, Kalman S, Keating D, Kiss E, Komp C, Lelaure V, Masuy D, Palm C, Peck MC, Pohl TM, Portetelle D, Purnelle B, Ramsperger U, Surzycki R, Thebault P, Vandenbol M, Vorholter FJ, Weidner S, Wells DH, Wong K, Yeh KC, Batut J (2001) The composite genome of the legume symbiont *Sinorhizobium meliloti*. Science 293:668–672

Göttfert M, Röthlisberger S, Kündig C, Beck C, Marty R, Hennecke H (2001) Potential symbiosis-specific genes uncovered sequencing a 410-kilobase DNA region of the *Bradyrhizobium japonicum* chromosome. J Bacteriol 183:1405–1412

Handberg K, Stougaard J (1992) *Lotus japonicus*, diploid legume species for classical and molecular genetics. Plant J 2:487–496

Hattori Y, Omori H, Hanyu M, Kaseda N, Mishima E, Kaneko T, Tabata S, Saeki K (2002) Ordered cosmid library of the *Mesorhizobium loti* MAFF303099 genome for systematic gene disruption and complementation analysis. Plant Cell Physiol 43:1542–1557

Hayashi M, Imaizumi-Anraku H, Akao S, Kawaguchi M (2000) Nodule organogenesis in *Lotus japonicus*. J Plant Res 113:489–495

Jarvis BDW, Pankhurst CE, Patel JJ (1982) *Rhizobium loti*, a new species of legume root nodule bacteria. Int J Syst Bacteriol 32:378–380

Kaneko T, Nakamura Y, Sato S, Asamizu E, Kato T, Sasamoto S, Watanabe A, Idesawa K, Ishikawa A, Kawashima K, Kimura T, Kishida Y, Kiyokawa C, Kohara M, Matsumoto M, Matsuno A, Mochizuki Y, Nakayama S, Nakazaki N, Shimpo S, Sugimoto M, Takeuchi C, Yamada M, Tabata S (2000a) Complete genome structure of the nitrogen-fixing symbiotic bacterium *Mesorhizobium loti*. DNA Res 7:331–338

Kaneko T, Nakamura Y, Sato S, Asamizu E, Kato T, Sasamoto S, Watanabe A, Idesawa K, Ishikawa A, Kawashima K, Kimura T, Kishida Y, Kiyokawa C, Kohara M, Matsumoto M, Matsuno A, Mochizuki Y, Nakayama S, Nakazaki N, Shimpo S, Sugimoto M, Takeuchi C, Yamada M, Tabata S (2000b) Complete genome structure of the nitrogen-fixing symbiotic bacterium *Mesorhizobium loti* (Supplement). DNA Res 7:381–406

Kohara Y, Akiyama K, Isono K (1987) The physical map of the whole *E. coli* chromosome: application of a new strategy for rapid analysis and sorting of a large genomic library. Cell 50:495–508

Kullik I, Fritsche S, Knobel H, Sanjuan J, Hennecke H, Fischer HM (1991) *Bradyrhizobium japonicum* has two differentially regulated, functional homologs of the sigma 54 gene (*rpoN*). J Bacteriol 173:1125–1138

Lopez-Lara IM, van den Berg JD, Thomas-Oates JE, Glushka J, Lugtenberg BJJ, Spaink HP (1995) Structural identification of the lipo-chitin oligosaccharide nodulation signals of *Rhizobium loti*. Mol Microbiol 15:627–638

Marie C, Broughton WJ, Deakin WJ (2001) *Rhizobium* type III secretion systems: legume charmers or alarmers? Curr Opin Plant Biol 4:336–342

Minamisawa K, Mitsui H (2000) Genetic ecology of soybean bradyrhizobia. In: Bollag JM, Stotzky G (eds) Soil biochemistry, vol 10. Marcel Dekker, New York, pp 349–377

Niner BM, Hirsch AM (1998) How many *Rhizobium* genes, in addition to *nod*, *nif/fix*, and *exo*, are needed for nodule development and function? Symbiosis 24:51–102

Niwa S, Kawaguchi M, Imaizumi-Anraku H, Chechetka SA, Ishizawa M, Ikuta A, Kouchi H (2001) Responses of a model legume *Lotus japonicus* to lipochitin oligosaccharide nodulation factors purified from *Mesorhizobium loti* JRL501. Mol Plant-Microbe Interact 14:848–856

Perret X, Broughton WJ, Brenner S (1991) Canonical ordered cosmid library of the symbiotic plasmid of *Rhizobium* species NGR234. Proc Natl Acad Sci USA 88:1923–1927

Rostas K, Kondorosi E, Horvath B, Simoncsits A, Kondorosi A (1986) Conservation of extended promoter regions of nodulation genes in *Rhizobium*. Proc Natl Acad Sci USA 83:1757–1761

Saeki K, Kouchi H (2000) The *Lotus* symbiont, *Mesorhizobium loti*: molecular genetic techniques and application. J Plant Res 113:457–465

Schauser L, Roussis A, Stiller J, Stougaard J (1999) A plant regulator controlling development of symbiotic root nodules. Nature 402:191–195

Sonnhammer EL, Eddy SR, Birney E, Bateman A, Durbin R (1998) Pfam: multiple sequence alignments and HMM-profiles of protein domains. Nucleic Acids Res 26:320–322

Stiller J, Martirani L, Tuppale S, Chian R, Chiurazzi M, Gresshoff PM (1997) High frequency transformation and regeneration of transgenic plants in the model legume *Lotus japonicus*. J Exp Bot 48:1357–1365

Sullivan JT, Ronson CW (1998) Evolution of rhizobia by acquisition of a 500-kb symbiosis island that integrates into a phe-tRNA gene. Proc Natl Acad Sci USA 95:5145–5149 (with published erratum appears in Proc Natl Acad Sci USA 95:9059)

Sullivan JT, Trzebiatowski JR, de Bruijn FJ, Ronson CW (2000) The symbiosis island of a *Mesorhizobium* strain that nodulates *Lotus*. In: Triplett EW (ed) Prokaryotic nitrogen fixation: a model system for the analysis of a biological process. Horizon Scientific Press, Wymondham, pp 693–704

Sullivan JT, Trzebiaowski JR, Cruickshank RW, Gouzy J, Brown SD, Elliot RM, Fleetwood DJ, McCallum NG, Rossbach U, Stuart GS, Weaver JE, Webby RJ, de Bruijn FJ, Ronson CW (2002) Comparative sequences analysis of the symbiosis island of *Mesorhizobium loti* strain R7A. J Bacteriol 184:3086–3095

Tajima S, Takane K, Nomura M, Kouchi H (2000) Symbiotic nitrogen fixation at the late stage of nodules formation in *Lotus japonicus* and other legume plants. J Plant Res 113:467–473

van Berkum P, Eardy BD (1998) Molecular evolutionary systematics of the *Rhizobiaceae*. In: Spaink HP, Kondorosi A, Hooykaas PJJ (eds) The *Rhizobiaceae*: molecular biology of model plant-associated bacteria. Kluwer, Dordrecht, pp 155–172

Young JPY (1996) Phylogeny and taxonomy of rhizobia. Plant Sci 186:45–52

14 Molecular Linkage Map of the Model Legume *Medicago truncatula* Gaertn.

P. Thoquet[1], A. Kereszt[1], J.M. Prosperi[2], and T. Huguet[1]

1 Introduction

The need for a more sustainable and environmentally safe agriculture has reinforced interest in the cultivation of legumes, which include important agricultural species such as alfalfa, clover, pea, soybean, bean, peanut, and more. These species have the capacity to establish an atmospheric dinitrogen fixing symbiose with soil bacteria collectively named rhizobia, and to form symbiotic root mycorrhizae with soil fungi, thus facilitating their uptake of phosphate, water, and other soil nutrients (Albrecht et al. 1999). However, genetic analysis of these processes remains difficult in the major crop legumes due to features such as tetraploidy, large genomes and/or the lack of efficient methods for transgenesis. Since the model plant *Arabidopsis thaliana* (as indeed is true of other Cruciferae) is unable to establish either rhizobial or mycorrhizal symbioses, the need to establish a model legume to analyze plant physiological processes which cannot be satisfactorily studied in *A. thaliana* has been recognized for over a decade (Cook and Denarie 2000).

The alfalfa relative *Medicago truncatula* possesses a number of interesting characteristics for both molecular and classical genetics (Barker et al. 1990; Cook 1999). Key attributes of *M. truncatula* include diploidy and autogamous fertilization, a small genome (500–600 Mbp/1C; Blondon et al. 1994), a rapid reproductive cycle, a high level of biodiversity, a number of available cultivars (Prosperi et al. 2001), and a well-characterized nitrogen-fixing symbiont, *Sinorhizobium meliloti*, whose genome has been recently sequenced (Galibert et al. 2001). *M. truncatula* is also being used as a model plant for studying mycorrhizal interactions (Sagan et al. 1995). Several tools for molecular genetics and genomics have been recently developed, including mutant collections (Sagan et al. 1995; Penmetsa and Cook 2000), cDNA libraries (Cowitz et al. 1998; Györgyey et al. 2000), a large-insert BAC library (Nam et al. 1999), and efficient methods of transformation (Chabaud et al. 1996; Trinh et al. 1998; Trieu et al. 2000).

[1] Laboratoire de Biologie Moléculaire des Relations Plantes-Microorganismes, CNRS-INRA, BP27, 31326 Castanet-Tolosan Cedex, France
[2] Station de Génétique et d'Amélioration des Plantes, INRA, Domaine de Melgueil, 34130 Mauguio, France

Biotechnology in Agriculture and Forestry, Vol. 52
Nagata/Tabata (Eds.) Brassicas and Legumes

Furthermore, the genus *Medicago* is part of the Galegoid phylum and is therefore related to a number of important crop legumes in addition to alfalfa such as pea, faba bean, chickpea, lentil, and clover (Doyle et al. 1996). Members of this phylum are expected to show a high level of nucleotide sequence conservation and similar genetic organization, and hence the potential for relatively easy transfer of genome information between member species.

In the *Medicago* genus, diploid populations of *M. sativa* were used for the construction of a high-density genetic map (Kalo et al. 2000) in order to avoid the complicated tetrasomic inheritance and linkage relationships of tetraploid alfalfa. Regarding *M. truncatula,* the first report of a genetic map has been published recently based on a polymorphic F2 population (Thoquet et al. 2002). The existence of a synteny between the model legume *M. truncatula* and allogamous *Medicago sativa* (alfalfa) opens the way to the identification and mapping of a number of orthologous genes and quantitative traits of interest. The use of this new genetic map has been illustrated recently (Thoquet et al. 2002) by the mapping of two genes of *M. truncatula*: the *Mtsym6* gene (Tirichine et al. 2000) involved in the strain × cultivar specificity of nitrogen fixation, and the *SPC* gene, which determines the direction of pod coiling (Lilienfeld and Kihara 1956).

2 Parental Lines

Two *M. truncatula* populations were selected on the basis of their phenotypic and molecular differences: the cultivar Jemalong and the Algerian natural population DZA315.16. Since *M. truncatula* is a self-fertilizing plant whose natural populations are composed of a variable number of different homozygous genotypes (Bonnin et al. 1996, 2001), a single seed from each population was selfed at least twice in order to ensure maximal homozygosity of the *M. truncatula* line. These two parental lines (Jemalong 6 and DZA315.16) were named after their respective population or cultivar, followed by a number that identifies the particular line.

These two lines can be easily distinguished from each other by leaf pigmentation pattern: Jemalong 6 shows a typical pattern on the adaxial leaf surface whereas DZA315.16 leaves have only infrequent and randomly distributed spots (Fig. 1A). Penmetsa and Cook (2000) reported the existence of a single gene responsible for the leaf spots between the Jemalong and A20 lines. Another striking difference lies in pod shape: Jemalong 6 pods have a spiny barrel shape characteristic of the *M. truncatula* ssp. *truncatula* pods described by Lesins and Lesins (1979), whereas DZA315.16 forms shorter and smaller pods and spines (Fig. 1B). Finally, the two lines show opposite coiling: anticlockwise for Jemalong 6 and clockwise for DZA315.16 (Fig. 1C).

The molecular diversity between the two lines was estimated by using anonymous amplified fragment length polymorphism (AFLP) banding patterns. We observed that 32% of AFLP bands are polymorphic between

Fig. 1. Phenotypic patterns which distinguish Jemalong 6 and DZA315.16 lines. *Bars* = 5 mm. **A** Typical leaf pigmentation found on adaxial leaf surface of Jemalong 6 (*left*) and DZA315.16 (*right*). **B** Pod shape of Jemalong 6 (*above*) and DZA315.16 (*below*). **C** Anticlockwise pod coiling of Jemalong 6 (*above*). Clockwise pod coiling of DZA315.16 (*below*). (Thoquet et al. 2002)

Jemalong 6 and DZA315.16, which is similar to the 30 to 34% observed among three ecotypes of *Arabidopsis thaliana* (Alonso-Blanco et al. 1998).

The cultivar Jemalong was the first *M. truncatula* cultivar to be released (Andrew and Hudson 1954) and is widely used in ley farming (Oram 1990). The cultivar Jemalong was initially selected for molecular studies because of its transformation/regeneration characteristics (Nolan et al. 1989; Chabaud et al. 1996). So far, most of the *Medicago* programs including mutagenesis, the construction of large-scale cDNA and genomic libraries, DNA sequencing, and genetic resources (Frugoli and Harris 2001) have made use of Jemalong lines.

3 Methodology

3.1 Plant Material

Individual seeds of the *M. truncatula* cultivar Jemalong and the Algerian natural population DZA315 (JMP, unpubl. results) were selfed at least twice and then crossed manually, Jemalong being the female parent (Thoquet et al. 2002). A population of 124 F2 plants, derived from a single F1 plant, was used for genetic mapping and is designated hereafter as the “mapping population”. The hybrid nature of the F1 plant was confirmed with molecular markers. More than 95% of the F2 seeds germinated, out of which about 6% gave albino plantlets that were discarded.

3.2 DNA Isolation and Marker Typing

DNA isolation, RAPD, and AFLP amplification and coding were as described by Thoquet et al. (2002). The intensity of the bands was not taken into account. Restriction fragment length polymorphism (RFLP) markers (Table 1) were analyzed as described by Thoquet et al. (2002). The polymorphism of 17 genes was identified after PCR amplification with primers anchored either on both

Table 1. List of genes and isoenzymes used

Markers	Nature	Function	RE[a]	LG[b]
ENBP1	SSR[c]	Chloroplastic protein		1
ENOD8	CAPS[d]	Nodulin	*Ssp*I	1
MtPT-1	CAPS	Phosphate transporter	*Rsa*I	1
PEPC	CAPS	PhosphoEnolPyruvate Carboxylase	*Rsa*I	2
Z-F1	RFLP	Zn-Finger1	*Eco*RI	3
GS.B	PCR[e]	Glutamine synthetase		3
Mtlec2	SSR	Lectine 2		3
ENOD12	PCR	Nodulin		3
ATPase	SSR	Vacuolar ATPase		4
MtST-1	PCR	Sucrose transporter		4
PGM	Isoenz[f]	PhosphoGlucoMutase		4
Gs.C	CAPS	Glutamine synthetase	*Vsp*I	5
*Lb*I	CAPS	Leghemoglobin I	*Ssp*I	5
rDNA	CAPS	45 s Ribosomal DNA	*Xba*I	5
ENOD40	CAPS	Nodulin	*Alu*I	5
GS.A	PCR	Glutamine synthetase		6
MtN1	PCR	Nodulin		6
GSHS	SSR	Glutathione synthetase		7
VR	CAPS	Vestitone reductase	*Dde*I	7
PGD	Isoenz	PhosphoGlucoDeshydrogenase		7
ENOD16	RFLP	Nodulin	*Eco*RV	8

All the references for these markers are from Thoquet et al. (2002); all markers, except for MtPT-1 and Mtlec2, have been used for synteny studies.

[a] RE, restriction enzyme.

[b] LG, *M. truncatula* linkage group.

[c] SSR, simple sequence repeat: polymorphism is revealed by the length of the amplified fragment.

[d] CAPS, cleaved amplified polymorphic sequence: polymorphism is revealed after digestion of the amplified fragment with restriction enzymes.

[e] PCR, polymerase chain reaction: polymorphism is revealed by the length of the amplified fragment.

[f] Isoenz, isoenzyme.

sides of a microsatellite motif (SSR) or within exons and crossing introns (data not shown). RFLP, CAPS, and PCR markers were scored as codominant. Isozyme markers were typed as described in Thoquet et al. (2002).

3.3 Map Construction

The genetic map was constructed using the MapMaker/Exp V3.0 software (Lander et al. 1987) with the Kosambi map function. The segregation data were tested for deviation from the expected Mendelian ratio using a Chi-square test. The markers showing significant deviation at $\alpha = 0.05$ were nevertheless used to build the linkage map.

3.4 Gene Mapping

Mtsym6 and *SPC* genes were analyzed using F3 populations and mapped as described in Thoquet et al. (2002).

4 Construction of an F2 Genetic Map

The 124 individuals of the mapping population were genotyped for 298 markers comprising 262 dominant anonymous markers (62 RAPD and 200 AFLP), 5 pairs of RAPD and 10 pairs of AFLP scored as codominant markers, 19 genes with known functions, and 2 codominant isoenzyme markers. The RAPD and AFLP primer generated, on average and respectively, 4 and 17 markers per primer combination.

The *M. truncatula* genetic map was built with a minimum (log of odd ratios) LOD score value of 5. The resulting *M. truncatula* genetic map is composed of 289 markers forming 8 linkage groups and 9 markers unlinked and spans 1225 Kosambi cM, this give an average of 470 kb/cM (Fig. 2).

The distribution of markers is approximately similar whatever the linkage groups (Table 2). The 8 linkage groups of *M. truncatula* differ in genetic size by a factor of 2.4, from 92 to 219 cM (Table 2), whereas no obvious correlation could be found between genetic size of linkage groups and pachytene chromosome length measurements (Kulikova et al. 2001).

About 27% of the markers do not show a typical Mendelian segregation ratio at $\alpha = 0.05$. Interestingly, the distorted markers are not scattered over the *M. truncatula* genome but concentrated on linkage groups I, II, and III, representing 86% of all the distorted markers with high χ^2 values for 1:3 segregation. Furthermore, these distorted markers are not randomly distributed throughout chromosomes: for example, a nearly linear gradient of distortion can be observed on linkage group III (Fig. 3), ranging from Mendelian equality of male and female markers to a fivefold predominance of male alleles. On

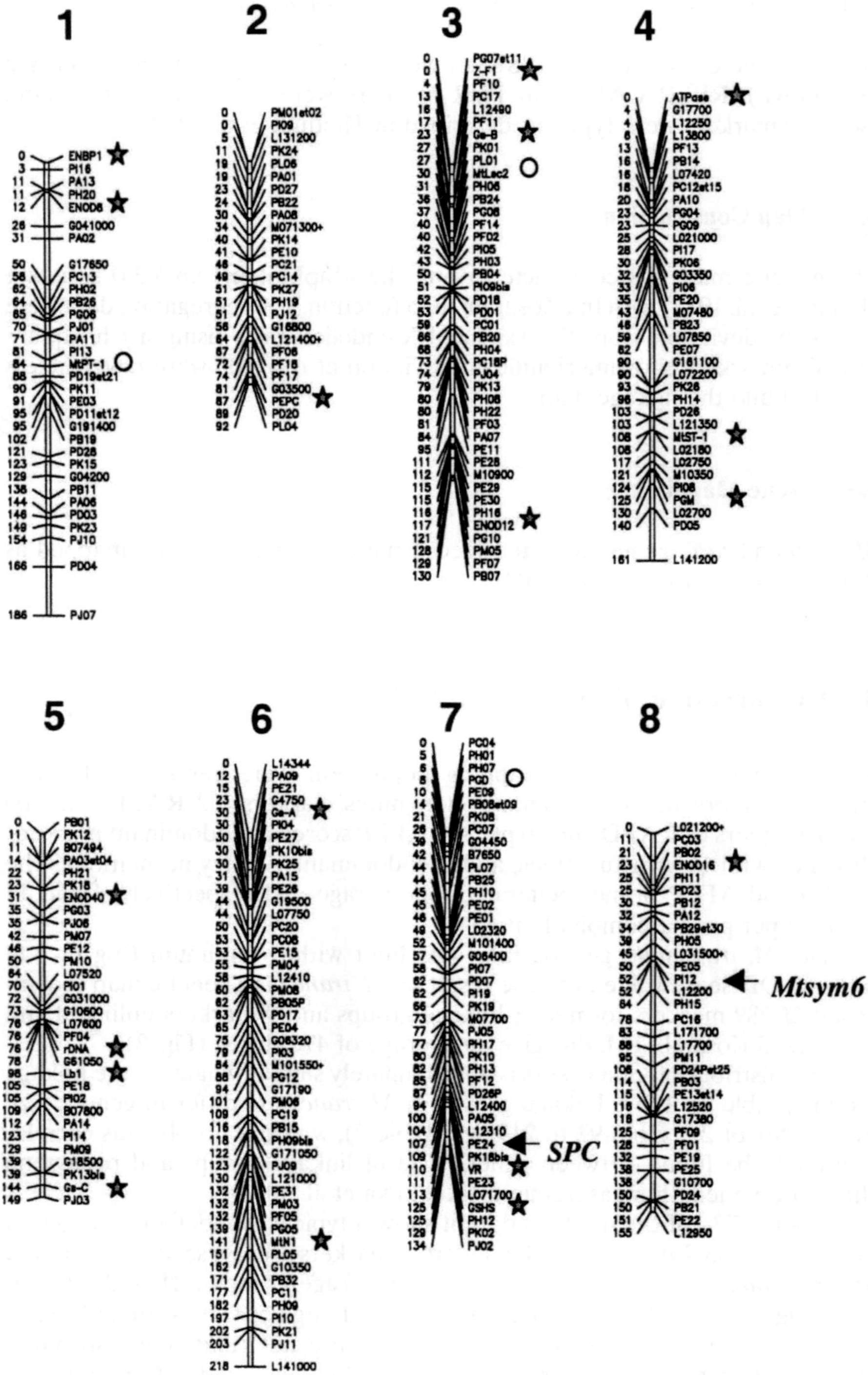

Fig. 2. Global F2 genetic map of *Medicago truncatula*. The *number* above the linkage groups refers to the homologous linkage group in *M. sativa* (Kalo et al. 2000). The *code* to the right of the linkage groups refers to the marker name. The *numbers* to the left of the linkage groups refer to the genetic distances (Kosambi cM) from the top and have been rounded for clarity. *Stars* refer to known genes used for synteny studies. *Circles* refer to known genes not used for synteny studies. (Thoquet et al. 2002)

Table 2. Distribution of markers according to the linkage group

Linkage group	No. of markers	cM[a]	Average distance (cM)	SD[b] (cM)
I	32	186	5.6	5.1
II	26	92	3.7	2.5
III	42	130	3.2	3.4
IV	36	161	4.6	5.7
V	31	149	4.6	4.4
VI	48	218	4.6	3.9
VII	41	134	3.3	2.7
VIII	33	155	5.1	4.2
	289	**1225**	**4.4**	**4.3**

[a] cM, Kosambi centimorgan.
[b] SD, standard deviation.

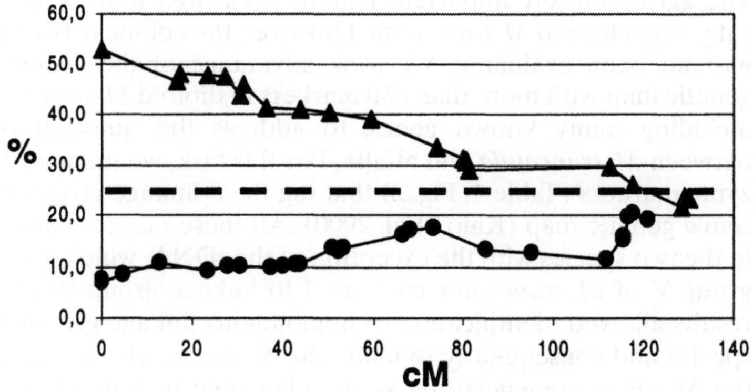

Fig. 3. Segregation distortion of the female and male markers along linkage group III of the F2 genetic map of *M. truncatula*. *Circles* and *triangles* refer to female and male alleles, respectively. *x-axis* Genetic distance from the top of the linkage group in Kosambi cM; *y-axis* frequency (%) of segregation of female and male alleles in the mapping population. A dominant marker is considered to be a male marker if the recessive allelic form is male (and the same for the female markers). If no distortion occurs, the segregation value should be close to 25% for both male and female markers (*dashed line*). (Thoquet et al. 2002)

a whole genome basis, the frequency of distorted male compared to female markers appears to be similar, 22 and 29%, respectively. From the analysis carried out by Jenczewski et al. (1997), it is clear that variations from expected Mendelian ratios are common within both inter- and intraspecific crosses. Although we cannot establish if deleterious recessive alleles, self-incompatibility alleles, or structural rearrangements are the distorting factors involved in this cross, it is noteworthy that Jemalong 6 and DZA315.16 have a similar overall DNA content (1.16 pg/2C). However, the gradient in distortion in favor of male alleles along linkage group III suggests the presence of specific gene(s) interfering with alleles' segregation. It would be of interest to know if the segregation distortion frequency depends on the lines involved in the cross and if it is always limited to the same three linkage groups.

The reliability and value of this genetic map is demonstrated in the following ways: (1) identical number of linkage groups and chromosomes (2n = 16) as well as the small number of unlinked markers [two random amplified polymorphic DNA (RAPD) and seven AFLP] at a minimal LOD 5; (2) no discrepancies between the global F2 map and both male and female maps; (3) only a limited clustering of markers on a few linkage groups. The excellent distribution of markers is illustrated by the fact that 90% of the markers are located within an interval of less than 10 cM from adjacent markers.

5 Alignment of *M. sativa* and *M. truncatula* Genetic Maps

The agronomically important legume *Medicago sativa* (alfalfa) is taxonomically very close to *M. truncatula*. However, the collinearity of the two genomes has not been evaluated. We took advantage of the availability of a dense genetic map with more than 850 markers in diploid *M. sativa* (Kalo et al. 2000), including many known genes, to address the question of macrosynteny between *M. truncatula* and alfalfa. For this task, we mapped 18 gene or isoenzyme markers (Table 1, Fig. 2) that tag the 8 linkage groups of the diploid *M. sativa* genetic map (Kalo et al. 2000). All these markers were similarly linked in the two species with the exception of the rDNA, which was found on linkage group V of *M. truncatula*, compared to linkage group VI of *M. sativa*. These results allowed identification of homologous linkage groups between the two species and consequently to name the *M. truncatula* linkage groups following the *M. sativa* nomenclature, as already stated in Kalo et al. (2000). This procedure of identification has been further extended to chromosome nomenclature (Kulikova et al. 2001). The estimation of the degree of genome collinearity between these two closely related legume species (as well as with more distant legumes) is presently in progress. However, several questions remain open: (1) what is the extent of macro- and microsynteny between *M. truncatula* and diploid alfalfa?; (2) are diploid and tetraploid alfalfa syntenic?; and (3) is synteny general within the *Medicago* genus, and more generally within legumes?

6 Genetic Mapping of Traits of Interest

Both parental lines, Jemalong 6 and DZA315.16, are efficiently nodulated by the reference *Sinorhizobium meliloti* strains 2011 (data not shown). However, we have recently identified in our laboratory one particular gene (*Mtsym6*) that is implicated in the inefficient symbiosis between Jemalong and the wild type *S. meliloti* strain A145 (Tirichine et al. 2000). In order to map this gene, we determined the nitrogen-fixing genotype of 73 F2 plants of the mapping population by inoculating their F3 progeny. In total, 15 F3 families were found to be Fix^-, 19 were Fix^+, and 39 displayed both Fix^+ and Fix^-, thus indicating that their F2 parents were heterozygous. Symbiotic typing data together with molecular marker segregation allowed us to map the *Mtsym6* gene on linkage group VIII (Fig. 2). This is the first report of the genetic mapping of a symbiotic gene in the model legume *M. truncatula*. The map-based cloning of *Mtsym6* is currently underway in the laboratory.

The clockwise (DZA315.16) or anticlockwise (Jemalong 6) polymorphism observed on parental pods (Fig. 1) was treated as a morphological marker. From 110 F3 families, we observed that 28 (25.5%) had anticlockwise and 82 had clockwise pod coiling. In agreement with Lilienfeld and Kihara (1956) and Bena et al. (1998), we reached the conclusion that pod coiling sense has a monogenic determinism with clockwise turning dominant. We named this gene *SPC* (for *S*ense of *P*od *C*oiling) and mapped it to linkage group VII. The map-based cloning of *SPC* is currently in progress in the laboratory.

7 Conclusion and Future Perspectives

The identification of polymorphic Jemalong 6 and DZA315.16 lines together with the making of an F2 genetic map and the identification of synteny with alfalfa, as described in this report, are likely to provide a powerful tool for both fundamental and applied approaches, such as studying the conservation of synteny between genomes of legumes and other model species, and the positional cloning of agriculturally important genes. We are currently creating a population of recombinant inbred lines (RILs) based on the same *M. truncatula* cross. These lines will be shared among the scientific community.

Acknowledgements. This work was supported in part by grants from INRA/CNRS for the *Medicago truncatula* genome project (1998–2000). PT and AK were recipients of postdoctoral fellowships from INRA (Santé des Plantes et Environnement Department).

References

Albrecht C, Geurts R, Bisseling T (1999) Legume nodulation and mycorrhizae formation: two extremes in host specificity meet. EMBO J 18:281–288

Alonso-Blanco C, Peeters AJM, Koorneef M, Lister C, Dean C, van den Bosch N, Pot J, Kuiper MTR (1998) Development of an AFLP based linkage map of L*er*, Col and Cvi *Arabidopsis thaliana* ecotypes and construction of a L*er*/Cvi recombinant inbred line population. Plant J 14:259–271

Andrew WD, Hudson WJ (1954) A superior strain of medic- Barrel medic 173. Agric Gaz Feb: 76–80

Barker DG, Bianchi S, Blondon F, Dattée Y, Duc G, Flament P, Gallusci P, Génier G, Guy P, Muel X, Tourneur J, Dénarié J, Huguet T (1990) *Medicago truncatula*, a model plant for studying the molecular genetics of the *Rhizobium*-legume symbiosis. Plant Mol Biol Rep 8:40–49

Bena G, Prosperi JM, Lejeune B, Olivieri I (1998) Evolution of annual species of the genus *Medicago*: a molecular phylogenetic approach. Mol Phylogenet Evol 9:552–559

Blondon F, Marie D, Brown S, Kondorosi A (1994) Genome size and base composition in *Medicago sativa* and *M. truncatula* species. Genome 37:264–270

Bonnin I, Huguet T, Ghérardi M, Prosperi JM, Olivieri I (1996) High level of polymorphism and spatial structure in a selfing plant species *Medicago truncatula* Gaertn. using RAPDs markers. Am J Bot 83:843–855

Bonnin I, Ronfort J, Wozniak F, Olivieri I (2001) Spatial effects and rare outcrossing events in *Medicago truncatula* (Fabaceae). Mol Ecol 10:1371–1384

Chabaud M, Larsonnaud C, Marmouget C, Huguet T (1996) Transformation of Barrel Medic (*Medicago truncatula* Gaertn.) by *Agrobacterium tumefaciens* and regeneration via somatic embryogenesis of transgenic plants with the MtENOD12 nodulin promoter fused to the GUS gene. Plant Cell Rep 15:305–310

Cook D (1999) *Medicago truncatula*: a model in the making! Curr Opin Plant Biol 2:301–304

Cook D, Dénarié J (2000) Progress in the genomics of *Medicago truncatula* and the promise of its application to grain legume crops. Grain Legumes 28:12–13

Cowitz PA, Smith LS, Long SR (1998) Expressed sequence tags from a root-hair enriched *Medicago truncatula* cDNA library. Plant Physiol 117:1325–1332

Doyle JJ, Doyle JL, Ballenger JA, Palmer JD (1996) The distribution and phylogenetic significance of a 50 kb chloroplast DNA inversion in the flowering plant family Leguminosae. Mol Phylogenet Evol 5:429–438

Frugoli J, Harris JM (2001) *Medicago* on the move. Plant Cell 13:458–63

Galibert F, Finan TM, Long SR, Puhler A, Abola P, Ampe F, Barloy-Hubler F, Barnett MJ, Becker A, Boistard P, Bothe G, Boutry M, Bowser L, Buhrmester J, Cadieu E, Capela D, Chain P, Cowie A, Davis RW, Dreano S, Federspiel NA, Fisher RF, Gloux S, Godrie T, Goffeau A, Golding B, Gouzy J, Gurjal M, Hernandez-Lucas I, Hong A, Huizar L, Hyman RW, Jones T, Kahn D, Kahn ML, Kalman S, Keating DH, Kiss E, Komp C, Lelaure V, Masuy D, Palm C, Peck MC, Pohl TM, Portetelle D, Purnelle B, Ramsperger U, Surzycki R, Thebault P, Vandenbol M, Vorholter FJ, Weidner S, Wells DH, Wong K, Yeh KC, Batut J (2001) The composite genome of the legume symbiont *Sinorhizobium meliloti*. Science 293:668–672

Gamas P, Niebel FD, Lescure N, Cullimore JV (1996) Use of a subtractive hybridization approach to identify new *Medicago truncatula* genes induced during root nodule development. Mol Plant Microbe Interact 9:233–242

Györgyey J, Vaubert D, Jiménez-Zurdo JI, Charon C, Troussard L, Kondorosi A, Kondorosi E (2000) Analysis of *Medicago truncatula* nodule expressed tags. Mol Plant Microbe Interact 13:62–71

Hoffmann B, Trinh TH, Leung J, Kondorosi A, Kondorosi E (1997) A new *Medicago truncatula* line with superior in vitro regeneration, transformation and symbiotic properties isolated through cell culture selection. Mol Plant Microbe Interact 10:307–315

Jenczewski E, Ghérardi M, Bonnin I, Prosperi JM, Olivieri I, Huguet T (1997) Insight on segregation distortions in two intraspecific crosses between annual species of *Medicago* (Leguminosae). Theor Appl Genet 94:682–691

Kalo P, Endre G, Zimanyi L, Csanadi G, Kiss GB (2000) Construction of an improved linkage map of diploid alfalfa (*Medicago sativa*). Theor Appl Genet 100:641–657

Kulikova O, Gualtieri G, Geurts R, Kim D-J, Cook D, Huguet T, de Jong H, Fransz P, Bisseling T (2001) Integration of the FISH-pachytene and genetic maps of *Medicago truncatula*. Plant J 27:49–58

Lander ES, Green P, Abrahamson J, Barlow A, Daly MJ, Lincoln SE, Newburg L (1987) MAPMAKER: an interactive computer package for constructing primary genetic linkage maps of experimental and natural populations. Genomics 1:174–181

Lilienfeld FA, Kihara H (1956) Dextrality and sinistrality in plants. Proc Jpn Acad 32:620–632

Nam YW, Penmetsa RV, Endre G, Uribe P, Kim DJ, Cook DR (1999) Construction of a bacterial artificial chromosome library of *Medicago truncatula* and identification of clones containing ethylene-response genes. Theor Appl Genet 98:638–646

Nolan KE, Rose RJ, Gorst JR (1989) Regeneration of *Medicago truncatula* from tissue culture: increase somatic embryogenesis using explants from regenerated plants. Plant Cell Rep 8:278–281

Oram RN (1990) Register of Australian herbage plant cultivars, 3rd edn. CSIRO, Clayton South, VIC, Australia

Penmetsa RV, Cook DR (2000) Production and characterization of diverse developmental mutants of *Medicago truncatula*. Plant Physiol 123:1387–1397

Prosperi JM, Auricht G, Génier G, Johnson R (2001) Medics (*Medicago* L.). In: Maxted N, Bennett SJ (eds) Plant genetic resources of legume in the Mediterranean. Kluwer, Dordrecht, pp 99–114

Rose RJ, Nolan KE, Bicego L (1999) The development of the highly regenerable seed line Jemalong 2HA for transformation of *Medicago truncatula* – implications for regenerability via somatic embryogenesis. J Plant Physiol 155:788–791

Sagan M, Morandi D, Tarenghi E, Duc G (1995) Selection of nodulation and mycorrhizal mutants in the model plant *Medicago truncatula* Gaertn after gamma rays mutagenesis. Plant Sci 111:63–71

Thoquet P, Ghérardi M, Journet E-P, Kereszt A, Ané J-M, Prosperi J-M, Huguet T (2002) The molecular genetic linkage map of the model legume *Medicago truncatula*: an essential tool for comparative legume genomics and the isolation of agronomically important genes. BMC Plant Biol 2 [http://www.biomedcentral.com/1471–2229/2/1]

Tirichine L, De Billy F, Huguet T (2000) *Mtsym6*, a gene conditioning *Sinorhizobium* strain-specific nitrogen fixation in *Medicago truncatula*. Plant Physiol 123:845–851

Trieu AT, Burleigh SH, Kardailsky IV, Maldonado-Mendoza IE, Versaw WK, Blaylock LA, Shin H, Chiou TJ, Katagi H, Dewbre GR, Weigel D, Harrison MJ (2000) Transformation of *Medicago truncatula* via infiltration of seedlings or flowering plants with *Agrobacterium*. Plant J 22:531–541

Trinh TH, Ratet P, Kondorosi E, Durand P, Kamate K, Bauer P, Kondorosi A (1998) Rapid and efficient transformation of diploid *Medicago truncatula* and *Medicago sativa* ssp. *falcata* lines improved in somatic embryogenesis. Plant Cell Rep 17:345–355

15 Genetic Mapping of Seed and Nodule Protein Markers in Diploid Alfalfa (*Medicago sativa*)

P. KALÓ, K. FELFÖLDI, A. SERES, G. ENDRE, and G.B. KISS

1 Introduction

The genetic map of diploid alfalfa was generated based on the genetic analyses of genotypes of markers collected for the individuals of an F2 population originating from the cross of two diploid alfalfa subspecies, *M. sativa* ssp. *quasifalcata* and *M. sativa* ssp. *coerulea* (Kiss et al. 1993). The constructed linkage map composed of eight linkage groups which contain more than 1800 morphological, allozyme, and DNA-based markers, of which more than 300 represent loci with known function or sequences in the genome (Kaló et al. 2000). The markers cover approximately 760 cM genetic distance. The average correlation between the physical and the genetic distance is about 1000 kb/cM. This genetic system was used to map protein markers originating from nodule and seed tissue.

The dominant protein species in seeds are the storage proteins that serve as nitrogen and sulfur sources for germinating seeds. Seed storage proteins of dicotyledonous plants are salt-soluble globulins coded by the individual members of multigene families (for review, see Higgins 1984). The storage proteins are often synthesized as pre-proteins followed by cleavage process yielding mature terminal products (subunits). The major storage protein families of leguminous plants are the low molecular weight (LMW) 2S albumins, the legumins (11S globulins), and the vicilins (7S globulins) (Higgins 1984). These three major classes of storage proteins comprise 20, 30, and 10%, respectively, of the total extractable proteins from mature seeds of alfalfa (Krochko and Bewley 2000). The legumin-like medicagins, the vicilin-like alfins, and the 2S albumins in alfalfa were characterized previously in detail by Stuart and Nelsen (1988), Krochko and Bewley (1990, 2000), Krochko et al. (1990), and Coulter and Bewley (1990). The storage proteins coded by multiple copy genes and their alleles differ in their amino acid composition. The slight variations in the amino acid composition can result in proteins with diverse molecular weights and/or isoelectric points that can be visualized when they are separated in two-dimensional polyacrylamide gels (2-D PAGE). When the protein spots are analyzed from individuals of a segregating population, the proteins with minor variations in their electro-

Institute of Genetics, Biological Research Center, Hungarian Academy of Sciences, Szeged, Hungary

Biotechnology in Agriculture and Forestry, Vol. 52
Nagata/Tabata (Eds.) Brassicas and Legumes

phoretic mobility can serve as genetic markers (Leonardi et al. 1988; Gerber et al. 1993).

Here we describe the characterization of the 7S (alfin) and 11S (medicagin) storage proteins in diploid alfalfa and their mapping based on the allelic variations in our segregating alfalfa population. Besides the 2-D PAGE analysis of the storage protein, other seed proteins were also surveyed for allelism and mapped when showing polymorphism. Similarly, the same technique was used to identify possible variations in the proteins isolated from root nodules that are capable of nitrogen fixation in symbiosis with the soil bacteria *Sinorhizobum meliloti*. The mapping of seed- and nodule-specific markers in correlation with already mapped molecular DNA markers allows marker-assisted breeding of storage proteins and other protein-based characters.

2 Results and Discussion

2.1 Characterization of Storage Proteins in Diploid Alfalfa by 2-D PAGE

The proteins for two-dimensional polyacrylamide gel electrophoresis (2-D PAGE) were isolated from seeds produced on the F2 individuals of the diploid alfalfa mapping population. Three different types of extraction were used as described in the Protocols (Sect. 4). The separation of the proteins in the first dimension was according to their isoelectric points (isoelectric focusing, IEF) followed by the second dimension according to their size in a denaturing polyacrylamide gel (SDS-PAGE). The protein spots were visualized by double staining, first with silver then by Coomassie BB (see Sect. 4.4). According to the staining behavior, protein spots could be classified into three categories: (1) bluish spots (Coomassie BB staining dominated); (2) brownish spots with different intensities (silver staining dominated); and (3) black spots (staining with both silver and Coomassie BB). After 2-D PAGE and staining, about 700 spots could be identified between pH 5.0 and 8.0 and between 18 and 95 kDa molecular mass. Figure 1a shows the 2-D gel of a protein sample extracted in O'Farrell lysis buffer, while Fig. 2a shows the separation of proteins from the S-1, or low-salt protein fraction, on the other hand, Fig. 2b that of the S-2, or high-salt protein fraction.

The major portion of the proteins in seeds is composed of storage proteins, and they thus appear as larger spots with higher staining intensity than do the seeds' nonstorage proteins. On the other hand, the nonstorage proteins represent far more protein species, and therefore they appear as many smaller spots. Since storage proteins are high molecular weight oligomeric proteins composed of subunits each possessing slightly different isoelectric points, they appear as a "string of bead-like" multiple spots in 2-D PAGE (e.g., medicagin subunits in Fig. 2a, b and Fig. 3). The vicilin-type 7S alfins and other abundant

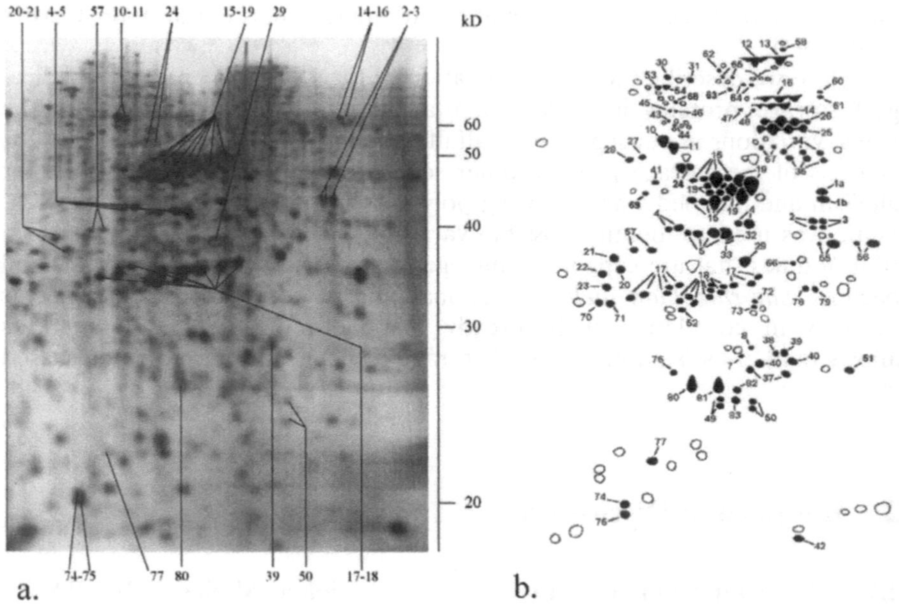

Fig. 1. **a** Two-dimensional gel of seed proteins extracted in O'Farrell lysis buffer from 12 seeds of the F1/1 *Medicago* plant. Some representative protein spots are indicated for orientation. **b** Schematic diagram of the protein spots mapped on the genetic map of alfalfa

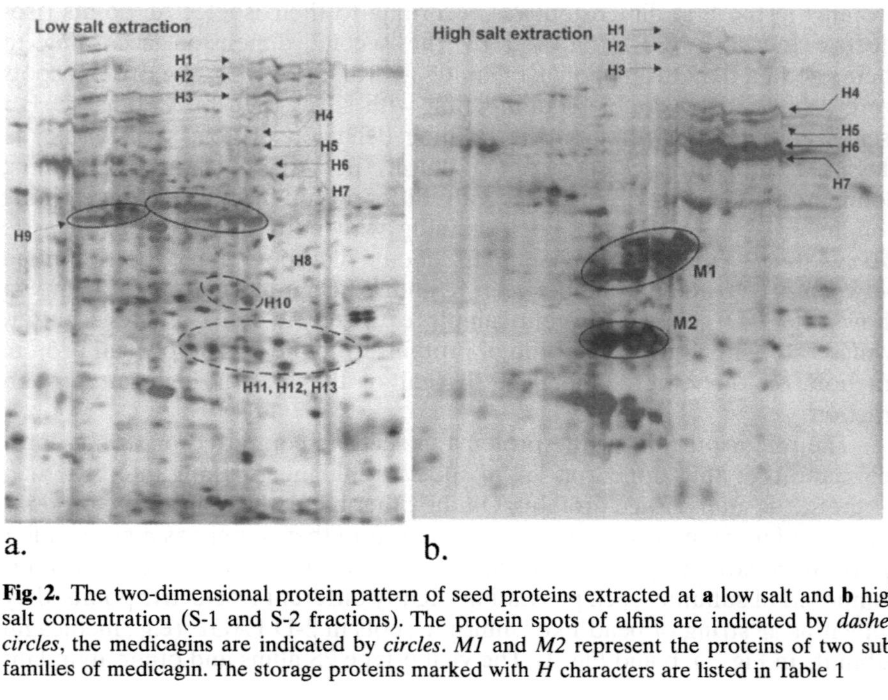

Fig. 2. The two-dimensional protein pattern of seed proteins extracted at **a** low salt and **b** high salt concentration (S-1 and S-2 fractions). The protein spots of alfins are indicated by *dashed circles*, the medicagins are indicated by *circles*. *M1* and *M2* represent the proteins of two subfamilies of medicagin. The storage proteins marked with *H* characters are listed in Table 1

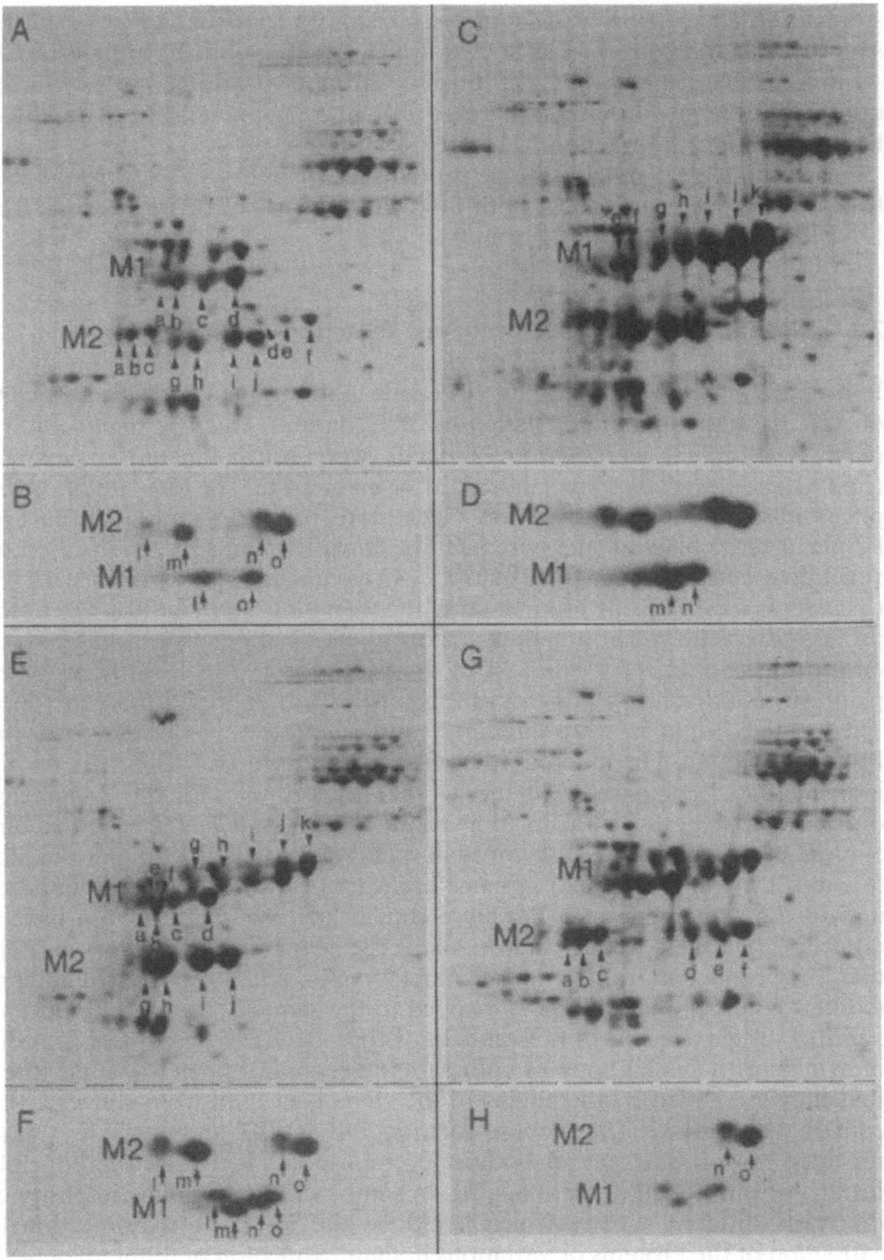

Fig. 3. The protein pattern of the acidic and basic subunits of medicagins in alfalfa. **A**, **C**, **E**, and **G** show the protein spots corresponding to the acidic subunits; **B**, **D**, **F**, and **H** present the basic subunits of the two type of medicagins, *M1* and *M2*. The protein spots belonging to the "string of bead-like" multiple spots of medicagin 1 (*M1*) and 2 (*M2*) are marked with small letters: *a, b, c, d, e, f, g, h, i, j, k, l, m, n,* and *o*

storage proteins, as well as most of the nonstorage proteins, can be extracted as a low-salt fraction (S-1) as shown in Fig. 2a. The alfins have light-brown staining properties and they also form a "string of bead-like" spots. On the other hand, the medicagin-type storage protein complexes can be extracted at high-salt buffer (S-2 fraction) (Figs. 2b, 3).

High molecular weight storage proteins also showing the "string of bead-like" feature were present in both low- and high-salt protein fractions (see H1-H9 in Fig. 2a, b).

2.2 Genetic Mapping Strategy of Seed Proteins

In order to map the seed proteins as genetic markers, the F2 individual plants of the basic mapping population of the diploid *Medicago sativa* cross (Kiss et al. 1993) were used. F2 plants were self-pollinated by manual tripping and the seeds were collected after maturation. The long life of the F2 plants made it possible to collect seeds more than 15 years after the first F2 plants were planted (the perennial *M. sativa* F2 plants were refreshed by vegetative cuttings from time to time). In general, 12 seeds from each F2 individual were mixed, ground, and the resulting powder was used for protein extraction. The random mixture of 12 seeds was analyzed to ensure the correct representation of the genotype of the corresponding F2 plant, since individual seeds from F2 plants already represent F3 genotypes. The protein samples were prepared and separated as described in the Protocols section. For mapping purposes, the pattern of 2-D gel of the F2 individual was compared with each other as well as with the 2-D image of the heterozygote F1/1 plant that contained both alleles of each protein. Scoring of the genotypes of the protein spots was based on all-or-nothing differences, i.e., the presence or absence of a spot (the two allelic protein products of the same gene may display a similar staining character but have a different molecular weight and/or isoelectric point). Thereby every protein that showed polymorphism in gel mobility was genotyped as a dominant marker. When two nearby spots mapped to the same location as of opposite parental origin (one maternal and the other paternal) and there was no individual with missing spot for both parental genotypes, then it was supposed that they are coded by the alleles of the same gene, and were converted to codominant markers. The protein isolation, 2-D PAGE, and evaluation were repeated several times and only those protein spots which gave reproducible results were included in the mapping. In some cases, the staining intensity of the spots differed slightly, which could be the result of disproportionate extraction of the proteins or variations in their expression of different genetic backgrounds.

Altogether 76 protein spots could be mapped based on their segregation pattern, but they determined only 29 loci in the genetic map of alfalfa. The genetic markers corresponding to protein spots on the gels are designated with a "2D" prefix (Fig. 4), followed by a number corresponding to the number of

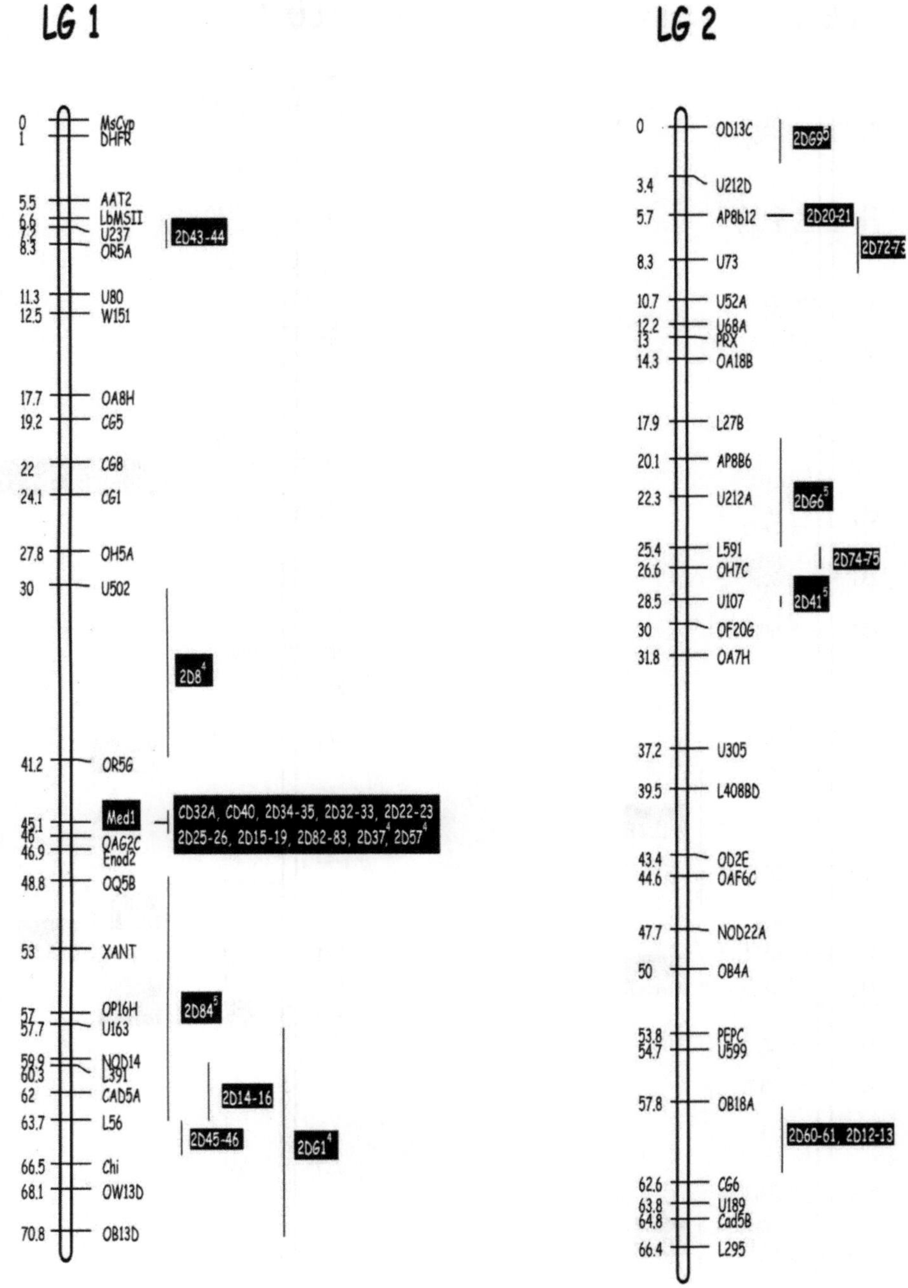

Fig. 4. The map position of the seed protein markers correlating to the core markers in the eight linkage groups of diploid alfalfa. The genetic markers corresponding to the protein spots on the gels are highlighted with a *black background* and a "*2D*" prefix followed by a *number* which corresponds to the number of the spot on the schematic diagram in Fig. 1b. *Superscripts 4* or *5* in the marker name of dominant markers refer to paternal and maternal origin, respectively. The seed protein markers that are linked to more than one core marker are located to the *right* of a *vertical bar*, indicating the genomic region to which they link

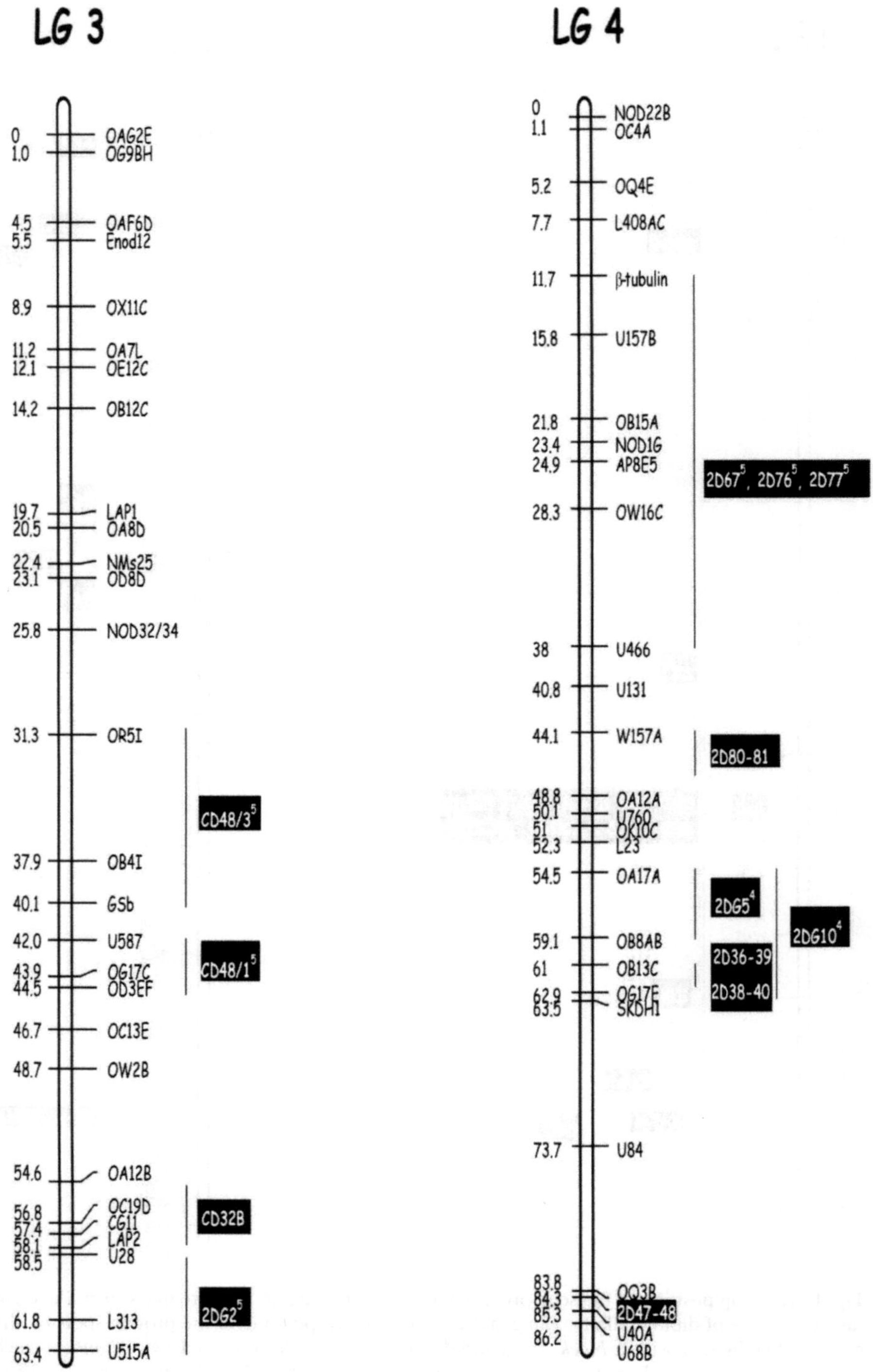

Fig. 4. *Continued*

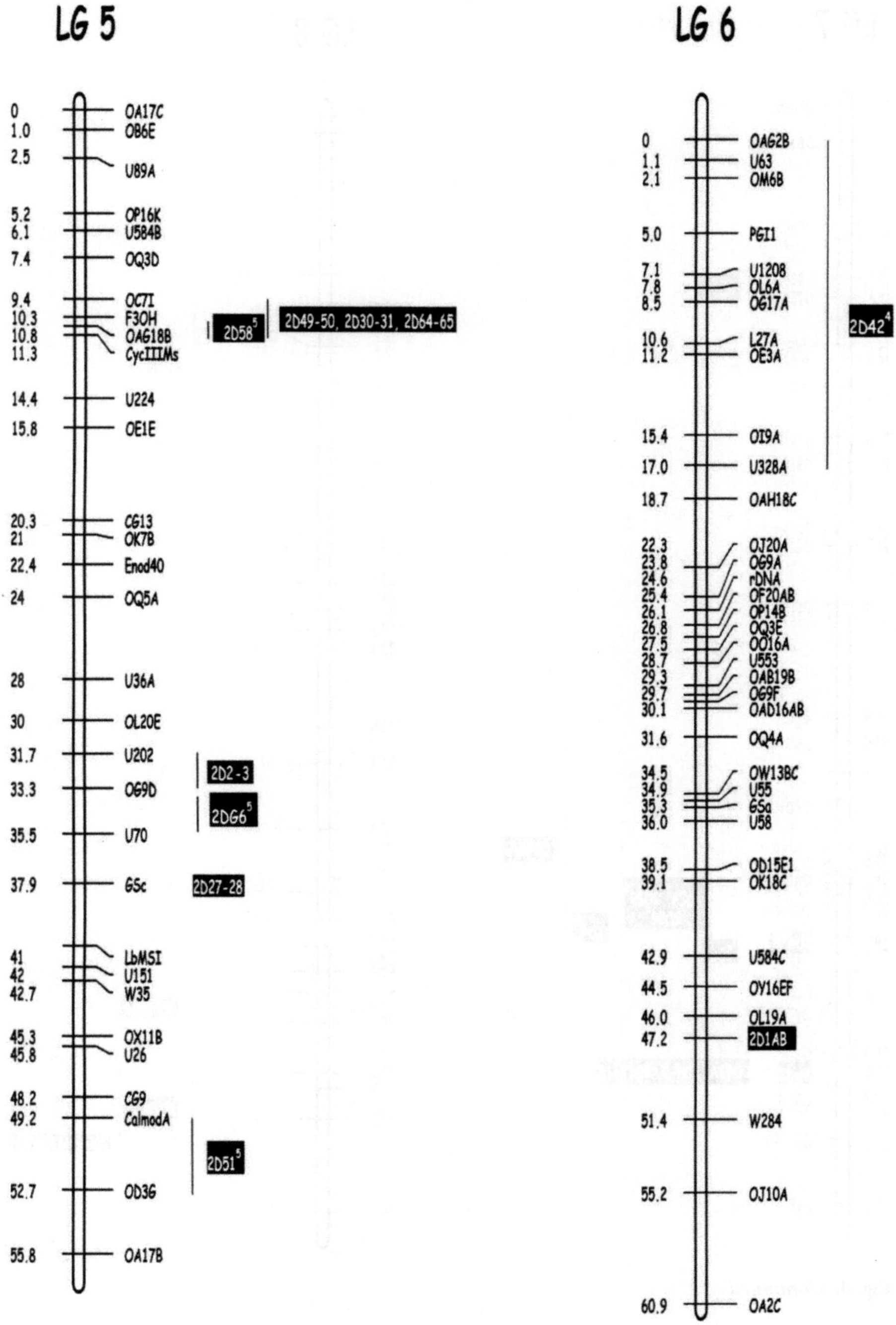

Fig. 4. *Continued*

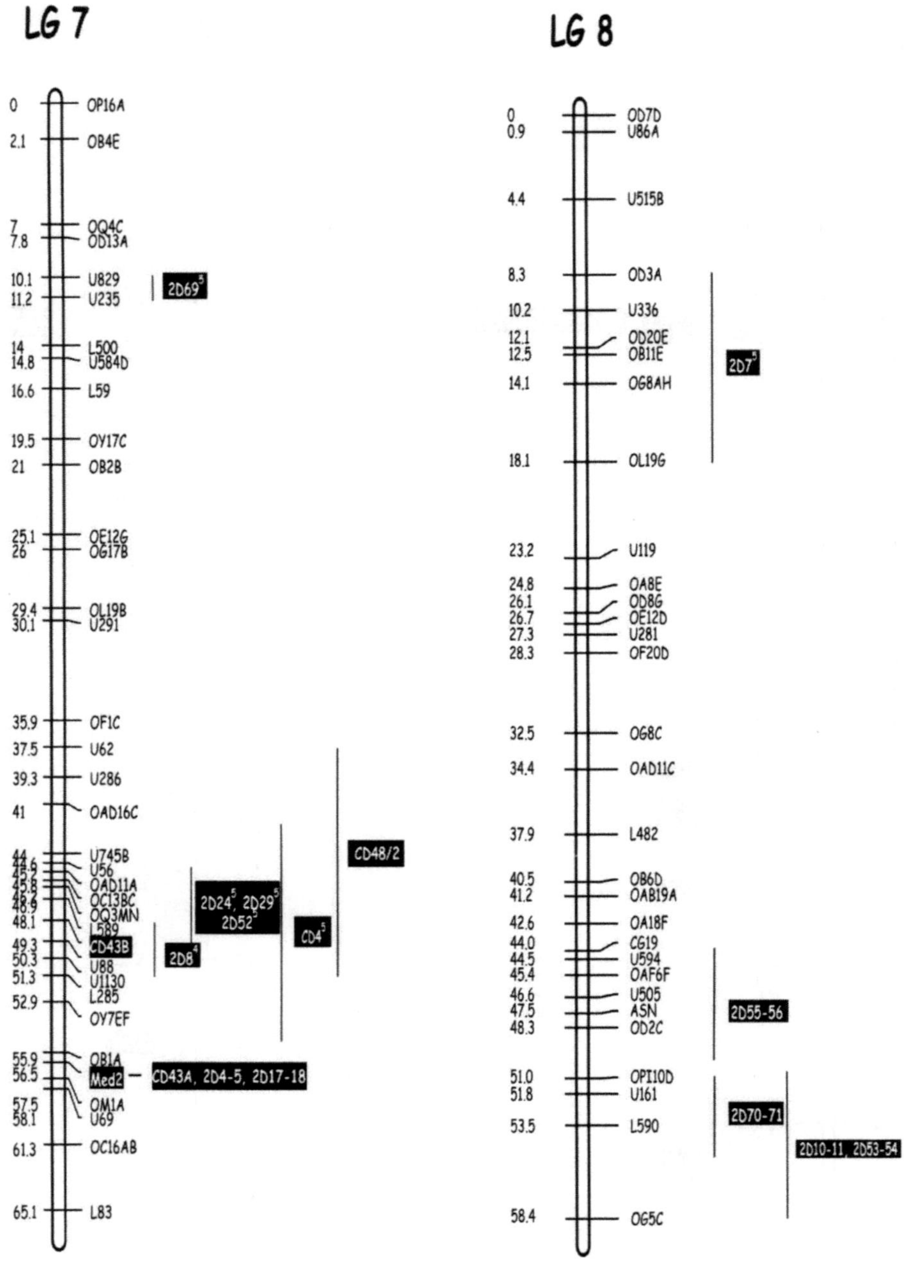

Fig. 4. *Continued*

the spot on the stained gel (Fig. 1a, b). As for codominant markers, they have two numbers separated by a hyphen (Fig. 4).

2.3 Genetic Mapping of the Medicagins

The medicagins have distinctive solubility in high NaCl solution (Krochko and Bewley (1990), and this character was exploited to identify and analyze them. The native medicagins have a Svedberg constant of 11S and can be divided into larger and smaller subunits that are acidic (A) and basic (B), respectively (Krochko and Bewley 1990; Krochko et al. 1990). After high-salt extraction the proteins were separated in two ways: the first division was done either by IEF to analyze the acidic subunits (Fig. 3A, C, E, G), or by nonequilibrium pH gradient between pH 5 and 8 (NEPHGE) for the basic subunits (Fig. 3 B, D, F, H). The two groups of protein spots correspond to the two subfamilies of the medicagin complexes and designated as M1 and M2 in a range between 40 and 49 kDa (Fig. 3A, C, E, G), representing the acidic subunits, as well as in the 23/24 and the 20 kDa range (Fig. 3B, D, F, H), representing the basic subunits of medicagins. The large subunits of the M1 medicagin are larger than those of the M2 medicagins, while the smaller subunits of M1 are smaller than those contributing to the M2 medicagins (Fig. 3). The representative spots of the large and small subunits of the two medicagin subfamilies M1 and M2 are indicated by small letters (a, b, c, d, e, f, g, h, i, j, k, l, m, n, and o) in Fig. 3.

After the genetic segregation of the large subunits of the two medicagins, M1 and M2, the collective appearance and disappearance of the spots were observed. The individual F2 plant represented in Fig. 3A was homozygous for the paternal allele displaying only the spots a, b, c, and d of the M1 medicagin. The pattern in Fig. 3C shows the genotype of a homozygous plant for the opposite, maternal allele (spots e, f, g, h, i, j, and k). The heterozygous configuration of M1 can be seen in Fig. 3E, G. Large subunits of the M2 medicagin could be genotyped similarly by identifying the cosegregating spots. The paternal and maternal homozygous configurations are shown in Fig. 3E, G, respectively, and the heterozygous configuration in Fig. 3A, C. The small subunits of medicagins followed similar patterns except that fewer spots segregated, and the M2 could only be evaluated as a dominant trait (see spots marked by i, m, n, and o in Fig. 3B, D, F, H). The segregation patterns of the large and small subunits reconfirmed the conception that the larger acidic subunits are associated with the smaller basic medicagin subunits of both M1 and M2 (Krochko and Bewley 1990, 2000; Krochko et al. 1990). After determining the genotypes of all seed samples tested the large and small subunits belonging to the M1 medicagin mapped together in linkage group 1 (LG1) and that of M2 mapped in LG7 (indicated as Med1 and Med2 in Fig. 4).

In addition to the genetic mapping of the medicagin subunits based on protein polymorphisms on 2-D gels, we were interested in mapping the cloned cDNA sequences encoding storage protein genes. Coding for cDNAs for legumin-type storage proteins had been cloned from pea (Domoney and

Table 1. Major storage proteins identified in the S-1 and S-2 fraction extracted from mature seeds of diploid alfalfa. The protein spots are presented in Fig. 2a, b

Protein spot	Molecular weight (approx. kDa)	Correlation to previously identified seed proteins
H1	89	86 kDa S-1 protein (Stuart and Nelsen 1988)
H2	86	
H3	80	
H4	70	68 kDa S-2 protein (Stuart and Nelsen 1988) and HMW3 (Krochko and Bewley 1990)
H5	68	
H6	63	HMW2 (Krochko and Bewley 1990)
H7	61	
H8	56	HMW1 (Krochko and Bewley 1990)
H9	54	
H10	52	7S protein (Stuart and Nelsen 1988) and HMW4; alfin (Krochko and Bewley 1990)
H11	36	
H12	35	
H13	32	

Casey 1983, 1985) and were available for RFLP mapping in alfalfa (Table 2). According to this work, the cDNA clone pCD43 encoding type-A legumins were mapped to two locations (loci CD43A and CD43B) on LG7. Locus CD43A colocalized with the position of the subunits of medicagin M2 (markers 2D4–5 and 2D17–18 identifying Med2 locus), indicating the correspondence between the proteins of M2 and a transcriptionally active loci of the type A legumin gene. The map position of CD43B did not colocalize with any seed protein mapped. Type-B legumin clones of pea (pCD40 and pCD32) mapped at one (CD40 on LG1) and two loci (CD32A on LG1 and CD32B on LG3), respectively. The map position of CD40 and CD32A colocalized with medicagin M1 locus (Med1) on LG1 and with many other seed protein markers (2D14–16, 2D22–23, 2D25–26, and more). According to this result, it is suggested that at least two type-B legumin genes map at this locus of *Medicago* LG1. There were no corresponding protein markers to the position of markers CD32B and CD43B, which can be explained by the lack of expression or the lack of protein polymorphism under the conditions *Medicago* seed proteins were tested.

2.4 Genetic Mapping of Alfins and Other Storage Proteins of the S-1 and S-2 Fractions

The vicilin-like storage proteins (alfins) did not show considerable polymorphism, as only three spots could be mapped from them: 2D52, 2D24, and 2D29, the last of which is presumed to be a degradation product.

The map position of two cDNA clones encoding vicilin in pea (pCD4 and pCD48; Ellis et al. 1986) was also determined by RFLP (Table 2). Marker

Table 2. The map position and the number of loci of pea cDNA clones encoding storage proteins in diploid alfalfa (*Medicago sativa*)

cDNA clone	Gene product	No. of mapped loci in alfalfa	LG of alfalfa	Reference
CD4	Vicilin	1	7	Ellis et al. (1986)
CD32	B-type legumin	2	1, 3	Domoney and Casey (1985)
CD40	B-type legumin	1	1	Domoney and Casey (1985)
CD43	A-type legumin	2	7	Domoney and Casey (1985)
CD48	Vicilin	3	3, 7	Domoney and Casey (1983)

CD48 mapped at three loci, CD48/1 and CD48/3 mapped at LG3, while CD48/2 mapped at LG7 together with the CD4 seed protein marker. No colocalization with protein markers was found for CD48/1 and CD48/2 on LG3. CD48/2 and CD4 mapped at LG7 within a region where the 7S-type alfin protein markers (the 50 kDa 2D24, the 36.5 kDa 2D29, and the 32 kDa 2D52) mapped. It has to be added, however, that the map location for markers CD4, CD48/2, and CD43B could not be determined exactly because of missing genotypes and distorted segregation in that region; therefore, a firm conclusion could not be drawn for colocalization.

The other abundant storage proteins present in both S-1 and S-2 fractions (Table 2), H1, H3, H8, and H9 (Fig. 2a, b) could not be mapped because of a lack of detectable polymorphism. The protein spots of H2 and H4-H5 (Fig. 2), corresponding to the 2D12–13 and 2D14–16 markers (from 2-D gels of protein samples extracted by O'Farrell lysis buffer), respectively, could be mapped as codominant markers in LG2 and LG1. The H6-H7 proteins (2D25–26 marker) are cosegregated with the protein spots of the medicagin 1 (M1) locus (Fig. 4).

2.5 Genetic Mapping of Nonstorage Seed Proteins

In addition to the storage protein markers, 55 dominant protein markers were mapped of which 44 became codominant markers. The mapped nonstorage seed protein markers are presented in Fig. 1a, b and their map position is shown in Fig. 4.

2.6 Mapping of Nodule Proteins

Similarly to the seed proteins, nodule proteins from the F2 individuals of the diploid alfalfa mapping population were extracted (in O'Farrell lysis buffer) and analyzed by 2-D PAGE (data not shown). Pairwise comparison of the protein patterns of the stained gel was used to detect the presence or absence of individual protein spots, which in turn were mapped as dominant markers. Seven nodule-specific proteins could be identified in this way and mapped on the genetic map. The position of these markers is highlighted in Fig. 4 as 2DG markers.

3 Conclusion and Future Perspectives

The pattern of seed storage proteins of individuals in a diploid alfalfa F2 segregation population was studied by two-dimensional polyacrylamide gel electrophoresis (2-D PAGE). The previously described major alfalfa storage proteins, medicagins and alfins, were identified. The analysis of the inheritance of the seed protein spots showed the segregation of 76 of the roughly 700 protein spots identified on the 2-D gels. The map position of these segregating seed protein markers was determined on the genetic map of diploid alfalfa that includes two loci of the medicagins. Some seed storage protein genes were also mapped as RFLP markers using heterologous cDNA clones from pea, and their map position was analyzed in comparison with the alfalfa storage protein markers. In addition to the seed proteins, the symbiotic nitrogen-fixing root nodules were also analyzed for segregating protein spots by 2-D PAGE. In the studied alfalfa mapping population, seven loci could be mapped based on the nodule protein patterns.

The analyses of 2-D gels revealed differences in the polypeptide composition of seed proteins in the individuals of the segregation population of diploid alfalfa. These results revealed that the 2-D PAGE technique is applicable for analyzing the protein compound of plant organs and for genetic mapping. Molecular DNA markers tightly linked to any protein markers can be used for marker-assisted selection in *Medicago sativa* and most probably in related species as well. The other significant outcome of the genetic mapping of seed storage proteins is that they can serve as common markers for analyzing the collinearity of the gene order between diploid alfalfa and pea (*Pisum sativum*). The study of macrosynteny between the two legumes is in progress. The comparative genome analyses enable us to transfer information from one legume to another.

4 Protocols

4.1 Plant Material

Mature and dry seeds of the individuals of the diploid ($2n = 2\times = 16$) alfalfa F2 mapping population (Kiss et al. 1993) were used for protein analyses. The segregation population was the progeny of the self-mated F1/1 plant derived from the cross between a diploid yellow-flowered *M. sativa* ssp. *quasifalcata* individual and a blue-flowered *M. sativa* ssp. *coerulea* individual. Since the seeds developed on the F2 plants represented F3 individuals, 12 seeds were pooled from each F2 plants before protein extraction to represent the genotype of each F2 individual. Since alfalfa is an outcrossing species, the seed production was reduced significantly after the second self-mating of some individuals of the F2 mapping population. Since only 72 out of 137 individuals of the mapping population produced the sufficient 12 seeds during the survey, the seed protein patterns of these 72 plants were analyzed.

Mature, symbiotic nitrogen-fixing pink nodules were collected from roots of F2 individuals upon infection with *Sinorhizobium meliloti* bacteria Rm41 (Szende and Ördögh 1960).

4.2 Protein Extraction from Alfalfa Seeds

To analyze the total protein content of seeds of diploid alfalfa, the dry seeds were homogenized with quartz sand in 1.5 ml plastic tubes on ice with 25 μl/seed of modified [5.0 M urea, 4% Nonidet P-40 (NP-40), 0.1 M dithiotreitol (DTT), 2% ampholine (pH 3.5–10, Pharmacia-LKB, Uppsala, Sweden) and 0.1% SDS] O'Farrell lysis buffer (O'Farrell 1975). After the homogenized material was centrifuged at 14,000 rpm for 10 min, the supernatant was used for further experiments. The samples were stored at –20 °C and subjected to a clearing spin after thawing.

Following the modified method of Krochko et al. (1990), two consecutive extraction steps were used to identify the different classes of storage proteins. The seeds were homogenized in 300 μl S-1 buffer (0.1 M NaCl in 25 mM potassium-phosphate buffer, pH 7.0) and after centrifugation (14,000 rpm, 10 min) the supernatant was collected (S-1, or low salt protein fraction). The pellet was resuspended in S-2 buffer (1.0 M NaCl in 25 mM potassium-phosphate buffer, pH 7.0) and the second supernatant was collected as before (S-2, or high salt protein fraction). Both extraction buffers contained 1 mM phenylmethylsulphonyl-fluoride (PMSF, Sigma-Aldrich, St. Louis, USA) as a protease inhibitor. The S-1 and S-2 protein samples were precipitated with ice-cold 80% acetone in the presence of 10 mM 2-mercaptoethanol. The proteins were allowed to precipitate for 1 h at –20 °C. After 10 min centrifugation the pellet was dried in SpeedVac (10 min) and resuspended in the modified O'Farrell lysis buffer.

4.3 Gel Electrophoresis of Protein Samples

Isoelectric Focusing (IEF). Isoelectric focusing of the protein samples was carried out in tube gels (12 cm length, 2.5 mm diameter) according to the method of O'Farrell (1975). The gel compound was 4% acrylamide/bisacrylamide (prepared from 30% stock solution containing 28.4% acrylamide and 1.6% bisacrylamide), 9.2 M urea, 2% NP-40 and 2% ampholine (0.4% pH 3.5–10 and 1.6% pH 5–8, Pharmacia). The tube gels were prefocused at 500 V for 1 h, and 20-μl samples were electrofocused at 500 V for 16 h and then at 900 V for an additional 4 h. The cathode solution (upper reservoir) was 0.02

M NaOH and the anode solution (lower reservoir) was 0.01 M phosphoric acid. The tube gels were equilibrated in SDS equilibration buffer (62.5 mM Tris-HCl, pH 6.8, 2% SDS, 50 mM DTT) for 30 min before the second-dimension electrophoresis.

Nonequilibrium pH Gel Electrophoresis (NEPHGE). The basic proteins were run in nonequilibrium pH gradient gels instead of IEF. The composition of the tube gels was identical to the IEF gels, but the protein samples were loaded onto the anodic end of the gel (the electrodes were reversed). Electrophoresis was performed at 400 V for 3 h without pre-running.

Two-Dimensional Gel Electrophoresis (2-D PAGE). After the proteins had been separated in the first dimension (IEF, NEPHGE), second-dimension electrophoresis was performed in denaturing SDS polyacrylamide gels (SDS-PAGE) according to the method of O'Farrell (1975). The measure of all the second-dimension gels was 120 × 180 × 1.00 mm in all cases, containing a 20-mm-high stacking gel (4% acrylamide/bisacrylamide) followed by the separating region (15% acrylamide/bisacrylamide).

4.4 Staining of the 2-D Gels

The protein spots separated on the 2-D gels were visualized by dual silver and Coomassie Brilliant Blue (Coomassie BB) staining. First, the gels were silver-stained by minor modification of the method published by Bloom et al. (1987). The gels were fixed in 50% methanol and 12% acetic acid for 1 h, then washed with 50% ethanol three times for 15 min and then rinsed in 0.2 g/l sodium-thiosulfate for 2 min. After rinsing the gels three times for 20 s in distilled water, they were incubated in 2 g/l silver nitrate solution containing 0.028% (v/v) formaldehyde for 20 min. After washing the gels in distilled water the protein pattern was developed with 60 g/l sodium-carbonate solution containing 0.018% (v/v) formaldehyde and 0.016 M sodium thiosulfate. The developing reaction was stopped by incubation of the gels in 50% (v/v) methanol and 12% (v/v) acetic acid for 10 min. The silver staining was followed by Coomassie BB staining in 0.1% (w/v) Coomassie BB, 50% (v/v) methanol, and 12% (v/v) acetic acid for 30 min. The gels were destained in 7% (v/v) acetic acid and 10% (v/v) methanol, and finally washed in 10% glycerol and dried between cellophane sheets (DrPack, Budaörs, Hungary).

4.5 DNA Isolation and Hybridization

The isolation of total plant DNA, the restriction digestion of the DNA, the transfer of the DNA and the DNA-DNA hybridization (RFLP) were performed as described previously by Kiss et al. (1993). The cDNA clones (pCD4, pCD32, pCD40, pCD43, and pCD48; Table 1) encoding the storage proteins of pea were kindly provided by R. Casey and N. Ellis (John Innes Centre, Norwich, UK).

4.6 Genetic Mapping of the Seed Proteins

The extraction and separation of the proteins on 2-D gels were performed at least twice. The 2-D protein patterns of the individuals of the mapping population were analyzed by visual comparison to each other. Only the reproducible protein spots with qualitative variations (presence or absence) were taken into account and used for genotyping and mapping. The genotypes were analyzed and the map positions of the markers were determined on the linkage map of diploid alfalfa using the colormap procedure (Kiss et al 1998). The 2-D protein markers were mapped comparing them to the core markers of the genetic map (Kaló et al. 2000) and presented graphically, taking into consideration the calculated relative genetic distances.

Acknowledgements. We are particularly indebted to S. Jenei, K. Lehoczky, Z. Liptay, and P. Somkúti for skillful technical assistance, and R. Casey and N. Ellis for pea storage protein encoding cDNA clones. This work was supported by the BRC, Szeged, Hungary, the Bástyai-Holczer Foundation, and by the following grants: AKP 96–360/62 and AKP 00–246/35 (MTA, Hungarian Academy of Sciences), OTKA F030408, T025467 (Hungarian Scientific Research Fund), OMFB EU-97-D8–063 (National Committee for Technical Development), NKFP Medicago Genomics grant No. 4/023/2001 (Ministry of Education), EuDicotMap grant No. BIO 4CT97 2170, and Medicago grant No. QLG2-CT-2000–30676 (European Union).

References

Bloom H, Beier H, Gross HS (1987) Improved silver staining of plant proteins, RNA and DNA in polyacrylamide gels. Electrophoresis 8:93–99

Coulter EM, Bewley JD (1990) Characterization of a small sulphur-rich storage albumin in seeds of alfalfa (*Medicago sativa* L.). J Exp Bot 41:1541–1547

Domoney C, Casey R (1983) Cloning and characterization of complementary DNA for vicilin, a major seed storage protein in *Pisum sativum* L. Planta 159:446–453

Domoney C, Casey R (1985) Measurement of gene number for seed storage proteins in *Pisum*. Nucleic Acid Res 13:687–699

Ellis THN, Domoney C, Castleton J, Cleary W, Davies DR (1986) Vicilin genes of *Pisum*. Mol Gen Genet 205:164–169

Gerber S, Rodolphe F, Bahrman N, Baradat P (1993) Seed-protein in maritime pine (*Pinus pinaster* Ait.) revealed by two-dimensional electrophoresis: genetic determinism and construction of a linkage map. Theor Appl Genet 85:521–528

Higgins TJV (1984) Synthesis and regulation of major proteins in seeds. Annu Rev Plant Physiol 35:191–221

Kaló P, Endre G, Zimányi L, Csanádi G, Kiss GB (2000) Construction of an improved linkage map of diploid alfalfa (*Medicago sativa*). Theor Appl Genet 100:641–657

Kiss GB, Csanádi G, Kálmán K, Kaló P, Ökrész L (1993) Construction of a basic genetic map for alfalfa using RFLP RAPD isozyme and morphological markers Mol Gen Genet 238:129–137

Kiss GB, Kereszt A, Kiss P, Endre G (1998) Colormapping: a non-mathematical procedure for genetic mapping. Acta Biol Hung 49:125–142

Krochko JE, Bewley JD (1990) Identification and characterization of the seed storage proteins from alfalfa. J Exp Bot 41:505–514

Krochko JE, Bewley JD (2000) Seed storage proteins in cultivars and subspecies of alfalfa (*Medicago sativa* L.). Seed Sci Res 10:423–434

Krochko JE, Charbonneau MR, Coulter KM, Bowley SR, Bewley JD (1990) A comparison of seed storage proteins in subspecies and cultivars of *Medicago sativa*. Can J Bot 68:940–948

Leonardi A, Damerval C, De Vienne D (1988) Organ-specific variability and inheritance of maize proteins revealed by two-dimensional electrophoresis. Genet Res Camb 52:97–103

O'Farrell PH (1975) High resolution two-dimensional electrophoresis of proteins. J Biol Chem 250:4007–4021

Stuart DA, Nelsen J (1988) Isolation and characterization of alfalfa 7S and 11S seed storage proteins. J Plant Physiol 132:129–133

Szende K, Ördögh F (1960) Die Lysogenie von *Rhizobium meliloti*. Naturwissenshaften 47: 404–405

16 Mapping the Chickpea (*Cicer arietinum* L.) Genome: Localization of Fungal Resistance Genes in Interspecific Crosses

P. Winter[1], S. Rakshit[2], M. Baum[3], and G. Kahl[1]

1 Introduction

1.1 Chickpea: a Major Player Worldwide

Chickpea (*Cicer arietinum* L.) is a grain legume of world importance (average annual production approximately 8,500,000 t) ranking third after pea and dry bean on the world market (Duke 1981; FAOSTAT 2000). Major production areas are the Indian subcontinent, West Asia and North Africa (WANA), European countries surrounding the Mediterranean basin, and Latin America. In these regions, chickpea belongs to the traditional diet of the people, and serves as animal fodder as well. In recent years chickpea production in the USA, Australia, and Canada has also increased and provides a growing surplus for export. In all these areas, chickpea is an important component of rain-fed, cereal-based intercropping systems.

1.2 Stagnant Chickpea Yields: a Case for DNA Markers

Despite considerable international investment in conventional breeding, productivity of chickpea has not yet been substantially improved. Yields stagnate, and prices in the most important consumer country, India, doubled from 1988 to 1995, and are still rising (FAOSTAT 2000). Many factors prevent increases in chickpea yield such as, e.g., inadequate amounts of fertilizers and pesticides, and some of the traditional methods of land preparation and seeding. Major constraints, however, are two fungal pathogens: *Fusarium oxysporum* f. sp. *ciceri*, and *Ascochyta rabiei*. Fusarium wilt, a vascular disease, is the main problem on the Indian subcontinent (Srivastava et al. 1984; Jiminéz-Diaz et al. 1993). Ascochyta blight, a foliar disease, affects all aerial parts of the plant and is most devastating in WANA countries, the USA, Australia, and Canada (Kaiser 1997; P. Taylor, pers. comm.). The tremendous potential to increase chickpea productivity by fungus-resistant cultivars is demonstrated by evi-

[1] Plant Molecular Biology, Biocentre, University of Frankfurt/Main, Marie-Curie-Strasse 9, 60439 Frankfurt/Main, Germany
[2] Indian Institute for Pulses Research, Kanpur, India
[3] International Centre for Agricultural Research in the Dry Areas (ICARDA), Aleppo, Syria

Biotechnology in Agriculture and Forestry, Vol. 52
Nagata/Tabata (Eds.) Brassicas and Legumes

dence that, for example, in the Mediterranean basin and adjacent WANA countries, by switching from traditional spring to winter sowing chickpea yields could be increased by 50 to 100%. This would extend the vegetative and reproductive growth period and allow the exploitation of winter rains in these areas. However, farmers still stick to their sowing habits to avert risks associated with ascochyta blight disease that spreads epidemically under favorable winter conditions (>150 mm rainfall, around 20 °C), eventually resulting in complete loss of the crop. Clearly, elite cultivars combining cold tolerance with durable ascochyta resistance would solve this problem. However, conventional breeding hardly has a chance to succeed in generating such cultivars, because new races of the pathogen evolve extremely fast. For example, conventional pathogenicity surveys in Syria revealed the presence of three pathotypes of *A. rabiei* of which a single pathotype III was predominant in all the chickpea-growing regions. This genotype H, which was identified by random amplified polymorphic DNA (RAPD) markers in all of the more recent pathogenicity surveys (1991–1995), probably evolved only recently, since it was not detected during the earlier survey of 1982 (Reddy and Kabbabeh 1985; Udupa et al. 1998). The abundance of this genotype has increased in all chickpea-growing regions of Syria (33% in 1991 to over 80% in 1995). It is highly aggressive and can completely kill variety Ghab 2 (ILC 3279) released for cultivation in Syria. Since the fungus can also occur in its perfect stage *Didymella* (formerly *Mycosphaerella*) *rabiei* (Kov.) V. Arx (review by Kaiser 1992), race evolution is not only fostered by somatic mutations but also by meiotic recombination.

Marker-assisted pyramiding of resistance genes (Robinson 1976) and/or transformation with complementary resistance genes are presently the only effective methods to achieve durable resistance.

As proven for tomato or rice (Tanksley and McCouch 1997), the application of modern biotechnological tools such as DNA markers and dense genetic maps can significantly increase the speed of crop improvement for traits as different as disease and insect resistance, quality factors facilitating marker-assisted selection, positional cloning of resistance genes, and mapping of quantitative trait loci (QTLs) of agronomic interest in many crops (Paterson et al. 1988; review in Winter and Kahl 1995). For chickpea, there is a need to develop a high-density DNA marker map for use in marker-assisted selection and cloning of important genes.

1.3 The Chickpea Genome: Monotony Rules

Chickpea is an annual, self-pollinated diploid (2n = 2× = 16) with a moderately sized genome of about 750 Mbp. Being a founder crop of modern agriculture, chickpea is probably of monophyletic origin and arose by selection from its wild ancestor *C. reticulatum* (Lev-Yadun et al. 2000). Genetic bottlenecks imposed during domestication, and by breeding practices, narrowed the genetic base of the crop as compared to its wild progenitors. Therefore, the many variations for qualitative and quantitative traits such as growth and podding behavior, flower and seed color, leaf form and color, disease resistance, or flowering time in chickpea is astonishing (review in Winter et al. 2001a). However, this great variety is probably not a result of large-scale intraspecific genetic variation, but instead caused by mutations in single genes.

Chickpea's genetic monotony hampered the application of molecular markers for genetic mapping and, consequently, researchers used interspecific crosses between *C. reticulatum* and the cultivar for linkage analysis of morphological and isozyme loci. Crosses between *C. reticulatum* and chickpea are superior to other wild species–cultivar combinations, as this combination displays nearly 100% pollen and plant fertility (review in Winter et al. 2001a). However, drawbacks of interspecific crosses have also been reported for rice (Wang et al. 1994) and tomato (Paran et al. 1995), where skewed segregation of markers and recombination suppression occur.

1.4 Microsatellites: Ideal Markers for an Invariable Genome

The availability of sufficient numbers of polymorphic markers is a prerequisite for successful linkage studies. Formerly used markers such as isozymes, restriction fragment length polymorphisms (RFLPs), and RAPDs did not provide sufficient polymorphism in chickpea. However, we could demonstrate considerable intraspecific variation by oligonucleotide fingerprinting with short, synthetic microsatellite motifs (Sharma et al. 1995a,b; reviews in Winter et al. 2001a,b).

Microsatellites, or simple sequence repeats (SSRs) are tandemly arranged repetitions of 1 to 5 nucleotide units with repeat numbers ranging from 2 to several hundreds or even thousands. They are ubiquitous and evenly distributed in all eukaryotic genomes. Microsatellites frequently change their length by replication slippage, unequal crossing-over, or other less understood mechanisms, thereby generating a wealth of highly polymorphic alleles (reviews in Gupta and Varshney 2000; Udupa and Baum 2001; Winter et al. 2001b).

The variability of microsatellites is exploited by so-called sequence-tagged microsatellite site markers (STMS, Beckmann and Soller 1990), a polymerase chain reaction (PCR)-based technique that uses microsatellite-flanking sequences as primers to amplify the microsatellites in between. The resulting locus-specific amplification products often exhibit considerable length differences due to the variable number of tandem repeats within the microsatellite. STMS are widely applied to plant genome analysis (reviews in Gupta and Varshney 2000; Winter et al. 2001b) because they are single-locus, codominant, easy to use and reliable markers with high polymorphic information content and having a potential for automated, nonradioactive detection (Mansfield et al. 1994).

In chickpea microsatellites, loci harboring the motifs $(GA)_n$, $(GAA)_n$, and $(TAA)_n$ are abundant and highly polymorphic even between closely related accessions. Microsatellite-containing genomic fragments were cloned, sequenced, and primers designed against the microsatellite-flanking sequences to generate approximately 250 STMS markers (Hüttel et al. 1999; Winter et al. 1999; P. Winter, unpubl.).

1.5 Recombinant Inbred Lines (RILs): Mapping Populations of Choice for Chickpea

To integrate dominant and codominant markers in one map, offspring populations with highly homologous genomes are desirable, since dominant markers cannot distinguish between hetero- and homozygous alleles. Therefore, this type of markers is not suitable for mapping in ordinary F2s where 25% of all loci are expected to be heterozygous. This problem may be solved by recombinant inbred lines (RILs, Burr et al. 1988) which are generated by single-seed descent from the F_2 to the F_6 and beyond, or by the pedigree method, until their genomes are virtually homozygous. From that point on the RILs are increased to provide sufficient seed for application of various kinds of markers, and allow a more precise evaluation of phenotypic variation as compared to F_2 populations. Since genetically invariable RILs can be tested over the years at several locations also genotype × environment effects can be determined.

2 The Genetic Map of Chickpea: Markers for Ascochyta Blight and Fusarium Wilt Resistance

Here, we report on the first extended genome map of chickpea, generated by the application of different DNA marker techniques to RIL populations from interspecific crosses between the cultigen and *C. reticulatum*. These populations segregate for resistance to ascochyta blight and fusarium wilt and other traits, and were used to position these loci on the genomic map.

2.1 The STMS Progenitor Map

Recently, more than 250 STMS markers have been developed for chickpea (Hüttel et al. 1999; Winter et al. 1999; Winter et al., unpubl.). The segregation of 120 of them was studied in a population of 90 RILs from an interspecies cross between *C. reticulatum* accession P.I. 489777 and the cultivar ICC 4958. As demonstrated in Fig. 1, STMS segregation patterns in RILs are robust, clear-cut and easy to evaluate, which makes these markers ideal starter landmarks for further mapping efforts.

The resulting map contained 112 markers in 11 linkage groups covering 613 cM. This STMS marker map provided a low-density core map (Winter et al. 1999), into which dominant markers such as RAPD (Williams et al. 1990), DNA amplification fingerprinting (DAF; Caetano-Anollés et al. 1991), AFLP) (Vos et al. 1995), and dominant microsatellite-based marker types such as inter-simple sequence repeat (ISSR; Gupta et al. 1994; Vogel and Scolnik 1997) markers could easily be integrated.

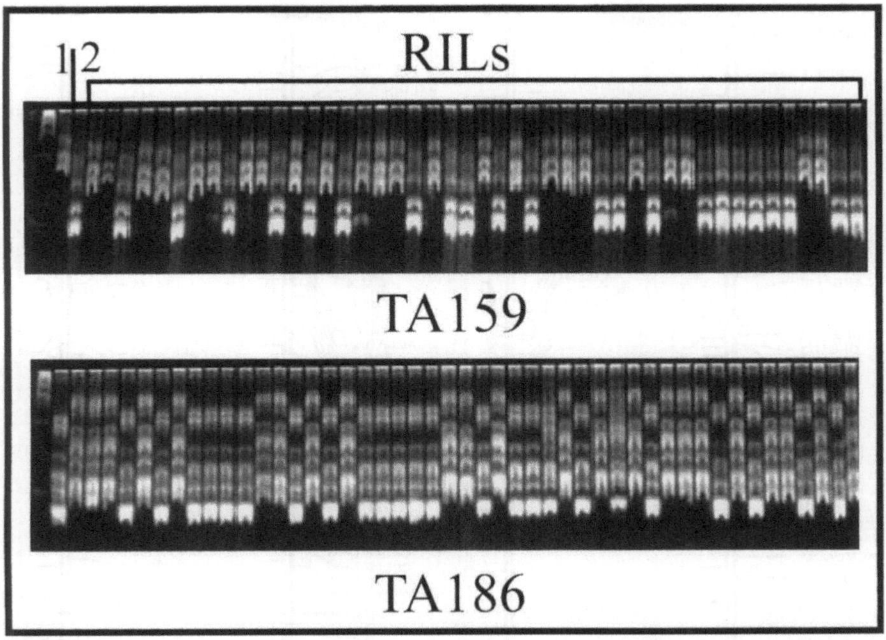

Fig. 1. Segregation of STMS markers *TA159* and *TA186* in RILs from the cross ILC 4958 ×*C. reticulatum*. Amplification products from the parental lines are separated in *lanes 1* and *2*. Note that the multiple bands observed per lane all segregate the same way, and thus are obviously derived from the same locus

2.2 The Most Extended Map of the Chickpea Genome: Mapping of Resistances to *Fusarium* Races 4 and 5

Building on the core map, 130 RILs from the above cross were used to extend the STMS map by adding isozyme, RFLP, DAF, ISSR, and AFLP markers. Since the parental line ICC4958 is resistant to races 0, 4, and 5 of *Fusarium oxysporum,* and *Cicer reticulatum* accession P.I. 489777 is susceptible, the segregation of the underlying resistance loci could also be followed.

2.2.1 The Genome Map of Chickpea: General Features

The most extended molecular marker map of the chickpea genome available to date is shown in Fig. 2. Of the 354 markers used for its generation, 303 are coupled in 16 linkage groups (LGs) spanning a total of 2077.9 cM. At a log of odds (LOD) score of 4, 53 (14.9%) markers are unlinked. The largest group (LG 1) comprises 39 markers and spans 373.9 cM, whereas the smallest LG (16) is made up of 2 AFLP markers linked without detectable recombination. There are 8 LGs larger than 150 cM (LGs 1 to 8), probably representing the 8 chickpea chromosomes.

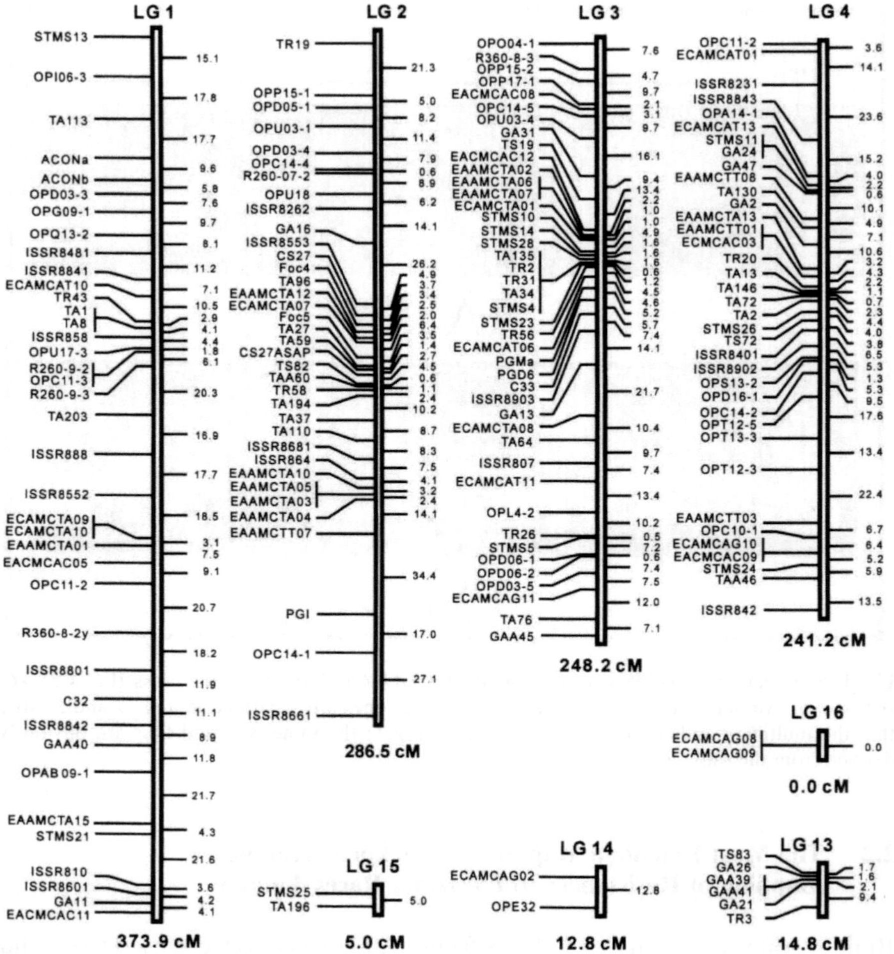

Fig. 2. Genetic map of the chickpea genome, comprising different marker types and loci for resistance genes against *Fusarium* races 4 and 5 (Foc4 and Foc5) in linkage group 2. For more details, see Winter et al. (2000)

The size of an LG does not necessarily reflect the number of linked markers. For instance, LG 1 with its 373 cM contains 39 markers, whereas in LG 5, 42 markers span only 225 cM. The average marker density in LGs is 6.8 cM. This implies that 1 cM represents approximately 360 kbp of the 750 Mbp chickpea genome (Arumuganathan and Earle 1991). Nevertheless, most mapped regions deviate from average spacing between the markers, because of the clustering and nonrandom distribution of markers. Though the presented map is quite extended as compared to its predecessors, we conclude from the fact that 51 markers are still unlinked, that large regions of the genome are still not covered by markers.

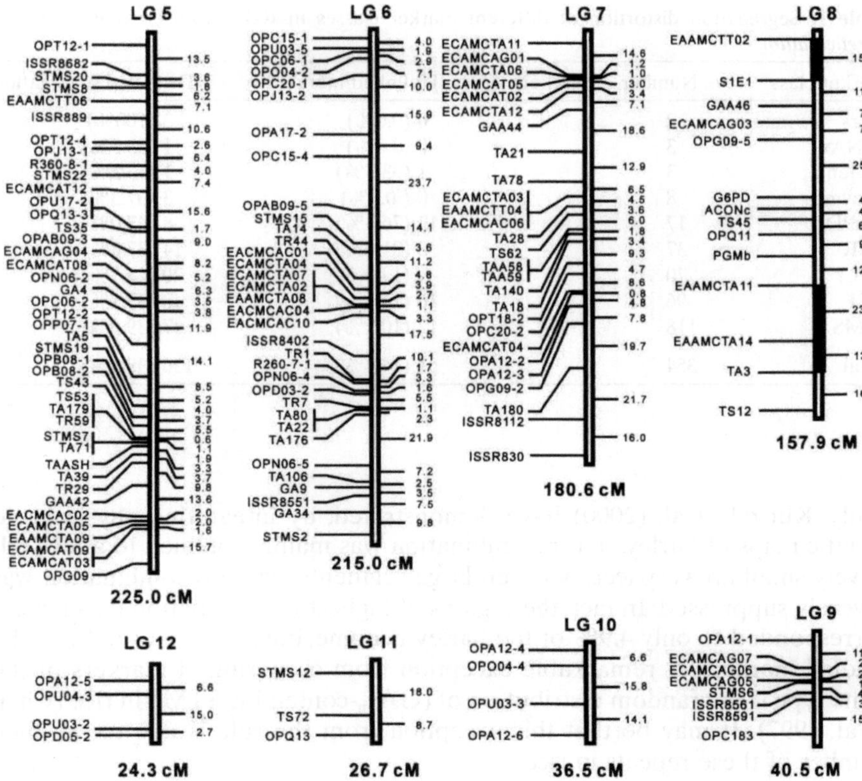

Fig. 2. *Continued*

2.2.2 *The Distribution of Markers Is Not Random*

There is a tendency for clustering of markers, which cannot be attributed to segregation distortion, because these clusters are visible to a similar extent also in normally segregating parts of the genome. Clustering of markers is observed at three levels: (1) all marker types (except RAPD) tend to cluster in specific genomic regions, (2) a tendency for clustering of markers of the same type exists, and (3) marker subtypes such as AFLPs derived from amplification of Eco RI/Mse I restriction fragments with a primer that contains a particular combination of specific bases, or STMS markers containing a particular microsatellite motif tend to cluster together.

Similar overall clustering and marker type-specific subclustering were observed in genetic maps of barley (Becker et al. 1995) and soybean (Young et al. 1999). Tanksley et al. (1992) explained these observations, also made in more advanced tomato and potato maps, reasoning that "centromeres and centromeric heterochromatin and in some instances telomeres experience up to tenfold less recombination than other areas of the genome." In a more recent

Table 1. Segregation distortion of different marker classes in RILs from cross ICC 4958 × *C. reticulatum*

Marker class	Number of markers	Unlinked markers	Distorted segregation
Scars	2	0 (0.0%)	2 (100%)
cDNAs	3	0 (0.0%)	1 (33.3%)
R-Gene	3	1 (33.3%)	1 (33.3%)
Isozymes	8	0 (0.0%)	3 (37.5%)
RAPD	17	13 (76.4%)	8 (47.0%)
ISSR	37	8 (21.6%)	14 (37.8%)
AFLP	70	1 (1.4%)	20 (28.5%)
DAF	96	16 (16.6%)	40 (42.1%)
STMS	118	12 (10.1%)	47 (39.8%)
Total	354	51 (14.4%)	136 (38.3%)

study, Künzel et al. (2000) have demonstrated, by integrating physical and genetic maps of barley, that recombination was mainly confined to a few, relatively small areas spaced between large segments where recombination was severely suppressed. In fact, the regions of highest recombination frequencies corresponded to only 4.9% of the barley genome, but harbored 47.3% of the studied markers. A remarkable exception from clustering of markers of the same type is the random distribution of $(GA)_n$-containing STMS in rice (Chen et al. 1997). It may be that this exception from the rule is due to the high number of these repeats in rice.

2.2.3 The Relation of Linked to Unlinked Markers Varies for Different Marker Types

The percentage of unlinked markers of a particular type varies for the different marker types. As summarized in Table 1, at LOD score 4, only 1.4% of AFLP markers, but 76% of RAPD markers are unlinked, whereas of three fusarium resistance loci, only the locus for resistance to fusarium wilt race 0 is unlinked. However, if the LOD score is reduced to 2.9 and 2.6, respectively, this locus can be placed at the end of LG 4, linking it to markers STMS 24 and ECAMCAC09. Lowering the LOD score, as a criterion for linkage, from 4 to 3, reduces the number of unlinked markers from 14.4 to 10.4%.

2.2.4 Segregation Distortion and Residual Heterozygosity

Theoretically, RILs in F_6 to F_7 harbor around 1–2% residual heterozygous loci detectable by codominant markers such as STMS. Indeed, heterozygotes were detected at a frequency of around 1–2%.

In the interspecific RILs, a large number of markers showed skewed segregation, generally favoring *C. reticulatum* alleles. In 68% of cases, the offspring contained the wild parent's locus, whereas the *C. arietinum* allele was present at 31% of loci. Only six markers showed the perfect Mendelian inheritance of 1:1. As judged from χ^2 tests, ($P < 0.05$) 136 (38.4%) of the 354 marker loci segregated distortedly. This value is close to the average segregation distortion reported for RILs (39.4 ± 2.5%) by Xu et al. (1997), whereas all other types of populations displayed around 20% biased segregation on average. The relatively large amount of segregation distortion in RILs is explained by the higher number of meioses (12–14) experienced by RILs as compared to the two meioses leading from F_1 to F_2. It has been shown for tomato RILs from an interspecific cross that loci revealing only skewed segregation in the F_2 deviated significantly from Mendelian inheritance in RILs. Distortedly segregating parental alleles in the F_2 were found more likely to produce gametes with distorted alleles, subsequently resulting in more distortion in the next generation (Paran et al. 1995).

Segregation distortion, eventually caused by preferential fitness of particular gametes or recombination suppression at specific genomic regions, may cause severe drawbacks for the map-based cloning of resistance and other genes, since it may lead to underestimation of the physical distance between the markers and the gene of interest. For example, the positional cloning of the Foc4 resistance gene on LG 2 from the most closely linked markers could be severely hampered by segregation distortion affecting the whole region surrounding this locus (Fig. 2). Therefore, the physical distance between the gene and the next markers could be much larger than expected from the average relation of 360 kbp/cM calculated for the whole genome. Consequently, it would be wise to first selectively increase the marker density in this region by bulked segregant analysis (Michelmore et al. 1991) and, secondly, to determine the physical distance between the markers by pulsed-field gel electrophoresis.

2.2.5 Tagging of Fusarium Wilt Resistance Genes

One important goal of genetic mapping in chickpea is the tagging of genes for resistance to important fungal pathogens such as *Ascochyta rabiei* and *Fusarium oxysporum* f. sp. *ciceri*. The RILs used here segregated for resistance to races 0, 4, and 5 of the pathogen, and hence, two of the respective resistance genes (Foc4 and -5) could be integrated into the current map. The Foc0 locus, however, was not linked to any marker at an acceptable LOD score, and thus could not be placed on the map. In a previous study, Mayer et al. (1997) developed two allele-specific associated primer (ASAP) pairs, namely, UBC170 and CS27, from RAPD bands that were linked at 7% recombination to the locus for resistance to fusarium wilt race 1. CS27 also segregated in the RILs used here, allowing us to infer the position of this locus on LG2 of the present map. Since CS27 is also linked to the genes for resistance to races 4 and 5, segregating in our RILs at distances of 3.7 and 18 cM, respectively, we conclude that the genes for resistance to fusarium wilt races 1, 4, and 5 form a cluster in chickpea. Moreover, the resistance gene for race 2 is probably also part of this cluster, since Tullu (1996) already demonstrated close linkage between resistance genes for races 1, 2, 4, and 5 in a narrow cross.

Clustering of resistance genes for different races of a pathogen and also for different pathogens has been demonstrated in several plant species. It may be the result of gene duplication, exon shuffling, and recombination processes that may have generated the different resistance genes from one or a few ancient progenitor genes (reviews in Michelmore 1996; Hammond-Kosack and Jones 1997; Hutcheson 1998; Young 2000). It is interesting that the resistance loci and CS27 are flanked by several microsatellite-based STMS and ISSR markers, because close linkage between microsatellite markers and resistance genes has also been documented for rice bacterial leaf blight (Blair and McCouch 1997). It is tempting to speculate that microsatellites, which are suspected to be involved in recombination processes, may also drive the evolution of resistance gene clusters in chickpea. For practical application in chickpea breeding, and especially marker-assisted selection, the close linkage of genes between fusarium wilt resistance and several STMS markers is very important as it allows the use of at least one of these highly polymorphic markers for analysis of segregation of wilt-resistance genes in a wide range of germplasm.

2.3 Mapping of QTL-1 for Ascochyta Blight Resistance: STMS Are Transferable Among Populations

Resistant chickpea varieties react to invasion by *Ascochyta rabiei* with a hypersensitive response (HR; Lamb et al. 1989; Dixon and Lamb 1990) accompanied by a rapid oxidative burst and other fast defense reactions, whereas susceptible lines do not (Höhl et al. 1990; Barz et al. 1993; Otte and Barz 1996). These observations suggest that ascochyta resistance in chickpea, like various biotic stress resistances in other plants, is governed by genes encoding receptor-like protein kinases serving as receptors for pathogen-derived avirulence factors.

In spite of histological evidence for an HR, the genetics of *ascochyta* resistance as determined in field experiments was elusive for a long time, because of its quantitative nature. Only recently, Tekeoglu et al. (2000) have provided clear-cut evidence for the existence of two to three major recessive and complementary resistance genes together with several modifiers. The two major genes together accounted for 50.3 and 45.0% of the estimated phenotypic variation in two subsequent years as shown by QTL mapping (Santra et al. 2000). Absence of either one or two major genes led to susceptibility, whereas the modifiers determined the degree of resistance (Tekeoglu et al. 2000). For these studies three different RIL populations, one from an interspecific cross between *C. arietinum* FLIP 84–92C (resistant parent) and *C. reticulatum* P.I. 599072 (susceptible parent), were tested in the field for three subsequent years using natural inoculum. Especially the interspecific population was well suited for QTL mapping of ascochyta resistance, because the population, comprising 142 F5:8 RILs, is large, and also genetically diverse enough for the efficient application of DNA markers for linkage analysis.

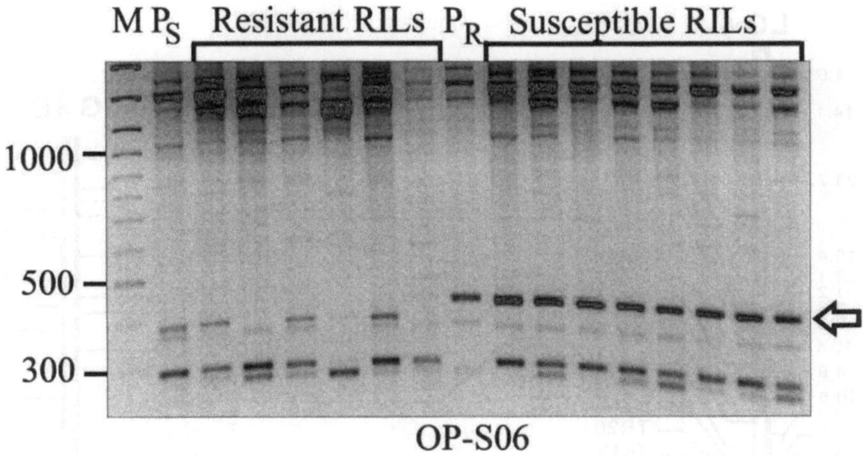

Fig. 3. Segregation of DAF marker OP-S06 in *ascochyta*-susceptible and resistant RILs from the cross FLP84–92C ×*C. reticulatum*. P_S Susceptible parent (*C. reticulatum*), P_R resistant parent (FLIP 84–92C). *Arrow* indicates the band indicative for QTL-1. Molecular weights of the 100-bp ladder (*M*) used as molecular weight standard is given in bp on the *left*

Santra et al. (2000) exploited this population for mapping of *A. rabiei* resistance loci. Interval mapping revealed two QTLs, QTL-1 and QTL-2, with LOD scores of 17.2 and 7.3, respectively. Individually, the two loci accounted for 42.5 and 19.9% of the resistance against *A. rabiei*. Two RAPD markers (UBC733b and UBC181a) flanked QTL-1 and were 10.9cM apart from each other, whereas QTL-2 was flanked by an ISSR and an isozyme marker 5.9cM apart. The two major QTLs were detected using resistance scoring data from two subsequent years. However, a third QTL-3 could only be detected in the data from a single year.

The density of markers around *ascochyta* resistance loci was still too low for map-based cloning of the underlying genes. Further, the positions of the resistance genes on the most extended genomic map of chickpea was still unknown. To solve these problems, we used 94 RILs from the cross of Santra et al. (2000) to identify more, and more closely linked markers for the major resistance locus QTL-1 (Rakshit et al., submitted), applying DAF in combination with bulked segregant analysis (BSA, Michelmore et al. 1991). Of 312 random 10mers, three produced five polymorphic bands between the bulks (Fig. 3). As can be drawn from Fig. 4, at LOD scores above 5, two of these markers (OPS06–1 and OPS03–1) were linked to markers UBC733B and UBC181a flanking the major *ascochyta* resistance locus. OPS06–1 was located between these two loci 4.4cM apart from marker UBC733B, and 9.5cM from UBC181a, and is probably the most closely linked marker for the major *A. rabiei* resistance locus available to date. OPS03–1 mapped about 25cM away from UBC181a on one flank of the resistance locus. The third DAF marker,

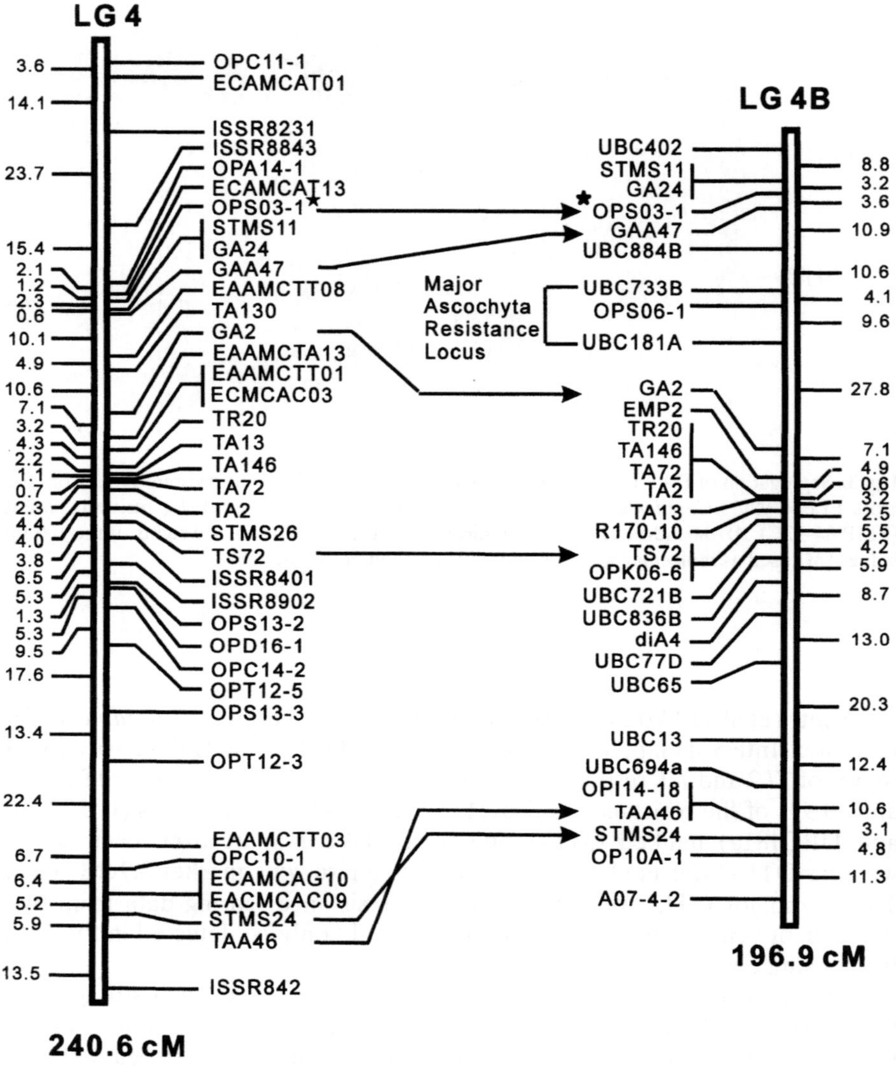

Fig. 4. Transferability of STMS markers between different mapping populations of chickpea. Genetic map surrounding the major *ascochyta* resistance locus on linkage group (*LG*) *4B* obtained from mapping the markers indicated in RILs from the cross FLIP 84–92C × *C. reticulatum*. The map position of several STMS markers (indicated by *arrows*) from *LG 4* is compared to their position on the map from another population shown in Fig. 2 (Winter et al. 2000). DAF marker OPS03–1, which could be mapped in both populations, is indicated by an *asterisk*. The position of QTL1 for *ascochyta* resistance is indicated

OPK06–5, was located on the other flank at the distal end of the linkage group approximately 30 cM away from UBC733b.

For the localization of QTL-1 on the map of Winter et al. (2000), OPS03–1 was transferred to the population on which this map is based, where it mapped to linkage group 4. To test the order of STMS markers in another population and to determine which STMS marker could be used for marker-assisted selection for ascochyta resistance, STMSs located on linkage group 4 were transferred to the population segregating for ascochyta resistance. Three of these markers, GAA47, GA24, and STMS 11 were linked to QTL-1 at distances of about 20–27 cM. Twelve out of 14 STMS markers could be used in both populations, and their order was essentially similar, though map distances differed between them (Fig. 4). The availability of STMS markers for the major *A. rabiei* resistance locus will elucidate the complex resistance against different races of this devastating fungus.

This study demonstrated that BSA can be used to tag at least QTL-1 with molecular markers, though this locus explains less than 50% of the resistance reaction. BSA then is a fast way to identify markers for qualitative and quantitative *ascochyta* resistances, and possibly also for other major QTLs. This should also help to identify markers for resistances against other pathotypes of the fungus and to clarify their genetic basis. However, since at least two loci are necessary to explain about 50% of the resistance (namely, QTL-1 and QTL-2), all individuals of the resistant bulk should possess both loci. Therefore, it was potentially possible to tag both loci with this approach. The fact that we tagged only one of these raises questions about the genetic background of the individuals rated as either resistant or susceptible. One possibility is that by chance none of the polymorphisms detected in this study was actually linked to QTL-2. Another possibility is that one or more of the individuals of the susceptible bulk (rating = 9, plants are dead) harbored QTL-2, but were nevertheless killed. In this case, polymorphism linked to QTL-2 would not be detected with BSA. However, DNA markers closely linked to this locus will be important for molecular breeding of ascochyta resistance, since this complex trait could be resolved into individual genetic components, which can be addressed with the efficacy of monogenic traits (Tanksley 1993).

The three STMS markers, GAA47, GA24 and STMS 11, linked distances of 20–27 cM to the resistance locus, are of appreciable value. Due to the high information content of STMS, at least one of the three markers will also identify polymorphisms in intraspecies crosses segregating for resistances against the different pathotypes of the fungus, and should allow one to determine if the same or another genomic region harbors major resistance genes against the other pathotypes.

One of the aims of the study was to evaluate the marker order in linkage group 4 of the genetic maps of Winter et al. (1999, 2000). Our results confirm the marker order obtained from the ICC 4958 × *C. reticulatum* population, though some minor inconsistencies exist, e.g., a few tightly linked markers changed places. Similar observations of variable distances between markers in different segregating chickpea populations were already reported by Kazan et al. (1993) and Simon and Muehlbauer (1997). These discrepancies can, however, be explained. Recombination frequencies for specific regions may change from one F_1 to another even in populations derived from crosses of the same parental lines. Therefore, fluctuations in map distances are not surprising. However, a realistic marker order will have to be determined in another – preferably intraspecific – population. Despite the monotony of the chickpea genome, such experiments are now possible, because highly polymorphic, codominant STMS markers are available that have again proven their value in this study.

Marker-assisted pyramiding of resistance genes could eventually lead to durable resistance. However, this approach requires the tagging of resistance genes by closely linked markers. Such markers are now available.

2.4 Conclusions and Perspectives: Toward Application of Marker Technology for Marker-Assisted Breeding

All published genetic maps of the chickpea genome, and also mapping of resistance loci for chickpea's most important fungal pathogens, are based on interspecies crosses between the cultigen and the wild relative *C. reticulatum*. These restrictions were dictated by the low level of genetic polymorphism between chickpea accessions. However, chickpea breeding normally applies intraspecific crosses for crop improvement. The present – especially STMS – marker technology can deal with the monotony of the chickpea genome and support routine breeding work with simple, fast, and inexpensive allelic tests.

However, the successful application of marker technologies requires at first a dense intraspecies map, linking traits of agronomic importance (yield, flowering time, cold, drought, and pest resistances) to highly polymorphic, codominant markers in sufficiently close vicinity to allow marker-assisted selection also in offspring from intraspecies crosses. Such an intraspecies map of chickpea will be available soon (J. Kumar, ICRISAT, pers. comm.).

Further, routine marker-assisted selection requires the tailoring of STMS markers and reactions and their combination with high-throughput screening techniques to speed up and simplify the application of marker technology. Successful tailoring of STMS markers for multiplex PCR and electrophoresis has been reported for soybean, where up to eight STMS markers can be amplified in a single PCR reaction and separated in a single lane of a sequencing gel. More than 70 of these tailored markers cover the whole genome (Narvel et al. 2000). However, this advancement was based on the availability of more than 600 STMS markers for soybean (Cregan et al. 1999). In chickpea, the number of STMS markers is still too low for such applications and needs to be increased.

3 Protocols

3.1 Plant Material and DNA Isolation

For the generation of RILs F_2 plants from crosses between the cultivated chickpea line ICC-4958 (fusarium wilt resistant) × *C. reticulatum* PI 498777 (susceptible) and *C. arietinum* FLIP 84–92C (resistant against *A. rabiei*) and *C. reticulatum* P.I. 599072 (susceptible), respectively, were advanced to the F_7 or F_8 using the single-seed descent method. Seeds of RILs from the interspecific mapping populations, were obtained from Dr. F.J. Muehlbauer, USDA, Pullman, USA. Plants were propagated in the greenhouse, and DNA was isolated from young leaflets using a modified cetyltrimethyl-ammoniumbromide (CTAB) protocol (Weising et al. 1995). Contaminating polysaccharides were precipitated according to Michaels et al. (1994).

3.2 Detection of Polymorphism

AFLP procedures were performed as outlined by Vos et al. (1995) using kits from Invitrogen (Karlsruhe, Germany), recommended for either large or small genomes. Initial screening for variability between parental lines with AFLP detected satisfying numbers of polymorphism with the small-genome kit. In addition to the primers provided with the kit, EcoRI primers containing two selective bases (CA, CC, CG, CT, and GA) at their 3′ ends were used for PCR. For radioactive detection of amplification products the *EcoRI* primer was labeled with ^{32}P, and PCR-generated fragments were separated on 5% denaturing polyacrylamide gels for 90 min. The gels were vacuum-dried and exposed to X-ray films (Kodak X-AR, Amersham, Freiburg, Germany) for 3–12 h.

DAF was performed essentially as described by Caetano-Anollés et al. (1991). PCR was carried out on a Perkin Elmer Geneamp 9700 thermal cycler using random 10mer primers procured from Eurogentec (Seraing, Belgium) or Roth (Karlsruhe, Germany), respectively. Each 15 μl PCR reaction contained 1.5 μl 10× PCR buffer (Eurogentec), 10 mM dNTPs, 0.4 U "Silverstar" DNA polymerase (Eurogentec), 40 pmol oligonucleotide primer, and 1 ng/μl template DNA. The DNA was first denatured for 2 min at 95 °C, followed by 40 cycles of 0.15 min denaturation at 95 °C, 1 min annealing at 35 °C and 2 min elongation at 72 °C, with a final elongation of 2 min at the same temperature. The reaction products were separated in 1.8% agarose gels, stained with ethidium bromide and viewed under ultraviolet light.

For RAPD analysis PCR was performed in a 20 μl mix consisting of 10 ng of template DNA, 6.5 pmol of random primer, (Operon Technologies, Alemeda, CA, USA), 0.4 mM spermidine, 200 mM of each dNTP, 1× *Taq* polymerase buffer, and 0.8 units of *Taq* DNA polymerase (Boehringer Mannheim, Germany). The amplification reactions were carried out on a Perkin Elmer thermocycler or PTC200 from MJ Research (Boston, MA, USA). An initial denaturation step at 94 °C for 4 min was followed by 5 cycles at 92 °C for 30 s, 36 °C for 2 min, and 72 °C for 90 s, then by 35 cycles at 92 °C for 5 s, 40 °C for 20 s, and 72 °C for 90 s. The final elongation was at 72 °C for 5 min. PCR products were separated on 2% agarose gels, stained with ethidium bromide and scored for the presence or absence of polymorphic bands.

For the detection of STMS polymorphisms, 35 cycles of PCR were performed with 50 ng DNA in 20 μl "Silverstar" reaction buffer containing 2 μM primers, 1.5 mM $MgCl_2$, 250 mM nucleotides, and 0.4 U "Silverstar" polymerase

(Eurogentec). The DNA was first denatured for 2 min at 94 °C, annealing was at 55 °C for 50 s, and elongation at 60 °C (TAA-repeats) or 72 °C (GA and GAA repeats) for 50 s. The reduced elongation temperature compared to the optimal 72 °C was necessary to stabilize the long (TAA) microsatellites (Su et al. 1996). Between 4–12 μl of the reaction mix was separated either on 2% agarose gels for a first analysis of amplification success, or on 8% polyacrylamide gels for determination of allele lengths and numbers. The detection of ISSR polymorphisms and corresponding primer sequences have been described by Ratnaparkhe et al. (1998).

Isozyme analysis was carried out as described by Tullu (1996). Abbreviations used here are: aconitase (ACO), phosphoglucomutase (PGM), phosphoglucose isomerase (PGI), 6-phosphogluconate dehydrogenase (6PGD), and glucose 6 phosphate dehydrogenase (P6GD).

3.2.1 Bulked Segregant Analysis (BSA) and Selective Genotyping

For detection of markers linked to the major ascochyta resistance locus, BSA was performed essentially as defined by Michelmore et al. (1991). Individual RILs displaying constant disease severity ratings of 1 to 3 over two subsequent years in the study of Tekeoglu et al. (2000) were grouped as resistant class, and lines with scores of 9 were grouped as susceptible class. Twelve such resistant and susceptible lines represented the resistant and susceptible bulks. For BSA, 1 ng of DNA of each of 12 individuals each from the resistant or the susceptible bulks were mixed and used as template.

Primers giving rise to polymorphic bands between the bulks were further tested for possible linkage to the resistance locus by analyzing the DNA of seven resistant and seven susceptible RILs. Here, DAF was performed on 15 ng of DNA of each individual RIL and the two parental lines. Bands present in all individuals of one group, but absent in all individuals of the other were assumed to be linked to one or the other resistance locus, and the respective primers were tested on the whole population.

3.3 Linkage Analysis

For linkage analysis, Mapmaker V2.0 (Lander et al. 1987) was used. Loci were first divided into linkage groups at a LOD score of 4 by two point analysis using the "group" command. Marker order within linkage groups was determined using the "try" command of the program. The order of STMS markers described by Winter et al. (1999) served as a starting order. Marker orders were scrutinized by multipoint analysis applying the "ripple" function. Markers that could not be assigned a unique placement for reasons other than lack of recombination were eliminated from the dataset. Final map distances were calculated by applying the "Kosambi" function provided by the program to account for the higher numbers of double crossovers experienced by the RILs.

Acknowledgements. Research of the authors was supported by grants from: Deutsche Forschungsgemeinschaft (grant Ka332/17.1–3), International Atomic Energy Agency, Breeding Section (grant CRP 302–02-GFR-8148), Bundesminsterium fuer Technische Zusammenarbeit (BMZ grant 89.7860–01.130), UNESCO (Paris, France), Fazit-Stiftung (Frankfurt, Germany), and Deutscher Akademischer Austauschdienst (DAAD). G. Kahl and P. Winter appreciate the hospitality of the Iwate Biotechnology Research Center (Director: Dr. Hitoshi Enei) at Kitakami, Japan, and STA support.

References

Arumuganathan K, Earle ED (1991) Nuclear DNA content of some important plant species. Plant Mol Biol Rep 9:208–218

Barz W, Bless W, Gunia W, Höhl B, Mackenbrock U, Meyer D, Tenhaken R, Vogelsang R (1993) Mechanism of resistance to fungal pathogens in cool-season food legumes. In: Singh KB, Saxena MC (eds) Breeding for stress tolerance in cool-season food legumes. Wiley, New York, pp 193–210

Becker J, Vos P, Kuiper M, Salamini F, Heun M (1995) Combined mapping of AFLP and RFLP markers in barley. Mol Gen Genet 249:65–73

Beckmann JS, Soller M (1990) Towards a unified approach to genetic mapping of eucaryotes based on sequenced tagged microsatellite sites. Bio/Technology 8:930–932

Blair MW, McCouch SR (1997) Microsatellite and sequence-tagged site markers diagnostic for the rice bacterial leaf blight resistance gene Xa-5. Theor Appl Genet 95:174–184

Burr B, Burr FA, Thompson KH, Albertsen MC, Stuber CW (1988) Gene mapping with recombinant inbreds in maize. Genetics 118:519–526

Caetano-Anollés G, Bassam BJ, Gresshoff PM (1991) DNA amplification fingerprinting: a strategy for genome analysis. Plant Mol Biol Rep 9:292–305

Chen X, Temnykh S, Xu Y, Cho YG, McCouch SR (1997) Development of a microsatellite framework map providing genome-wide coverage in rice. Theor Appl Genet 95:553–567

Cregan PB, Jarvik T, Bush AL, Shoemaker RC, Lark KG, Kahler AI, Kaya N, VanThoai TT, Lohnes DG, Chung J, Specht J (1999) An integrated genetic linkage map of the soybean genome. Crop Sci 39:211–217

Dixon RA, Lamb CJ (1990) Molecular communication in interactions between plants and microbial infection in plants. Annu Rev Plant Pathol Plant Mol Biol 41:339–367

Duke JA (1981) Handbook of legumes of world economic importance. Plenum Press, New York

FAOSTAT (2000) http.//apps.fao.org/

Gupta M, Chyi Y-S, Romero-Severson J, Owen JL (1994) Amplification of DNA markers from evolutionary diverse genomes using single primers of simple sequence repeats. Theor Appl Genet 89:998–1006

Gupta PK, Varshney RK (2000) The development and use of microsatellite markers for genetic analysis and plant breeding with emphasis on bread wheat. Euphytica 113:163–185

Hammond-Kosack KE, Jones JJ (1997) Plant disease resistance genes. Annu Rev Plant Phys Plant Mol Biol 48:575–607

Höhl B, Pfautsch M, Barz W (1990) Histology of disease development in resistant and susceptible cultivars of chickpea (*Cicer arietinum* L.) inoculated with spores of *Ascochyta rabiei*. J Phytopathol 129:31–45

Hüttel B, Winter P, Weising K, Choumane W, Weigand F, Kahl G (1999) Sequence-tagged microsatellite site markers for chickpea (*Cicer arietinum* L.). Genome 42:210–217

Hutcheson SW (1998) Current concepts of active defense in plants. Annu Rev Phytopathol 36:59–90

Jiminéz-Diaz RM, Crino P, Halila MH, Masconi C, Trapero-Casas AT (1993) Screening for resistance to fusarium wilt and ascochyta blight in chickpea. In: Singh KB, Saxena MC (eds) Breeding for stress tolerance in cool-season food legumes. Wiley, New York, pp 77–95

Kaiser WJ (1992) Epidemiology of *Ascochyta rabiei*. In: Singh KB, Saxena MC (eds) Breeding for stress tolerance in cool-season food legumes. Wiley, New York, pp 117–134
Kaiser WJ (1997) Inter- and intranational spread of *Ascochyta* pathogens of chickpea, faba bean, and lentil. Can J Plant Pathol 19:25–224
Kazan K, Muehlbauer FJ, Weeden NF, Ladizinsky G (1993) Inheritance and linkage relationships of morphological and isozyme loci in chickpea (*Cicer arietinum* L.). Theor Appl Genet 86: 417–426
Künzel G, Korzun L, MeisterA (2000) Cytologically integrated physical restriction fragment length polymorphism maps for the barley genome based on translocation breakpoints. Genetics 154:397–412
Lamb CJ, Lawton MA, Drou M, Dixon RA (1989) Signals and transduction mechanism for activation of plant defences against microbial attack. Cell 56:215–224
Lander ES, Green P, Abrahamson J, Barlow A, Daly J, Lincoln SE, Newsburg L (1987) MAPMAKER: an interactive computer package for constructing primary genetic linkage maps of experimental populations. Genomics 1:174–181
Lev-Yadun S, Gopher A, Abbo S (2000) The cradle of agriculture. Science 288:1602–1603
Mansfield DC, Brown AF, Green DK, Carothers AD, Morris SW, Evans HJ, Wright AF (1994) Automation of genetic linkage analysis using fluorescent microsatellite markers. Genomics 24:225–233
Mayer MS, Tullu A, Simon CJ, Kumar J, Kaiser WJ, Kraft JM, Muehlbauer FJ (1997) Development of a DNA marker for *Fusarium* wilt resistance in chickpea. Crop Sci 37:1625–1629
Michaels SD, John MC, Amasino RM (1994) Removal of polysaccharides from plant DNA by ethanol precipitation. Bio/Techniques 17:274–276
Michelmore R (1996) Flood warning – resistance genes unleashed. Nat Genet 14:376–378
Michelmore RW, Paran I, Kesseli RV (1991) Identification of markers linked to disease resistance genes by bulked segregant analysis: a rapid method to detect markers in specific genomic regions using segregating populations. Proc Natl Acad Sci USA 88:9828–9832
Narvel JM, Chu W-C, Fehr WR, Cregan PB, Shoemaker RC (2000) Development of multiple sets of simple sequence repeat DNA markers covering the soybean genome. Mol Breed 6:175–183
Otte O, Barz W (1996) The elicitor-induced oxidative burst in cultured chickpea cells drives the rapid insolubilization of two cell wall structural proteins. Planta 200:238–246
Paran I, Goldmann I, Tanksley SD, Zamir D (1995) Recombinant inbred lines for genetic mapping in tomato. Theor Appl Genet 90:542–548
Paterson A, Lander E, Hewitt JD, Peterson S, Lincoln SE, Tanksley SD (1988) Resolution of quantitative traits into Mendelian factors by using a complete linkage map of restriction fragment length polymorphisms. Nature 335:721–726
Ratnaparkhe MB, Santra DK, Tullu A, Muehlbauer FJ (1998) Inheritance of inter-simple-sequence polymorphisms and linkage with a fusarium wilt resistance gene in chickpea. Theor Appl Genet 96:348–353
Reddy MV, Kabbabeh S (1985). Pathogenic variability in *Ascochyta rabiei* (Pass) lab in Syria and Lebanon. Phytopathol Mediterr 24:265–266
Robinson RA (1976) Plant pathosystems. Springer, Berlin Heidelberg New York
Santra DK, Tekeoglu M, Kaiser WJ, Muehlbauer FJ (2000) Identification and mapping of QTLs conferring resistance to *Ascochyta* blight. Crop Sci 40:12606–12612
Sharma PC, Winter P, Bunger T, Huttel B, Weigand F, Weising K, Kahl G (1995a) Abundance and polymorphism of di-, tri- and tetranucleotide tandem repeats in chickpea (*Cicer arietinum* L.). Theor Appl Genet 90:90–96
Sharma PC, Huttel B, Winter P, Kahl G, Gardner RC, Weising K (1995b) The potential of microsatellites for hybridization- and PCR-based fingerprinting of chickpea (*Cicer arietinum* L.) and related species. Electrophoresis 16:1755–1761
Simon CJ, Muehlbauer FJ (1997) Construction of a chickpea linkage map and its comparison with maps of pea and lentil. J Hered 88:115–119
Srivastava SK, Singh SN, Khare MN (1984) Assessment of yield losses in some promising gram cultivars due to fusarium wilt. Indian J Plant Prot 12:125–126
Su X-Z, Wu Y, Sifri C D, Wellems TE (1996) Reduced extension temperatures required for PCR amplification of extremely A+T-rich DNA. Nucleic Acids Res 24:1574–1475
Tanksley SD (1993) Mapping polygenes. Annu Rev Genet 27:205–233

Tanksley SD, McCouch SR (1997) Seed banks and molecular maps: unlocking genetic potential from the wild. Science 277:1063–1066

Tanksley SD, Ganal MW, Price JP, de Vincente MC, Bonierbale MW, Broun P, Fulton TM, Giovannoni H, Grandillo S, Martin GB, Messeguer R, Miller JC, Miller L, Patterson AH, Pineda O, Röder M, Wing RA, Wu W, Young ND (1992) High density molecular linkage maps of tomato and potato genomes. Genetics 132:1141–1160

Tekeoglu M, Santra DK, Kaiser WJ, Muehlbauer FJ (2000) Ascochyta blight resistance inheritance in three chickpea recombinant inbred lines. Crop Sci 40:1215–1256

Tullu A (1996) Genetics of fusarium wilt resistance in chickpea. PhD Thesis, Crop and Soil Sciences Department, Washington State University, Pullman, WA, USA

Udupa SM, Baum M (2001) High mutation rates and mutational bias at (TAA)n microsatellite loci in chickpea (Cicer arietinum L.). Mol Gen Genomics 265:1097–1103

Udupa SM, Weigand F, Saxena MC, Kahl G (1998). Genotyping with RAPD and microsatellite markers resolves pathotype diversity in the ascochyta blight pathogen of chickpea. Theor Appl Genet 97:299–307

Vogel JM, Scolnik PA (1997) Direct amplification from microsatellites: detection of simple-sequence repeat-based polymorphism without cloning. In: Caetano-Anollés G, Gresshoff PM (eds) DNA markers: protocols, applications and overviews. Wiley, New York, pp 133–150

Vos P, Hogers R, Bleeker M, Reijans M, van de Lee T, Hoernes M, Frijters A, Pot J, Peleman J, Kuiper M, Zabeau M (1995) AFLP: a new technique for DNA fingerprinting. Nucleic Acids Res 23:4407–4414

Wang G, Mackill DJ, Bonman MJ, McCouch SR, Nelson RJ (1994) RFLP mapping of genes conferring complete and partial resistance to blast in a durable resistant rice cultivar. Genetics 136:1421–1434

Weeden NF, Ambrose M, Swiecicki W (1993) *Pisum sativum*, pea. In: O'Brian SJ (ed) Genetic maps, 6th edn. Cold Spring Harbor Laboratory Press, Cold Spring Harbor, NY, pp 624–634

Weising K, Nybom H, Wolff K, Meyer W (1995) DNA fingerprinting in plants and fungi. CRC Press, Boca Raton

Williams JGK, Kubelik AR, Livak KJ, Rafalski JA, Tingey SV (1990) DNA polymorphisms amplified by arbitrary primers are useful as genetic markers. Nucleic Acids Res 18:6531–6535

Winter P, Kahl G (1995) Molecular marker technologies for crop improvement. World J Microbiol Biotechnol 11:449–460

Winter P, Pfaff T, Udupa SM, Hüttel B, Sharma PC, Sahi S, Arreguin-Espinoza R, Weigand F, Muehlbauer FJ, Kahl G (1999) Characterization and mapping of sequence-tagged microsatellite sites in the chickpea (*Cicer arietinum* L) genome. Mol Gen Genet 262:90–101

Winter P, Benko-Iseppon A-M, Hüttel B, Ratnaparkhe M, Tullu A, Sonnante G, Pfaff T, Tekeoglu M, Santra D, Sant VJ, Rajesh PN, Kahl G, Muehlbauer FJ (2000) A linkage map of the chickpea (*Cicer arietinum* L.) genome based on recombinant inbred lines from a *C. arietinum* x *C. reticulatum* cross: Localization of resistance genes for *Fusarium* races 4 and 5. Theor Appl Genet 101:1155–1163

Winter P, Staginnus C, Sharma PC, Kahl G (2001a) Organisation and genetic mapping of the chickpea genome. In: Jaiwal PK, Singh RP (eds) Biotechnology for the improvement in legumes. Kluwer, Dordrecht, pp 285–360

Winter P, Hüttel B, Weising K, Kahl G (2001b) Microsatellites and molecular breeding: exploitation of microsatellite variability for the analysis of a monotonous genome. In: Jain SM, Brar DS, Ahloowalia BS (eds) Molecular techniques in crop improvement. Kluwer, Dordrecht

Xu Y, Zhu L, Xiao J, Huang N, McCouch SR (1997) Chromosomal regions associated with segregation distortion of molecular markers in F2 backcross, double haploid, and recombinant inbred populations in rice (*Oryza sativa* L). Mol Gen Genet 253:535–545

Young ND (2000) The genetic architecture of resistance. Curr Opin Plant Biol 3:285–290

Young WP, Schupp JM, Keim P (1999) DNA methylation and AFLP marker distribution. Theor Appl Genet 99:785–790

Subject Index

MIX
Papier aus verantwortungsvollen Quellen
Paper from responsible sources
FSC® C105338

If you have any concerns about our products,
you can contact us on
ProductSafety@springernature.com

In case Publisher is established outside the EU,
the EU authorized representative is:
Springer Nature Customer Service Center GmbH
Europaplatz 3, 69115 Heidelberg, Germany

Printed by Libri Plureos GmbH
in Hamburg, Germany